L. Cemič

Thermodynamics in Mineral Sciences

Ladislav Cemič

Thermodynamics in Mineral Sciences

An Introduction

With 82 Figures and 23 Tables

Springer

Ladislav Cemič
Institut für Geowissenschaften
der Christian-Albrechts-Universität
Abteilung Mineralogie
Olshausenstrasse 40
24098 Kiel
Germany

Email: lc@min.uni-kiel.de

ISBN-13 978-3-642-06363-3 e-ISBN-13 978-3-540-28808-4

Springer is a part of Springer Science+Business Media
Springeronline.com

Printed in The Netherlands

Cover design: E. Kirchner, Heidelberg

Printed on acid-free paper 30/3141/as 5 4 3 2 1 0

Preface

Thermodynamics is an important tool to interpreting the conditions at which natural geomaterial equilibrate. It allows one to determine, for example, the equilibrium pressures and temperatures and the nature and chemical composition of phases involved mineralogical and petrological processes. Simple chemical model systems, which are often studied in the laboratory in order to understand more complicated natural systems, generally consist of few chemical components. In order to use phase equilibrium results obtained from model systems for interpreting the conditions of formation of natural geologic materials, extrapolations in compositional space and other *P-T* conditions are often required. This can only be done using the mathematical formalism that is offered by thermodynamics.

An number of excellent books on thermodynamics with regards to the fields of mineralogy, petrology and geochemistry have been published over past 40 years. Many of them are, however, written for more advanced students and experienced researchers and it is often assumed that the reader already possesses some prior knowledge of the subject. Consequently, discussions and presentations of basic concepts, which are necessary for beginning students and others attempting to learn thermodynamics for the first time, are often given short shrift. Therefore, the aim of this book is to explain the basic principles of thermodynamics at an introductory level, while trying not to loose much of the mathematical rigor that is one of the most important and central aspects of this subject. Moreover, many students in geosciences are required to take thermodynamic courses in chemistry departments where they are mostly confronted with treatments of gases and fluids. Thereby, the connection to geological or mineralogical problems is often not perceived and, unfortunately, students do not come to understand the importance of thermodynamics for a number of different areas in the geosciences.Therefore, this introductory textbook was written with the aim of reaching beginning geoscience students. In order to demonstrate the usefulness and power of thermodynamics, various experimental phase equilibria results, calorimetric data, etc. from geological literature are used to demonstrate different types thermodynamic calculations. The problems at the end of each chapter are given in order that students can practice the more theoretical concepts that are presented in text.

As a note of thanks, I wish to acknowledge Charles A. Geiger who spent much time and effort correcting my English and for his critical comments on the text and presentation. Also, I would like to express my gratitude to my wife Trudel for her patience and understanding during the weekends that I spent working on my book.

Kiel, May 15. 2005

Lado Cemič

Contents

List of symbols

A	area
a_i^{α}	activity of component i in phase α
B	isothermal bulk modulus
C	number of components
c	molarity
c_p	heat capacity at constant pressure
c_v	heat capacity at constant volume
$\boldsymbol{C_p}$	molar heat capacity at constant pressure
$\boldsymbol{C_v}$	molar heat capacity at constant volume
$\Delta_{tr}\boldsymbol{C_p}$	heat capacity change associated with a phase transition
$\Delta_r\boldsymbol{C_p}$	heat capacity change associated with a chemical reaction
$\Delta_{ord}\boldsymbol{C_p}$	heat capacity change associated with ordering
E	extensive property
F	Helmholtz free energy
$\boldsymbol{f_i}$	fugacity of a pure non-ideal gas i
$\boldsymbol{f_o}$	standard fugacity
f^{id}	fugacity of a non-ideal gas in an 'ideal gas mixture'
f_i	partial fugacity of a non-ideal gas i
G	Gibbs free energy as an extensive property

$\boldsymbol{G}$	Gibbs free energy of a pure phase
G_i^{∞}	Gibbs free energy of a component i at infinite dilution
G^L	Landau's Gibbs free energy
$\overline{G}^{ss}$	molar Gibbs free energy of a homogeneous solution
$\overline{G}$	molar Gibbs free energy
$\overline{G}^{\mathrm{ord}}$	molar Gibbs free energy of ordering
$\Delta_m\overline{G}^{mm}$	molar Gibbs free energy of a mechanical mixture
$\Delta_m\overline{G}$	molar Gibbs free energy of mixing
$\Delta_m\overline{G}^{id}$	molar Gibbs free energy of an ideal solution
$\Delta_m\overline{G}^{\mathrm{ex}}$	excess Gibbs free energy of mixing
$\Delta_r G$	Gibbs free energy of reaction
$\Delta_r\boldsymbol{G}^{\mathrm{o}}$	standard Gibbs free energy of reaction
$\Delta_r\boldsymbol{G}^{rec}$	Gibbs free energy of a reciprocal reaction
$\Delta_f\boldsymbol{G}_{298}$	standard Gibbs free energy of formation
$\Delta_{dis}\overline{G}$	Gibbs free energy of disordering
H	enthalpy as an extensive property
$\overline{H}^{sol}$	molar enthalpy of solution
$\overline{H}^{mm}$	molar enthalpy of a mechanical mixture
$\overline{H}^{\mathrm{ord}}$	enthalpy of ordering
$\boldsymbol{H}_i$	molar enthalpy of a pure component i
H_{i}	partial molar enthalpy of a component i
H_i^{ex}	partial molar enthalpy of mixing (partial molar excess enthalpy)
$H_i^{ex\infty}$	partial molar excess enthalpy of a component i at infinite dilution
$\Delta_m\overline{H}^{ex}$	enthalpy of mixing (molar excess enthalpy)
$\Delta_r\boldsymbol{H}$	enthalpy of reaction between pure phases
$\Delta_r\boldsymbol{H}_{298}$	standard enthalpy of reaction
$\Delta_r\boldsymbol{H}^{exch}$	enthalpy of an exchange reaction
$\Delta_r\boldsymbol{H}^{rec}$	enthalpy of a reciprocal reaction
$\Delta_{tr}\boldsymbol{H}$	enthalpy of a phase transition

$\Delta_{ord}H$	enthalpy of ordering
$\Delta_{dis}\overline{\boldsymbol{H}}$	enthalpy of disordering
$\Delta_f\boldsymbol{H}$	enthalpy of formation
$\Delta_f\boldsymbol{H}_{298}$	standard enthalpy of formation
$\Delta_{sol}\boldsymbol{H}$	enthalpy of solution
$\Delta_{tr}\boldsymbol{H}$	enthalpy of a phase transition
$\Delta_{fus}\boldsymbol{H}$	enthalpy of fusion
I	intensive property
k	Boltzmann constant
$K_{p,T}$	thermodynamical equilibrium constant
K_D	distribution coefficient
M	molar mass
m	mass
m_i	molality
n_i	number of moles of component i
N	number of species
P	total pressure
$\boldsymbol{P}_i$	pressure of a pure gas i
P_i	partial pressure of a gas i
Q	order parameter, heat
q	number of atoms per formula unit
R	Gas constant; number of independent reactions
r	compositional parameter

S	entropy as an extensive property
$\boldsymbol{S}$	molar entropy of a pure phase
$\overline{S}$	molar entropy
$\boldsymbol{S}_{298}$	conventional standard entropy (third law entropy)
S_i^{ex}	partial molar excess entropy of a component i
$\overline{S}^{ord}$	molar entropy of ordering
$\Delta_m\overline{S}$	molar entropy of mixing
$\Delta_m\overline{S}^{ex}$	molar excess entropy of mixing
$\Delta_r\boldsymbol{S}$	entropy of reaction between pure phases
$\Delta_r\boldsymbol{S}^{exch}$	entropy of exchange reaction
$\Delta_r\boldsymbol{S}^{rec}$	entropy of a reciprocal reaction
$\Delta_{tr}\boldsymbol{S}$	entropy of a phase transition
$\Delta_{ord}S^{conf}$	configurational entropy associated with ordering
$\Delta_f\boldsymbol{S}$	entropy of formation
$\Delta_f\boldsymbol{S}_{298}$	standard entropy of formation
T	temperature in K
t	temperature in °C
U	internal energy as an extensive property
$\boldsymbol{U}$	molar internal energy of a pure phase
U_i^{∞}	internal energy of component i at infinite dilution
V	volume as an extensive property
V_E	volume of a unit cell
$\overline{V}$	molar volume
$\overline{V}^{ord}$	volume of ordering
$\boldsymbol{V}_i$	molar volume of a pure component i
$\boldsymbol{V}_{298}$	molar volume of a pure phase at room temperature
V_i	partial molar volume of a component i

$\Delta_m \overline{V}^{ex}$	molar volume of mixing (molar excess volume)
$\Delta_r V$	volume of reaction between pure phases
$\Delta_r V^{rec}$	volume of reciprocal reaction
$\Delta_r V^{exch}$	volume of exchange reaction
W	work
W^H	enthalpic interaction parameter
W^S	entropic interaction parameter
W^V	volumetric interaction parameter
W^G	Gibbs free energy interaction parameter
Z	partition function
z	number of formula units per unit cell
α	thermal expansion
β	compressibility coefficient
γ	activity coefficient
ϵ	energy quantum
$\boldsymbol{\varphi}_i$	fugacity coefficient of a pure non-ideal gas i
φ_{i}	partial fugacity coefficient of a non-ideal gas i in a non-ideal mixture
$\boldsymbol{\mu}_i$	chemical potential of a pure phase i
$\boldsymbol{\mu}^{\mathrm{o}}$	standard chemical potential referred to a pure phase
$\boldsymbol{\mu}^{0,\infty}$	standard chemical potential referred to a phase at infinite dilution
μ_{i}	chemical potential of a component in a solution (partial molar Gibbs free energy)
μ_i^{ex}	excess chemical potential of a component i
$\mu_i^{ex\infty}$	excess chemical potential of a component i at infinite dilution
ρ	density

ϕ	number of phases
Ω	thermodynamic probability
x_i	mole fraction of a component i

Chapter 1 Definition of thermodynamic terms

1.1 Systems

A thermodynamic system is defined as the part of the universe that is subject of consideration. The universe outside the chosen system is called the system's *surroundings*. The system and its surroundings are separated from each other by walls, which can be physical or in thought. A system may or may not interact with its surroundings. Whether or not an interaction takes place, depends on the nature of the walls surrounding the system. Three kinds of systems can be distinguished based on the permeability of the walls to energy and matter. They are:

a. Isolated systems,
b. Closed systems and
c. Open systems

Systems are referred to as *isolated*, if their boundaries or walls prevent any exchange of energy and matter between the system and its surroundings. This means, their walls are rigid, unmovable and perfectly insulating. Systems of this kind have constant energy and mass content. Along their boundaries discontinuities of matter and energy can exist (Fig. 1.1a).

Isolated systems are not observable. They do not exist in nature, because all walls allow some energy transfer. Whether or not a system can be considered isolated depends on the time scale of observation.

The walls of a *closed* system allow energy transfer, but are impermeable to matter. Hence, these systems posses a constant mass and variable energy content. But this does not exclude a change in internal composition caused by chemical reactions. Along the walls of these systems discontinuities of matter but not energy are possible. Closed systems play an important role in the treatment of thermodynamic processes (Fig. 1.1b).

Systems are called *open*, if energy and matter can pass into or out of the system. Along their boundaries or walls neither energy nor matter discontinuities can exist for long periods of time. Systems of this type are characteristic for the treatment of metasomatic processes (Fig. 1.1c).

In mineralogy the systems are often subdivided in different parts, which then belong to different types. For example, the solid phases of a rock can be considered to represent a closed system, but fluids like H_2O, CH_4 and CO_2 constitute an open system.

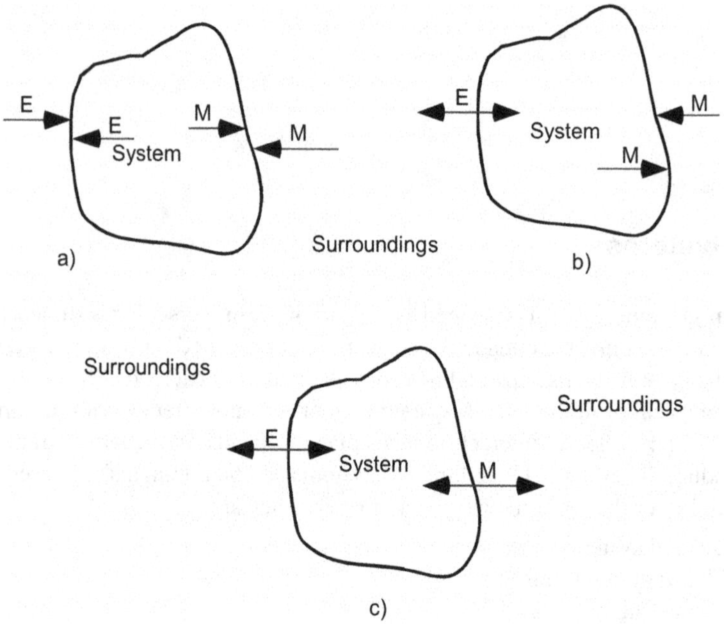

Fig. 1.1 Types of thermodynamic systems. a) isolated, b) closed, and c) open. Symbols: double arrow = exchange in both directions possible; single arrow = exchange not possible; E = energy, M = matter

1.2 Phase

A *phase* is defined as a chemically, physically and structurally homogeneous part of a system. In a system phases are separated from each other by phase boundaries. On the boundaries, the properties of the phases change discontinuously. A system can consist of one or more phases. If it consists of only one phase, it is called *homogenious*. Systems with more than one phase are referred to as *heterogeneous*. For example, a monomineralic rock (e.g. marble) represents a homogeneous and a polymineralic one (e.g. granite) a heterogeneous system.

Most substances can occur in at least three phases: solid, liquid and vapor. In addition to this, many solids exist in different forms having different crystal structures, and are called *polymorphs*. For example, at temperatures below 571°C SiO_2 crystallizes as low quartz, between 571°C and 870°C as high quartz, between 870 and 1470°C as tridymite and from 1470°C up to the melting point at 1713°C as cristobalite.

If the above definition of a phase would be taken in its strict sense, every feldspar grain, say, in a granite would represent a different phase, since every grain is separated from its neighbor by a phase boundary, whose physical properties differ greatly from those of the bulk. Such a construct would be very impractical. We, therefore, neglect the surfaces and consider that all minerals with the same composition and the same structure form just one phase.

1.3 Components

The chemical entities that are required to describe the composition of a system in equilibrium are called *components*. In contrast to the number of phases, which is rigorously determined by thermodynamics, the choice of chemical entities is arbitrary. It depends ultimately upon the nature of the actual thermodynamic problem.

The choice of a component in the case of one single phase with fixed composition is quite simple. The composition of the phase serves as the component. For example, Al_2SiO_5 describes the composition of all three polymorphs: kyanite, andalusite, and sillimanite. Similarly Mg_2SiO_4 represents the composition of orthorhombic olivine and the two high-pressure polymorphs, the β-phase and γ-phase. Systems consisting of only one component are referred to as *unary*.

Most natural olivines, $(Mg,Fe)_2SiO_4$, posses a variable composition. For their description two components, namely Mg_2SiO_4 and Fe_2SiO_4, are required. The same is true in case of the orthopyroxene solid solution, $(Mg,Fe)_2Si_2O_6$. The components here are $Mg_2Si_2O_6$ and $Fe_2Si_2O_6$. Systems involving two components are called *binary*.

The Mg-Fe exchange between orthorhombic olivine and orthopyroxene can be represented by the following reaction:

$$1/2Mg_2SiO_4 + FeSiO_3 = 1/2Fe_2SiO_4 + MgSiO_3. \tag{1.1}$$

The total number of components involved in reaction (1.1) is three, because only three components are independent. The fourth component can be formulated in terms of the other three, simply by rearrangement of Eq. (1.1). For example, the component $FeSiO_3$ is given by:

$$FeSiO_3 = 1/2Fe_2SiO_4 + MgSiO_3 - 1/2Mg_2SiO_4. \tag{1.2}$$

As a rule, the number of components, C, is given by the number of species, N, reduced by the number of independent reactions, R, among the species. That is:

$$C = N - R. \tag{1.3}$$

In the our example, the number of species is 4 and the number of independent reactions is 1, so that

$$C = 4 - 1 = 3. \tag{1.4}$$

A system consisting of three components is referred to as *ternary*.

Species are not necessarily identical with phases. In our case we have four species (Mg_2SiO_4, $FeSiO_3$, Fe_2SiO_4, $MgSiO_3$) but only two phases, namely olivine and orthopyroxene.

There is a straightforward algebraic approach for determining the number of independent reactions in a system for a given number of species. A discussion is beyond the scope of this book. For a detailed presentation of the method, the interested reader is referred to the extensive literature (e.g Spear 1993 and references therein).

Generally, components are real chemical entities, such as oxides or compositions of minerals. But this needs not always to be the case. It is also possible to define entities with negative masses of elements. Components of this kind are particularly advantageous when phases, as in the above example, exhibit variable compositions (see Thompson 1982). For example, the compositional variability of an Mg-Fe-orthopyroxene can be described with the components $MgSiO_3$ and $FeMg_{-1}^{opx}$. Mg_{-1}^{opx} represents -1 mole Mg^{2+}. The composition of ferrosilite can then be given by the following equation:

$$FeSiO_3 = MgSiO_3 + FeMg_{-1}^{opx}. \tag{1.5}$$

The same formalism can be used to describe the composition of fayalite, namely:

$$\frac{1}{2}Fe_2SiO_4 = \frac{1}{2}Mg_2SiO_4 + FeMg_{-1}^{ol}. \tag{1.6}$$

Using the relationships (1.5) and (1.6), Eq. (1.1) becomes:

$$FeMg_{-1}^{opx} = FeMg_{-1}^{ol}. \tag{1.7}$$

According to the above definition components designate chemical entities and have physical meaning only in reference to phases, which are part of the physical world. It should therefore, in principle, not bear a mineral name. Nonetheless this kind of component notation is found very often in mineralogical literature. This is not critical, as long as the considered component does not change its 'form of appearance', but it becomes problematic as soon, as the mineral whose name is used to describe the component undergoes some phase transition.

Example: In the binary system Mg_2SiO_4 - Fe_2SiO_4 the components are designated very often as forsterite and fayalite. This denotation causes no problems as long as processes at low and moderate pressures are considered. At high pressures, however, when olivine transforms first into β-phase and later into γ-phase, the usage of mineral names becomes confusing.

1.4 Functions and variables of state

Classical thermodynamics is concerned only with macroscopic properties of a system such as volume, pressure, temperature, electrical potential, etc. Microscopic properties, for example the distances between atoms in a crystal structure, are not considered. One distinguishes between two groups of properties:

Extensive properties are additive and mass dependent, that means, the property of a system represents simply the sum of the properties of its constituent parts. To this group belongs, for example, volume of a phase or system.

Intensive properties are not additive and do not depend on mass. Typical intensive properties are temperature, pressure, density etc.

If a closed system consists of a pure single phase, only two intensive properties determine completely the rest. For example, if the temperature and pressure of a melt are fixed, then the density and viscosity of this melt are also fixed. One can write:

$$I_k = f(I_1, I_2); \qquad k = 3, 4, \ldots, n. \tag{1.8}$$

Eq. (1.8) is called *function of state* and the arbitrarily chosen properties I_1 and I_2 are the *variables of state*. Let the density of a phase, ρ, be the dependent variable and temperature, T, and pressure, P, the variables of state. For this case, the state function, ρ, reads:

$$\rho = f(T, P). \tag{1.9}$$

The choice between dependent and independent variable of state is arbitrary. In Eq. (1.9) the density was chosen as the dependent variable. But it could as well be pressure or temperature. It is merely a question of convenience which variable is taken as the dependent and which as independent one. Usually, pressure and temperature act as independent variables, because they are relatively easy accessible in experiment.

In case of a mixture additional variables are required in order to specify which particular composition is under the consideration. For this purpose the molar proportions of components composing the mixture must be given. If there are k components present $k - 1$, molar fractions, x_i, are needed. Instead of Eq. (1.8), we now have

$$I_k = f(I_1, I_2, x_1, x_2, \ldots x_{k-1}). \tag{1.10}$$

For example, the density, ρ, of an Mg-Fe-olivine can be expressed as a function of temperature, pressure, and the molar fraction of Fe_2SiO_4 in olivine, $x^{ol}_{Fe_2SiO_4}$:

$$\rho = f(T, P, x^{ol}_{Fe_2SiO_4}). \tag{1.11}$$

An *extensive* property of a pure phase is determined by three variable, one of which is conveniently the mass. The other two are intensive variables.

$$E_k = m \times f(I_1, I_2); \qquad k = 3, 4, \ldots, n. \tag{1.12}$$

Example: The volume of a forsterite crystal is direct proportional to its mass. Beyond that, its volume depends also on temperature and pressure, that is

$$V^{ol} = m^{ol}[g] \times f(T, P). \tag{1.13}$$

To determine an extensive property of a mixture the mole fractions of the components must be introduced again. In place of Eq. (1.10) we obtain:

$$E_k = m \times f(I_1, I_2, x_1, x_2 \ldots x_{k-1}). \tag{1.14}$$

If an extensive property , E_k, is divided by the mass of a phase or system, m, a *specific property* is obtained. Multiplying the specific property by the molar mass yields a *molar property*.

In order to calculate the numerical value of a function of state, the algebraic link between the variables of state must be known. This is not an easy task, since classical thermodynamics does not offer any physical explanation describing the functional interrelationships between variables. The problem can be overcome by focusing the study on changes of the functions due to changes of state.

Changes in the functions are expressed mathematically by differentials, that is, a function is differentiated with respect to its independent variables. The obtained quantities are termed partial derivatives. They describe how changes of independent variables influence the function of state. The total change of a function is expressed by the *total differential*, that is, the sum of all partial derivatives multiplied by the respective infinitesimal increments of the independent variables. Applied to Eq. (1.8), the total differential, dI_k, reads:

$$dI_k = \left(\frac{\partial I_k}{\partial I_1}\right) dI_1 + \left(\frac{\partial I_k}{\partial I_2}\right) dI_2. \tag{1.15}$$

The differentiation of a continuous single-valued function yields a total differential, which can be integrated to obtain the original function. In this case the integral is always the same, no matter which path of integration was chosen. Expressions whose integrals are independent of path of integration are called *exact differentials*.

Using these relationships a set of mathematical formulas, which can be applied to predict the behavior of a system under all possible conditions, has been derived. The starting point for their derivation are two laws which represent everyday experience. These are the *first* and the *second law* of thermodynamics.

1.5 The concept of thermodynamic equilibrium

A state of a system is determined by the values of the state variables. A change in one or more variables causes time-dependent processes to take place. The processes end as soon as the new state, which has its own values of variables, is reached. We say, the system is at *equilibrium* again. It persists in this state, if no further distortions occur. A more strict definition of an equilibrium reads: *A system is at equilibrium if all variables such as pressure, temperature, volume, etc. remain constant, independently of the time of observation. After being disturbed, such a system will revert to its original state, after the disturbance comes to an end.*

Example: At some pressure and temperature a forsterite crystal has a given volume. If the pressure is increased while the temperature is kept constant, its volume decreases. However, as soon as the pressure is brought back to the original value, the volume of the forsterite crystal will be exactly the same as it was before the change in pressure.

1.6 Temperature

The term temperature originally stems from man's sense for cold and warm. In the field of thermodynamics it represents a property, whose value is fixed by two independent variables. In order to define temperature, we must come back to the concept of equilibrium. If two systems are brought in contact with each other, a number of processes such as volume and pressure changes, take place. This happens in both systems until an overall equilibrium is reached. We call this status *thermal equilibrium*. Experience shows that systems which are in equilibrium with a given system, are also in equilibrium with each other. That is, no processes take place if they are brought into contact with one another. This is the subject matter of the *zeroth law* of thermodynamics that reads: *Systems that are in thermal equilibrium with each other posses one common intensive property, namely temperature.*

The time required to reach thermal equilibrium depends particularly on the nature of shared walls. With respect to the permeability to heat, two types of walls can be distinguished. The first is called *diathermal*. If two systems with this kind of walls are brought into contact, their thermodynamic states can be influenced by the mutual heat exchange. The second type of wall is called *adiabatic*. A system that is surrounded by this kind of wall cannot be affected by heat transfer. It can only be influenced by moving the walls, shaking, or by other processes causing an internal motion. In nature, there are no true adiabatic walls, but dewars approach this condition closely. Changes that take place in a system with adiabatic walls, are called *adiabatic processes*. In contrast, processes taking place in a system surrounded by diathermal walls are referred to as *isothermal* if the heat exchange between the system under consideration and the neighboring system occur so quickly that the overall temperature remains constant. The necessary prerequisite is that the neighboring system is large enough so not to change its temperature, when heat is taken away or added to it. Systems of this kind are called *heat reservoirs*.

Before one can measure temperature, an empirical temperature scale has to be defined. For this purpose a suitable system is brought into contact with several other systems in order to equilibrate. The chosen system must be small in comparison to the system whose temperature is to be measured, so that it will not affect significantly the properties of the measured system. Thus the value of an appropriate variable of state (e.g. volume of a gas, expansion of a Hg-column in a glass capillary) is measured. It is assumed that this variable, x, is a linear function of temperature:

$$t(x) = ax + b. \tag{1.16}$$

The coefficients a and b in Eq. (1.16) are arbitrary, and are assigned to two easily reproducible fixed points. In the case of the *Celsius temperature scale*, these points are the freezing and boiling point of water at ambient pressure. Their values are 0 and 100, such that

$$ax_o + b = 0 \text{ and } ax_1 + b = 100. \tag{1.17}$$

The coefficients a and b are obtained by solving Eq. (1.17)

$$a = \frac{100}{x_1 - x_o} \text{ and } b = -\frac{100}{\left(\frac{x_1}{x_o} - 1\right)}. \tag{1.18}$$

Inserting a and b in Eq. (1.16) yields

$$t(x) = 100 \cdot \frac{x - x_o}{x_1 - x_o} [°C]. \tag{1.19}$$

Generally, the relationship given in Eq. (1.19) does not hold strictly. Different *thermometers* give slightly different temperatures. The differences depend on the temperature, the reference system (thermometer type, e.g. gas, liquid etc.) and on the state variable, x, that is measured. Temperatures determined by gas thermometers, where volume, V, or pressure, P, as a function of temperature, is measured, deviate least from each other.

$$t(V) = 100 \cdot \frac{V - V_o}{V_1 - V_o} [°C]; \quad P = \text{const}, \tag{1.20}$$

$$t(P) = 100 \cdot \frac{P - P_o}{P_1 - P_o} [°C]; \quad V = \text{const}. \tag{1.21}$$

At very low gas pressures Eq. (1.20) and Eq. (1.21) yield the same temperature for all gases. This fact is the basis of *thermodynamic temperature.* If the coefficient $100 \cdot \frac{P_o}{P_1 - P_o}$ for different values of P_0 is plotted versus P_0 and then extrapolated to $P_0 = 0$, a constant, T_0, is obtained that holds for all gases:

$$T_o = \lim_{P \to 0} 100 \cdot \frac{P_o}{P_1 - P_o}; \quad V = \text{const}. \tag{1.22}$$

The numerical value of T_0 is 273.16 K. It is the temperature where ice coexists with water and vapor (triple point) in Kelvin (K). This temperature equals 0.01 degree on the Celsius scale, whose zero is the freezing point of water at *ambient* pressure. Hence, the *Kelvin temperature* scale is shifted by 273.15 relative to the Celsius scale, such that:

$$T[K] = t[°C] + 273.15. \tag{1.23}$$

In thermodynamic calculations only the Kelvin temperature scale is used. It has the advantage of being always positive and independent of a particular substance.

In experimental mineralogy thermocouples are often used for measuring the temperature, where EMF vs. temperature curve is calibrated against the melting points of standard substances such as different halogenides, nobel metals, etc.

1.7 Pressure

Mineralogical processes inside the Earth often occur at high to very high pressures. Some minerals are thermodynamically stable only at high pressures. A well known example is ferrosilite, $FeSiO_3$. Thus, the variable pressure plays an important role in many mineralogical studies.

Pressure is defined as force per unit area. Its dimension is N/m^2 (Newtons per square meter) or Pa (Pascal). This is a small unit. In order to avoid large numbers, MPa (Megapascal = 10^6 Pa) or GPa (Gigapascal = 10^9 Pa) are used. In the older mineralogical literature pressure is often given in bar or kbar (10^3 bar). The numerical conversion between the two units is:

$$1 \text{ bar} = 10^5 \text{ Pa}. \tag{1.24}$$

1.8 Composition

In addition to temperature and pressure, *composition* is the third important variable. This variable is required to describe the state of a system consisting of more than one component. The simplest way to do this, is to give the number of moles of each constituent participating in the mixture. This procedure, however, is impractical. It is more convenient to use normalized quantities such as *mole fractions*, *weight percent, molarities,* and *molalities.*

The mole fraction of the *i*-th component in a solution is defined as the number of moles of *i* divided by the sum of all moles of all the components in the solution, that is:

$$x_i = \frac{\text{number of moles of i}}{\text{total number of moles}} = \frac{n_i}{\sum_{i=1}^{k} x_i}. \tag{1.25}$$

Example: The mole fraction of grossular in a $(Mg,Fe,Ca,Mn)_3Al_2Si_3O_{12}$ garnet is calculated as follows:

$$x^{grt}_{Ca_3Al_2Si_3O_{12}} = n^{grt}_{Ca_3Al_2Si_3O_{12}} / (n^{grt}_{Ca_3Al_2Si_3O_{12}} + n^{grt}_{Mg_3Al_2Si_3O_{12}} + n^{grt}_{Fe_3Al_2Si_3O_{12}} + n^{grt}_{Mn_3Al_2Si_3O_{12}}). \tag{1.26}$$

The sum of all mole fractions present in a phase is always one,

$$\sum_{i=1}^{k} x_i = 1. \tag{1.27}$$

The multiplication of the mole fraction by 100 yields the *mole percent*. This unit of measure is used in the graphical presentation of compositions and in phase diagrams. Mole percent is not suitable for thermodynamic calculations.

Ceramists often use *weight percent* (*wt%*) instead of mole percent to specify the composition of a phase or system. This unit is defined as the mass of the component under consideration divided by the total mass of the phase or system times 100, i.e.

$$wt\%(i) = \frac{\text{mass } i}{\text{total mass}} \times 100 = \frac{m_i}{\sum_{i=1}^{k} m_i} \times 100. \tag{1.28}$$

Molarity (c_i) gives the concentration of a component *i* expressed by the number of moles of solute per liter of solution, i.e.:

$$c_i = \frac{1000\, n_i}{V}, \tag{1.29}$$

where *V* is the volume of the solution.

Molality (m_i) gives the concentration of the *i*th-component in terms of numbers of moles, per kg solvent, that is

$$m_i = \frac{n_i}{\dfrac{n_{H_2O} \cdot M_{H_2O}}{1000}} = \frac{1000\, n_i}{n_{H_2O} \cdot M_{H_2O}}, \tag{1.30}$$

where M_{H_2O} is the molar mass of water.

Molarity and molality are mainly used to describe the concentrations of components in aqueous solutions.

1.8. 1. Graphic representation of composition

For the graphic representation of composition, mole fractions, mole percent, or weight percent can be used.

In a two-component system composition can be represented on a straight line. The length of a line depends on the type of compositional specification. If the composition is expressed by the mole fraction, the line starts at zero and ends at 1. In the case that mole percent or weight percent are used as the compositional units, the line starts also at zero, but it ends at 100. The components of the system are represented at the endings of the concentration line. If temperature and pressure are not specified, the compositions of all phases belonging to the system under consideration are represented on concentration line regardless their stability conditions. Polymorphs plot at the same compositional point (see Fig. 1.2).

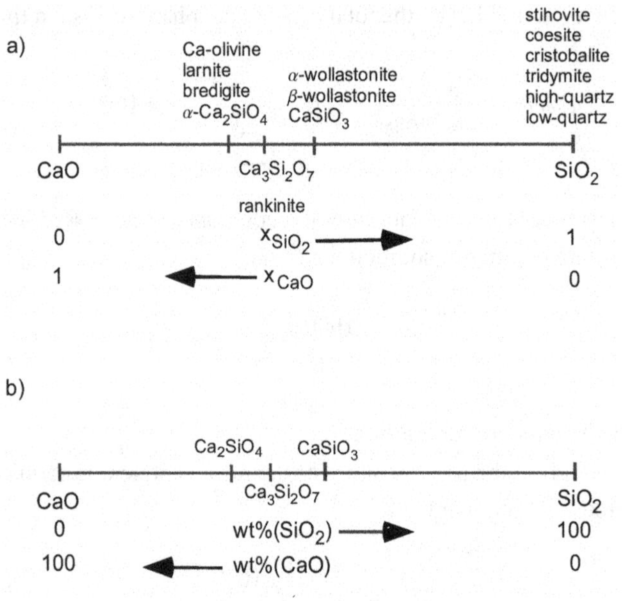

Fig. 1.2 Graphic representation of the composition in the binary system CaO-SiO_2. a) concentration in mol fractions, b) concentration in weight percent.

The molar ratio of phases that constitute a given bulk composition, can be calculated using the *lever rule*. According to this rule, the amount of a phase is directly proportional to the distance between the bulk composition and the composition of the coexisting phase, sitting on the other side of the bulk composition point. The validity of the lever rule can be demonstrated as follows:

Suppose a mixture consisting of the components A and B has the composition x_B^b. Suppose further that at some P and T, A and B react with each other forming the phases α and β, with compositions x_B^{α} and x_B^{β}, respectively. The question to be answered is: what are the molar proportions of the two phases.

According to Eq. (1.25), the mole fraction of component B in the bulk composition reads:

$$x_B^b = \frac{n_B^b}{n_A^b + n_B^b} \tag{1.31}$$

with n_A^b and n_B^b giving the number of moles for the components A and B, respectively.

The following relations hold between the numbers of moles of the components A and B in phases α and β and the bulk composition:

$$n_A^b = n_A^\alpha + n_A^\beta \text{ and } n_B^b = n_B^\alpha + n_B^\beta. \tag{1.32}$$

Inserting the expression of Eq. (1.32) into Eq. (1.31), one obtains

$$x_B^b = \frac{n_B^\alpha + n_B^\beta}{(n_A^\alpha + n_A^\beta) + (n_B^\alpha + n_B^\beta.)}. \tag{1.33}$$

From the definition of mole fraction,

$$x_B^\alpha = \frac{n_B^\alpha}{n_A^\alpha + n_B^\alpha} \text{ and } x_B^\beta = \frac{n_B^\beta}{n_A^\beta + n_B^\beta} \tag{1.34}$$

one derives the expressions for n_B^α and n_B^β, namely:

$$n_B^\alpha = x_B^\alpha(n_A^\alpha + n_B^\alpha) \text{ and } n_B^\beta = x_B^\beta(n_A^\beta + n_B^\beta). \tag{1.35}$$

Replacing the number of moles n_B^α and n_B^β in the numerator of Eq. (1.33) by the above expressions,

$$x_B^b(n_A^\alpha + n_A^\beta) + x_B^b(n_B^\alpha + n_B^\beta) = x_B^\alpha(n_A^\alpha + n_B^\alpha) + x_B^\beta(n_A^\beta + n_B^\beta) \tag{1.36}$$

is obtained.

A small rearrangement of terms leads to a proof of the lever rule, namely:

$$\frac{(n_A^{\beta} + n_B^{\beta})}{(n_A^{\alpha} + n_B^{\alpha})} = \frac{(x_B^b - x_B^{\alpha})}{(x_B^{\beta} - x_B^b)}. \tag{1.37}$$

The sums $(n_A^{\alpha} + n_B^{\alpha})$ and $(n_A^{\beta} + n_B^{\beta})$ give the total compositions of the phases α and β, respectively. The differences between the mole fractions, given on the right side of Eq. (1.37), can be read off the line in Fig. 1.3. The numerator equals the distance a and the denominator the distance b, respectively.

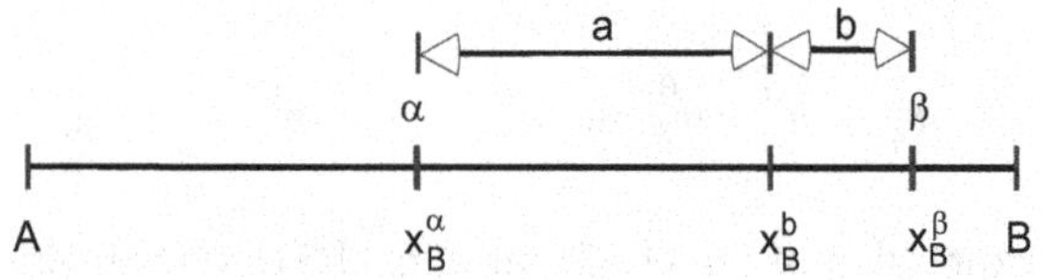

Fig. 1.3 Illustration of the lever rule. The distances a an b are directly proportional to the amount of the phases β and α, respectively. x_B^{α} gives the mole fraction of the component B in the phase α and x_B^{β} that in the phase β. x_B^b designates the mole fractions of the component B in the bulk composition.

Example: An analysis of a phase mixture consisting of rankinite, $Ca_3Si_2O_7$, and wollastonite, $CaSiO_3$, yielded 45.455 wt% SiO_2 and 54.545 wt% CaO. What are the molar proportions of rankinite and wollastonite in the mixture?

To solve this problem, first the given weight percents are to be converted into mole percents. In this conversion weight percents are taken as masses and the mole fractions are calculated according to Eq. (1.25) as follows:

$$x_{SiO_2} = \frac{\dfrac{wt(\%)(SiO_2)}{M_{SiO_2}}}{\dfrac{wt(\%)(SiO_2)}{M_{SiO_2}} + \dfrac{wt(\%)(CaO)}{M_{CaO}}} = \frac{\dfrac{45.455}{60.084}}{\dfrac{45.455}{60.084} + \dfrac{54.545}{56.077}} = 0.4375.$$

According to Eq. (1.27), the mole fraction of CaO can be calculated directly from the known mole fraction of SiO_2, namely:

$$x_{CaO} = 1 - 0.4375 = 0.5625.$$

A plot of x_{SiO_2} or x_{CaO} on the composition line shows that the bulk composition lies between rankinite and wollastonite (see Fig. 1.4).

If we now make use of the lever rule (see Eq. (1.37), the molar proportions of the phases present in the mixture are calculated as follows:

$$\frac{n_{CaO}^{wo} + n_{SiO_2}^{wo}}{n_{CaO}^{rnk} + n_{SiO_2}^{rnk}} = \frac{a}{b} = \frac{0.4375 - 0.4}{0.5 - 0.4375} = \frac{0.0375}{0.0625}.$$

Since the maximum concentrations of CaO and SiO_2 at the ends of the composition line are 1, the calculated number of moles of wollastonite is one half and that one of rankinite one fifth of the respective formula unit. In order to obtain the whole formula units, the right side of the above equation has to be multiplied by 5/2. The sums $n_{CaO}^{wo} + n_{SiO_2}^{wo}$ and $n_{CaO}^{rnk} + n_{SiO_2}^{rnk}$ give the total amount of wollastonite and rankinite, respectively and therefore

$$\frac{n_{CaSiO_3}^{wo}}{n_{Ca_3Si_2O_7}^{rnk}} = \frac{0.0375}{0.0625} \times \frac{5}{2} = \frac{3}{2}$$

or

$$n_{CaSiO_3}^{wo} = \frac{3}{2} \cdot n_{Ca_3Si_2O_3}^{rnk}.$$

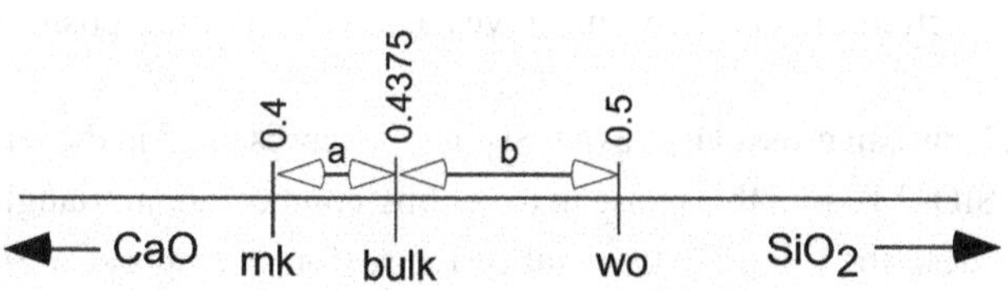

Fig. 1.4 The system CaO - SiO_2. In agreement with the lever rule, the distance a is directly proportional to the amount of wollastonite, $CaSiO_3$, in the bulk composition. Similarly, the distance b corresponds to the amount of rankinite, $Ca_3Si_2O_7$. The numbers give the mole fractions of SiO_2.

Thus, we are able to calculate the number of moles of rankinite for any arbitrarily chosen mole number of moles of wollastonite and vice versa. For example, if we assume wollastonite $n_{CaSiO_3}^{wo} = 3$, the number of moles of rankinite, $n_{Ca_3Si_2O_7}^{rnk}$, equals 2. The corresponding mole fractions are therefore:

$$x_{CaSiO_3}^{wo} = \frac{n_{CaSiO_3}^{wo}}{n_{Ca_3Si_2O_7}^{rnk} + n_{CaSiO_3}^{wo}} = \frac{3}{3+2} = \frac{3}{5} = 0.6$$

of wollastonite and

$$x_{Ca_3Si_2O_7}^{rnk} = 1 - x_{CaSiO_3}^{wo} = 1 - 0.6 = 0.4$$

of rankinite.

For the graphic representation of a ternary system Gibbs' compositional triangle is used. It is an equilateral triangle with the components given at the corners. The three sides represent the three binary subsystems. Compositions within the triangle can be plotted in different ways:

a) First, one component of the ternary system or phase under consideration is neglected and for the remaining two, the mole fractions are calculated and plotted on the proper side of the compositional triangle. Next, a line is drawn from the plotting position to the opposite corner, that is the corner with the component neglected in the calculation of the mole fractions. This line divides the triangle into two parts in constant proportions along its entire length. Thereafter, this procedure is repeated for another two components. The intersection of the two lines defines the plotting position of the ternary system or phase.

b) Another way to plot a ternary phase in a compositional triangle is to use the mole fractions of all three components present in it. Since each component is fixed by two other, each mole fraction appears on two sides of the triangle, such that a line connecting the two plotting positions runs parallel to the third side of the triangle. The intersection of such two lines gives again the plotting position of the ternary phase.

Example: How can grossular, $Ca_3Al_2Si_3O_{12}$, be represented in the ternary system CaO-Al_2O_3-SiO_2? To plot this phase in the Gibbs' compositional triangle, following the method described in a), two subsystems, for example CaO-Al_2O_3, and Al_2O_3-SiO_2 are chosen. Next, the mole fractions of the components in these two subsystems are calculated using the oxide based formula of grossular, $3CaO \cdot Al_2O_3 \cdot 3SiO_2$. The calculations yield the following results:

$$x_{CaO}^{lim\text{-}cor} = \frac{3}{3+1} = \frac{3}{4} = 0.75 \text{ and } x_{Al_2O_3}^{lim\text{-}cor} = 1 - x_{CaO}^{lim\text{-}cor} = 1 - 0.75 = 0.25$$

in the subsystem CaO-Al_2O_3 and

$$x_{Al_2O_3}^{cor\text{-}qtz} = \frac{1}{1+3} = 0.25 \text{ and } x_{SiO_2}^{cor\text{-}qtz} = 1 - 0.25 = 0.75$$

in the subsystem Al_2O_3-SiO_2.

The mineral abbreviations in the superscript signify the binary subsystem to which the mole fractions refer.

The mole fractions, as calculated above, are plotted on the corresponding sides of the Gibbs' compositional triangle. Then, lines are drawn to the SiO_2 and CaO corners, respectively. The intersection of these lines gives the plotting position of grossular (see Fig. 1.5).

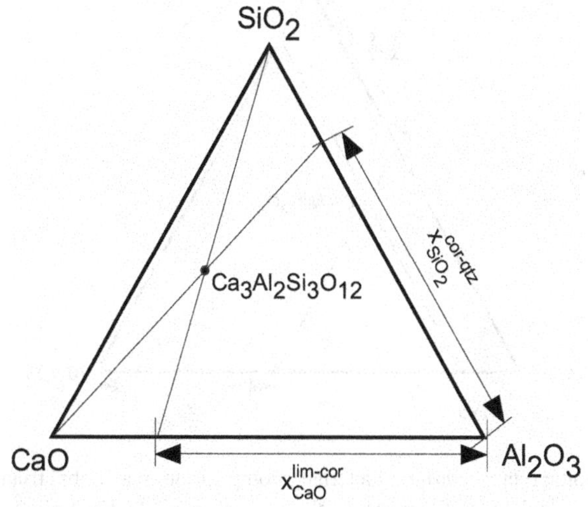

Fig. 1.5 Graphic representation of a ternary phase in Gibbs' compositional triangle. The plotting position of grossular, $Ca_3Al_2Si_3O_{12}$, is determined using the mole fractions in the subsystems CaO - Al_2O_3 and Al_2O_3 - SiO_2.

To plot grossular according to method b) the ternary mole fractions are to be calculated first. Using the oxide formula of grossular once again, we obtain:

$$x_{CaO}^{grt} = \frac{3}{3+1+3} = \frac{3}{7} = 0.429$$

and

$$x_{Al_2O_3}^{grt} = \frac{1}{3+1+3} = \frac{1}{7} = 0.143.$$

As shown previously, the mole fractions of any two components suffice. The use of the mole fraction of the third one is not necessary. If we plot the calculated mole fractions, each one on the *two* appropriate sides of the compositional triangle, and connect the two mole fractions with one another, a parallel line to the third side of the triangle is obtained. The intersection of two lines yields the plotting position of grossular (see Fig. 1.6).

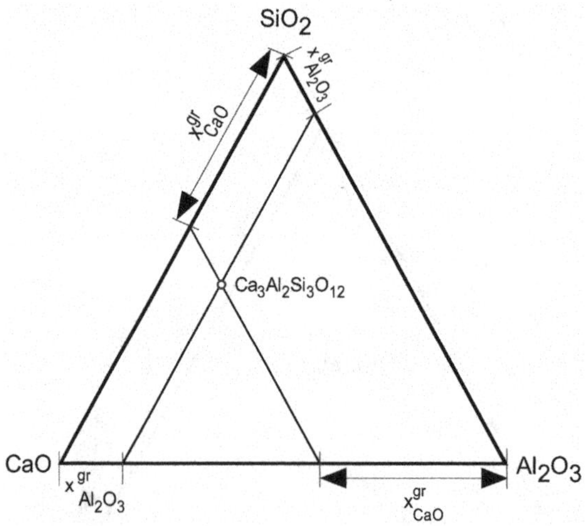

Fig. 1.6 Graphic representation of a ternary composition in a Gibbs' triangle. The mole fractions are calculated with respect to all three components. Each mole fraction is plotted on two sides of the triangle.

In order to read the mole fractions of a ternary composition from the Gibbs' compositional triangle, the procedure described above has to be reversed. Another way to arrive at the molar proportions is to drop perpendiculars from the plotting position on the three sides of the triangle. The lengths of the three perpendiculars are directly proportional to the three molar fractions. This method is demonstrated in Fig. 1.7.

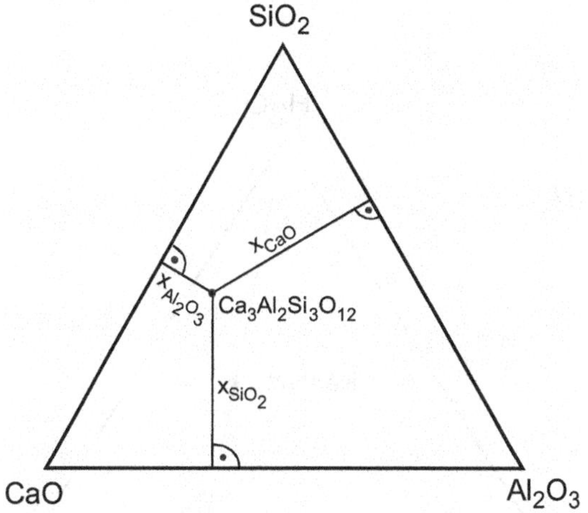

Fig. 1.7 Determination of mole fractions of a ternary phase from its position in a Gibbs' compositional triangle. The lengths of the perpendiculars: x^{grt}_{CaO}, $x^{grt}_{Al_2O_3}$, and $x^{grt}_{SiO_2}$ correspond to the mole fractions of the oxides: CaO, Al_2O_3, and SiO_2 respectively.

If compositions with more than three components are to be presented graphically, a procedure is applied that is called compositional *projection*. This means that the dimensionality of the full space is reduced by its projection onto a compositional subspace.

Example: Consider kaolinite, $Al_2Si_2O_5(OH)_4$. The composition of this mineral can be visualized in a ternary diagram with the components Al_2O_3, SiO_2 and H_2O. In addition, it can also be represented in a binary system. In this case, the plotting position of kaolinite in the triangle has to be projected onto one of the three sides of the Gibbs' triangle. Thereby, three dimensions are reduced to two. For example, if it is to be represented on the line Al_2O_3-SiO_2 the projection is carried out by drawing a line from the H_2O corner of the triangle through the plotting position of kaolinite to the basis line. The intersection of this line with basis of the triangle marks the composition of kaolinite in the projection. This projection procedure is demonstrated in Fig. 1.8.

In a multi component systems with $C \geq 4$ projections are carried out in the same way. For example, a phase consisting of four components can be represented in a Gibbs' compositional triangle as a projection from one of the four corners of a tetrahedron. Of course, it is not necessary to draw the three dimensional diagram first. The plotting position in the projection is found simply by using the mole fractions calculated disregarding the component from which the projection is to be carried out.

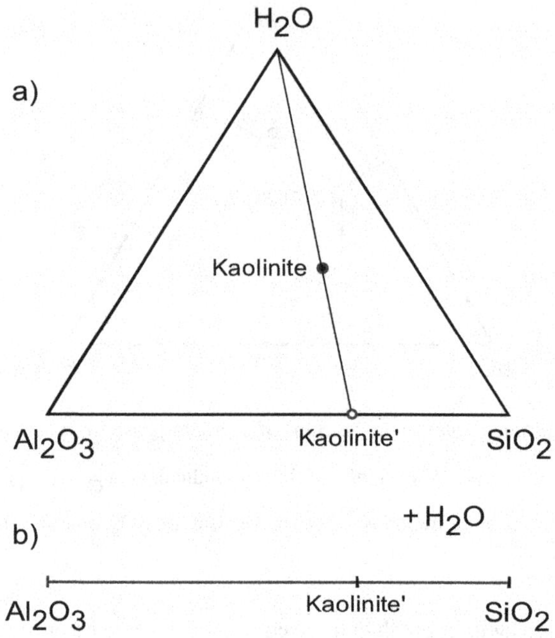

Fig. 1.8 Projection of kaolinite composition from the H_2O apex onto the Al_2O_3-SiO_2 base line. a) The plotting position of kaolinite, $Al_2Si_2O_5(OH)_4$, in the compositional triangle, b) kaolinite in projection.

Example: The four-component phase muscovite, $KAl_2[AlSi_3O_{10}](OH)_2$, can be represented as a projection from the H_2O apex onto the base of the tetrahedron, defined by K_2O-Al_2O_3-SiO_2. In order to do this, the molar fractions of the components are calculated as though water was not part of the system. The following values for the mol fractions are obtained:

$$x_{K_2O}^{mu} = \frac{1}{1+3+6} = \frac{1}{10} = 0.1,$$

$$x_{Al_2O_3}^{mu} = \frac{3}{1+3+6} = \frac{3}{10} = 0.3$$

and

$$x_{SiO_2}^{mu} = \frac{6}{1+3+6} = 0.6.$$

The result of this projection is demonstrated in Fig. 1.9.

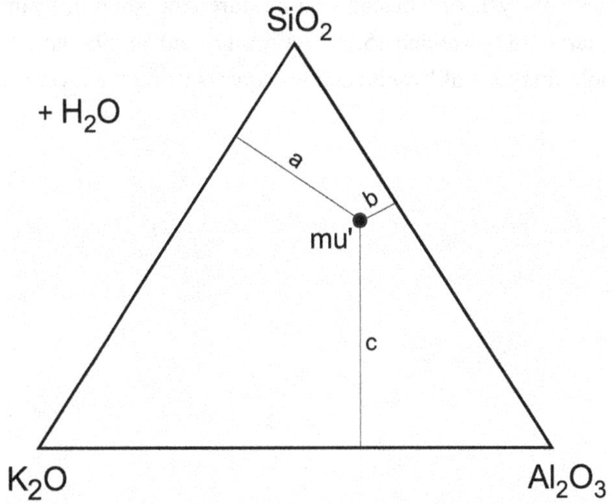

Fig. 1.9 Muscovite (mu') after projection from the H_2O apex onto the base of the triangle K_2O - Al_2O_3 - SiO_2. The distances *a, b*, and *c* correspond to the mole fraction in the ternary subsystem as follows: $a \propto (Al_2O_3)/(K_2O + Al_2O_3 + SiO_2)$, $b \propto (Al_2O_3)/(K_2O + Al_2O_3 + SiO_2)$ and $c \propto (SiO_2)/(K_2O + Al_2O_3 + SiO_2)$.

1.9 Problems

1. Show that the function $z = ax/y$ possesses a total differential.
2. Calculate the mole fractions of MgO and SiO_2 for forsterite and enstatite.
3. Plot the forsterite and enstatite composition on the line MgO-SiO_2.
4. Calculate the mole fractions of MgO, Al_2O_3 and SiO_2 for a water-free cordierite, $Mg_2Al_3AlSi_5O_8$.

- Convert the mole fractions of MgO, Al_2O_3 and SiO_2 into wt%.

5. Plot the following compositions in the Gibbs compositional triangle MgO-Al_2O_3-SiO_2: enstatite, $MgSiO_3$, kyanite, Al_2SiO_5, spinel, $MgAl_2O_4$, and sapphirine, $Mg_2Al_4O_6(SiO_4)$.

6. Plot the tremolite composition, $Ca_2Mg_5Si_8O_{22}(OH)_2$, as a projection from the H_2O apex onto the base of the MgO-CaO-SiO_2-H_2O tetrahedron.

7. The chemical analysis of a mechanical mixture consisting of kyanite, Al_2SiO_5, and low quartz, SiO_2, yielded 45.707 wt% SiO_2 and 54.293 wt% Al_2O_3. Calculate the mole fractions of kyanite and low quartz using the lever rule.

Chapter 2 Volume as a state function

For geochemical and geophysical calculations volume of minerals and fluids plays an important role. It determines basically the direction in which reactions proceed during the pressure changes. It belongs to one of the few relatively easily conceivable thermodynamic properties.

The volume of a solid is normally given in m^3 or cm^3. In older literature additional volume dimensions such as $cal\,bar^{-1}$ (calorie pro bar) or $J\,bar^{-1}$ (Joule pro bar) are often found. The following relationships exist between different dimension specifications:

$$1\ \text{bar} = 10^5\ \text{Nm} = 10^5\ \text{Pa (Pascal)}$$

$$1\,\text{J}\,\text{bar}^{-1} = \frac{\text{N} \times \text{m}}{\left(\frac{\text{N}}{\text{m}^2}\right) \times 10^5} = 10^{-5}\,\text{m}^3 = 10\,\text{cm}^3.$$

Taking into account that

$$1\ \text{cal} = 4.1844\ \text{J}$$

the relationship

$$V[\text{cal bar}^{-1}] = 10 \text{ x } 4.8144 = 41.844\ \text{cm}^3 = 41.844 \text{ x } 10^{-6}\ \text{m}^3$$

is obtained.

The volume given in $[cal\,bar^{-1}]$ or $[J\,bar^{-1}]$ is occasionally referred to as *volume coefficient.*

2.1 Volume of pure phases

In the case where the mass of a pure phase is kept constant, its volume is determined completely by temperature and pressure. Changes in pressure result in definite changes in temperature. The relationship between pressure, temperature, and volume is given by the *equation of state*:

$$V = f(P, T). \tag{2.1}$$

In a three-dimensional diagram with the coordinates temperature, T, pressure, P and volume, V, this function represents a surface (see Fig. 2.1).

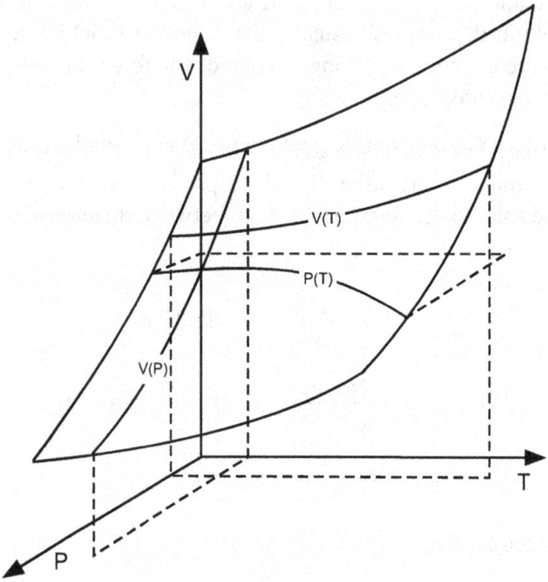

Fig. 2.1 Schematic diagram of the volume of a pure phase as a function of state $V = f(P,T)$. $V(T)$ = isotherm, $V(P)$ = isobar, and $P(T)$ = isochore.

Experiments aimed at studying the relationships between the variables must be designed such that all but one of the variables are kept constant. In respect to which variable is changed and which one is kept constant, the following partial functions of volume are obtained: $V = f(T)_P$ and $V = f(P)_T$. The curves given by the first function are called *isobars*. They represent the temperature dependence of volume at constant pressure. The second function renders the *isotherms*. These curves represent the pressure dependence of the volume at constant temperature. In addition to these two functions, a third additional one is possible. One can also keep volume constant while the temperature changes. In this case changes in temperature cause definite changes in pressure and one obtains the following partial function: $P = f(T)_V$. The resulting curve is referred to as an *isochore*. The three functions are depicted in Fig. 2.1.

2.1.1 Thermal expansion and compressibility

Fig. 2.2 shows the volume of the unit cell, V_E, of diopside, $CaMgSi_2O_6$, as a function of temperature.

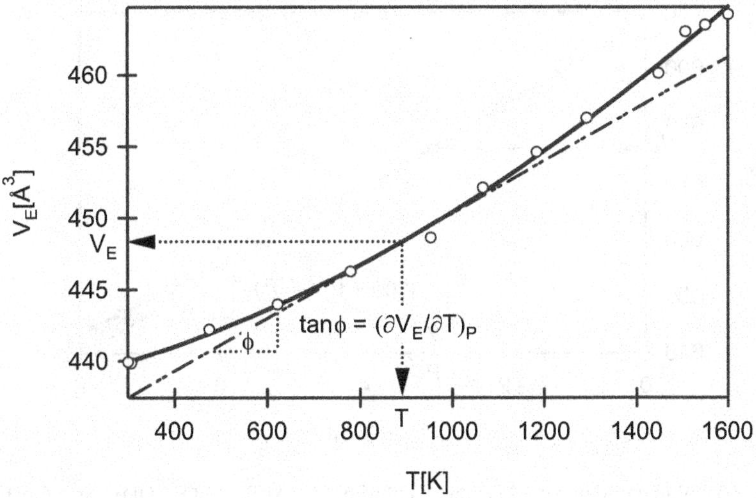

Fig. 2.2 The volume of the unit cell of diopside, $CaMgSi_2O_6$, as a function of temperature at 0.1 MPa. (After Richet et al. 1998).

The volume of the unit cell increases with increasing temperature non-linearly. The slope of the volume vs. temperature curve, which is given by the derivative $(\partial V_E/\partial T)_P$, is smaller at low temperatures and becomes greater at higher temperatures.

Fig. 2.3 shows the volume of the unit cell of deuterated synthetic chlorite, $Mg_5Al(Si_3Al)O_{10}(OD)_8$, as a function of pressure at room temperature. Increasing pressures results in a decrease in the volume of the unit cell. The slope of the volume vs. pressure curve is given by the derivative $(\partial V_E/\partial P)_T$. It decreases with increasing pressure and approaches zero at very high pressures.

If both temperature and pressure change, the volume change is given by the total differential:

$$dV = \left(\frac{\partial V}{\partial T}\right)_P dT + \left(\frac{\partial V}{\partial P}\right)_T dP. \tag{2.2}$$

In Eq. (2.2) the subscript E is dropped, because the relationship holds not only for the volume of an unit cell but, generally, for the volume of an arbitrary mass of a pure phase. The only precondition that must be fulfilled is that the mass of the phase remains constant in the course of a thermodynamic process.

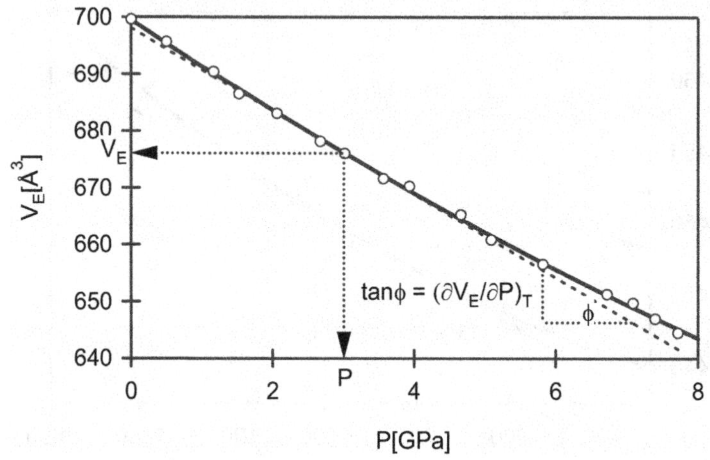

Fig. 2.3 Volume of deuterated synthetic chlorite, $Mg_5Al(Si_3Al)O_{10}(OD)_8$, as a function of pressure at 298 K. (After Welch and Crichton 2002).

The fractional increase of volume with temperature at constant pressure is referred to as the *coefficient of thermal expansion*, α, that is

$$\alpha = \frac{1}{V}\left(\frac{\partial V}{\partial T}\right)_P. \tag{2.3}$$

In a similar way the *compressibility coefficient,* β, is defined as the negative fractional change of volume with pressure at constant temperature:

$$\beta = -\frac{1}{V}\left(\frac{\partial V}{\partial P}\right)_T. \tag{2.4}$$

The negative sign is taken to make the coefficient a positive number, because increasing pressure leads to a decrease in volume. Strictly, V in the term $1/V$ is the volume of the phase at pressure P, that is at the point of differentiation. For the practical use, however, it is often replaced by V_0, that is the volume in some reference state, e.g. at ambient pressure and temperature. If the substance under consideration is a solid phase, and thus has relatively small changes with both temperature and pressure, little error is introduced by this simplification.

In the mineralogical literature, the *isothermal bulk modulus* is often used instead of the compressibility coefficient. It is defined as the reciprocal of the compressibility, that is:

$$B = \frac{1}{\beta} = -V\left(\frac{\partial P}{\partial V}\right)_T. \tag{2.5}$$

Generally, both thermal expansion and compressibility are pressure and temperature dependent. In the case of solids, however, these dependences are relatively small. At low and moderate pressures and temperatures both coefficients are often considered to be constant. Moreover, the fact that pressure and temperature act in opposite directions lowers the error, because many mineralogical processes take place at elevated temperatures *and* pressures.

In order to describe the behavior of volume at very high pressures (above 1 GPa) the Birch-Murnaghan equation is frequently used. This equation of state has the following form:

$$P = \frac{3}{2}B_o\left[\left(\frac{V_o}{V_P}\right)^{7/3} - \left(\frac{V_o}{V_P}\right)^{5/3}\right]\left\{1 + \left(\frac{3}{4}B' - 4\right)\left[\left(\frac{V_o}{V_p}\right)^{2/3} - 1\right]\right\}, \tag{2.6}$$

where V_0 designates the volume at ambient pressure. B_o is the bulk modulus and B' its pressure derivative.

Using the definitions of thermal expansion and compressibility, Eq. (2.2) can be rewritten as follows:

$$dV = \alpha V dT - \beta V dP. \tag{2.7}$$

In isochoric processes the volume remains constant, that is $dV = 0$. Inserting this in Eq. (2.7) yields:

$$\alpha V dT - \beta V dP = 0. \tag{2.8}$$

From Eq. (2.8) follows:

$$\left(\frac{\partial P}{\partial T}\right)_V = \frac{\alpha}{\beta}. \tag{2.9}$$

Eq. (2.9) describes the pressure change per unit temperature in the case where the volume is kept constant.

In the examples presented above, we used the volume of one unit cell. For thermodynamic calculations, however, the *molar volume*, ***V***, is the more appropriate quantity. It is defined as the volume that is occupied by one mole of formula units of the substance under consideration. The spatial arrangement of atoms is predetermined by the crystal structure of the substance.

Examples of calculated thermal expansion and compressibility

Example 1: Tab. 2.1 gives and Fig. 2.4 shows the molar volume of α-eucryptite, $LiAlSiO_4$, as a function of temperature at a constant pressure of 1.94 GPa. The data are taken from Zhang et al. (2002), who measured the lattice constants in the temperature range 298-1073 K.

Table 2.1 Volume of α-eucryptite, $LiAlSiO_4$, as a function of temperature at 1.94 GPa (Zhang et al. 2002)

T [K]	V [$cm^3 mole^{-1}$]
300	209.083
373	209.357
473	209.622
573	210.107
673	210.592
773	211.243
873	211.731
923	212.014
973	212.250
1073	213.046

Although the molar volume clearly increases non-linearly, a straight line is first fitted to the data points. The result is shown by the dotted line in Fig. 2.4. Of course, this fit is only a rough approximation. Nonetheless, this procedure is frequently carried out especially in the case where the average thermal expansion over a restricted temperature interval is required. In our example the linear regression yields:

$$V\,[\mathrm{cm^3 mol^{-1}}] = 207.35 + 5.0739 \times 10^{-3} T\,[\mathrm{K}].$$

Because of the assumed linear relationship between the volume and temperature, the thermal expansion is temperature independent. It is calculated as follows:

$$\bar{\alpha} = \frac{1}{V_{298}} \frac{(V_T - V_{298})}{(T - 298.15)}, \tag{2.10}$$

where $\bar{\alpha}$ is the average thermal expansion and V_{298} the volume at 298.15 K.

Using the coefficients determined by the regression procedure, one obtains:

$$V_{298} = 207.35 + 5.0739 \times 10^{-3} \times 298.15 = \underline{208.863\ \mathrm{cm^3 mol^{-1}}}$$

for the volume at 298.15 K and

$$V_T = V_{700} = 207.35 + 5.0739 \times 10^{-3} \times 700 = \underline{210.902\ \mathrm{cm^3 mol^{-1}}}$$

for the molar volume at 700 K.

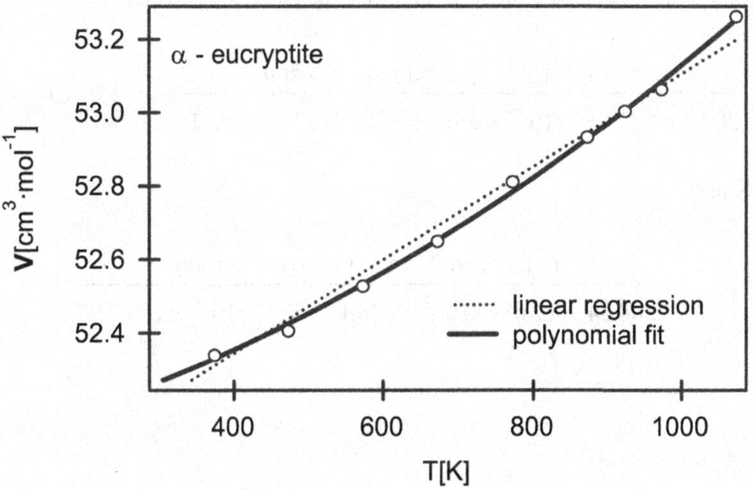

Fig. 2.4 Molar volume of α-eucryptite as a function of temperature at a constant pressure of 2.4 GPa (Zhang et al. 2002). A straight line and a second-order polynomial are fitted to the data points.

In this example 700 K is taken as the upper temperature limit. Because of the assumed linear behavior, the choice of upper temperature is entirely arbitrary. Hence, any other temperature within the measured interval could have been used as well.

The average thermal expansion coefficient, $\bar{\alpha}$, for α-eucryptite is then

$$\bar{\alpha} = \frac{1}{208.863 \text{ cm}^3\text{mol}^{-1}} \times \frac{(210.902 \text{ cm}^3\text{mol}^{-1} - 208.863 \text{ cm}^3\text{mol}^{-1})}{(700 \text{ K} - 298.15 \text{ K})}$$

$$= \underline{2.43 \times 10^{-5}\text{K}^{-1}}.$$

If one uses a second-order polynomial of the form $V = a + bT + cT^2$, instead of a straight line, a much better fit is obtained. It yields the following coefficients:

$a = 208.34$,
$b = 17.087 \times 10^{-4}$ and
$c = 2.47 \times 10^{-6}$.

With these data, the thermal expansion is then calculated according to Eq. (2.3):

$$\alpha = \frac{1}{(a + bT + cT^2)}(b + 2cT). \tag{2.11}$$

Using the volume at 298.15 K as a reference the calculation reads:

$$\alpha_{500} = \frac{17.087\times10^{-4} + 2 \times 2.47\times10^{-6} \times 500}{208.34 + 17.087\times10^{-4} \times 298.15 + 2.47\times10^{-6} \times 298.15^2} = \underline{2.0\times10^{-5}\text{K}^{-1}}$$

for 500 K and

$$\alpha_{1000} = \frac{17.087\times10^{-4} + 2 \times 2.47\times10^{-6} \times 1000}{208.34 + 17.087\times10^{-4} \times 298.15 + 2.47\times10^{-6} \times 298.15^2}$$

$$= \underline{3.18\times10^{-5}\text{K}^{-1}}$$

for 1000 K, respectively.

A comparison of the two results reveals appreciable differences. As a general rule it holds that higher temperatures give rise to greater thermal expansion coefficients. It is necessary, therefore, to specify the temperature region over which a given thermal expansion coefficient holds.

Example 2: Yang et al. (1997) determined the unit-cell dimensions of kyanite, Al_2SiO_5, at room temperature and at various pressures up to 4.56 GPa. Their results can be used to calculate the molar volume of kyanite as a function of pressure. The pressures and the molar volumes are presented in Table 2.2 and displayed in Fig. 2.5.

Table 2.2 Molar volume of kyanite, Al_2SiO_5, as a function of pressure (Yang et al. 1997)

Pressure [GPa]	V[$cm^3 mol^{-1}$]
0.00	44.1572
0.68	44.0186
1.35	43.8525
1.98	43.7175
2.54	43.6113
3.10	43.4725
3.73	43.3386
4.32	43.2212
4.56	43.1674

A plot of the data exhibits a linear relationship between the molar volume and pressure (see Fig. 2.5) that can be described by the equation:

$$V\,[\mathrm{cm^3 mol^{-1}}] = 44.156 - 0.218P\,[\mathrm{GPa}].$$

The linear behavior suggests a constant compressibility over the entire pressure regime.

Using the volume at 0.1 MPa ($V_o = 44.156\ \mathrm{cm^3 mol^{-1}}$) as a reference and that at 4.0 GPa ($V_{4.0} = 43.285\ \mathrm{cm^3 mol^{-1}}$) as the upper limit, the compressibility coefficient is:

$$\beta = -\frac{1}{44.156\ \mathrm{cm^3 mol^{-1}}} \times \frac{43.285\ \mathrm{cm^3 mol^{-1}} - 44.156\ \mathrm{cm^3 mol^{-1}}}{4.0\times10^9\ \mathrm{Pa}}$$

$$= \underline{4.93\times10^{-12}\ \mathrm{Pa^{-1}}}$$

and the bulk modulus

$$B = \frac{1}{\beta} = \frac{1}{4.93\times10^{-12}\ \mathrm{Pa^{-1}}} = \underline{202.84\ \mathrm{GPa}}.$$

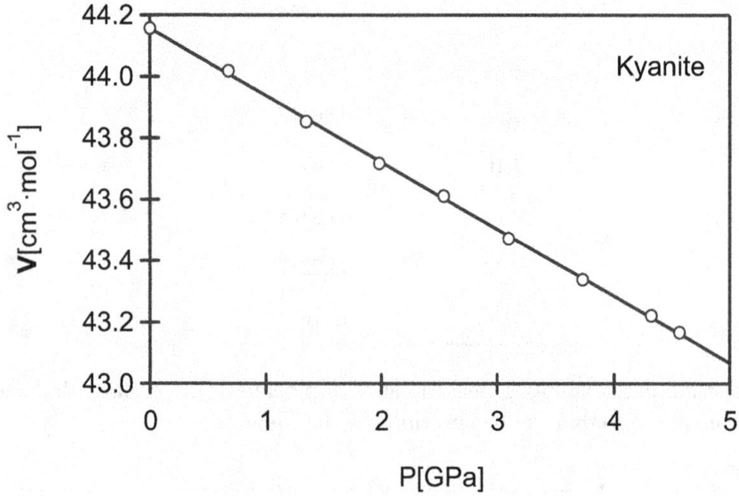

Fig. 2.5 Volume of kyanite as a function of pressure at room temperature (Yang et al. 1997).

Example 3: The volume of the unit-cell versus pressure curve for chlorite (see Fig. 2.3) shows a significant curvature. This means that the compressibility coefficient is pressure dependent within regime under consideration. In order to account for this behavior, a polynomial of the second-order is fitted to the data. A least squares best fit yields:

$$V_E[\text{Å}^3] = 699.56 - 8.2814P + 0.1601P^2; \qquad P[\mathrm{GPa}].$$

Using this polynomium, the compressibility coefficient can be calculated at any a pressure within the pressure range over which the volume-pressure relationship was measured. Using the volume of the unit-cell at 0.1 MPa (699.56 Å^3) as a reference, the compressibility coefficient at 1 GPa is calculated according to Eq. (2.4) as follows:

$$\beta_{1GPa} = -\frac{1}{699.56}(-8.2814 + 2 \times 0.1601 \times 1)$$
$$= 11.38\times10^{-3}\ \text{GPa}^{-1} = \underline{11.38\times10^{-12}\ \text{Pa}^{-1}}.$$

An analogous calculation for $P = 7$ GPa yields:

$$\beta_{7GPa} = \underline{8.63\times10^{-12}\ \text{Pa}^{-1}}.$$

As expected, different compressibility coefficients are obtained depending upon the pressure. The higher the pressure, the smaller the compressibility coefficient.

In Fig. 2.6 the calculated compressibility coefficient of chlorite is shown as a function of pressure. Because a second-order polynomial is used to fit the experimental data, a linear relationship between compressibility and pressure results.

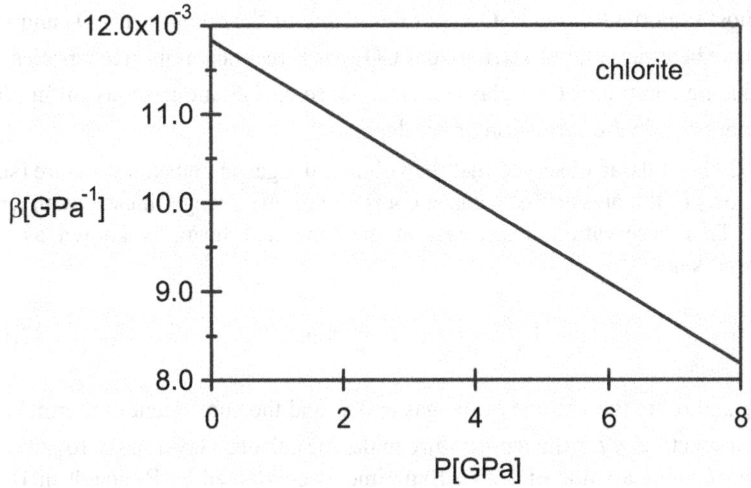

Fig. 2.6 Compressibility coefficient for chlorite as a function of pressure at room temperature. The second-order polynomial is used to describe the volume vs. pressure relationship. (Experimental data are taken from Welch and Crichton 2002)

In thermodynamic tables used by geoscientists the compressibility is occasionally given as a function of pressure, i.e. $\beta_P = \beta_o - \beta_1 P$. In our case for chlorite β_o = 11.83 x10^{-3} GPa^{-1} and β_1 = 45.76 x10^{-6} GPa^{-2} is obtained. If the thermal expansion is also given as a function of temperature in such tables, equations of this kind read: $\alpha_T = \alpha_o + \alpha_1 T$.

In this connection, it is necessary to emphasize that equations describing the pressure and temperature behavior of compressibility and thermal expansion respectively are purely, empirical in nature. The second-order polynomials that are used to fit the experimental data, have no physical meaning. Therefore, any extrapolations outside the experimental data base should be viewed cautiously.

A better, although, empirical approach to describe the compressibility of solids up to very high pressures offers the Birch-Murnaghan equation (see Eq. (2.6)). If this equation is plotted to the experimental data of Welch and Crichton (2002), disregarding the given uncertainties, a bulk modulus, B_0, of 82.9 GPa is obtained. Its derivative B' amounts to 4.7.

2.1.2 Volume of ideal gases as a state function

A mineralogist who is mainly concerned with solid rocks might think gases don't play such an important role in geosciences. However, a closer look at the subject unveils the immense importance of this group of substances. Any equilibria involving solids involve also gases exerting a vapor pressure over the solids. In oxidation/reduction processes gases determine the direction in which reactions proceed.

The most important gas is H_2O vapor consisting of the components O_2 and H_2. In carbonate bearing mineral assemblages CO_2 gas is the most important species. In more reducing conditions CH_4 plays an essential role. H_2S and gaseous sulfur play an important role in the formation of ore deposits.

In 1802 Gay-Lussac observed that the volume of a gas at constant pressure (sufficiently low) or the pressure of a gas at constant volume change linearly with temperature. This observation, expressed in mathematical form, is known as the Guy-Lussac's law:

$$V = V_0(1 + \alpha t); \quad P = \text{const}, \tag{2.12}$$

where V_0 and α are the volume of the gas at 0°C and the coefficient of thermal expansion, respectively. t is the temperature in degree celsius. Gay-Lussac found α to be constant having a value of 1/267. Experiments conducted by Regnault in 1847 yielded a value for α of 1/273. Later more precise measurements showed that some gases obey Gay-Lussac better than others. However, the deviation from the law was found to become smaller with decreasing pressure for all gases. If P in Eq. (2.12) is set to ambient pressure (101325 Pa), α equals 1/273.15. With this value the

Gay-Lussac law reads:

$$V = V_0\left(1 + \frac{t}{273.15}\right) = V_0\left(\frac{273.15 + t}{273.15}\right) = V_0\frac{T}{T_0} \propto T. \tag{2.13}$$

In Eq. (2.13) T designates the thermodynamic temperature as defined in Eq. (1.23).

According to the Gay-Lussac's law a gas would have zero volume at 0 K but this is impossible, therefore, the absolute zero is physically not attainable. Hence, the Gay-Lussac's law is a limiting law and it holds strictly only for a *hypothetical* gas consisting of molecules having negligible volume and no intermolecular interaction. Such a gas is called *ideal* or *perfect*. Real gases approach this state at very low pressures and high temperatures.

In the seventeenth century the British physicist Robert Boyle discovered that the product between volume and pressure for the air is (nearly) constant. Somewhat later a French physicist Mariotte, who independently repeated Boyle's experiments, added the very important restriction that the inverse relationship between volume and pressure holds only if the temperature of the gas is kept constant. The law, that is called Boyle's law (in the Anglo-Saxon part of the world) or Mariotte's law (in France) or Boyle-Mariotte's law (in Germany), holds strictly only for the ideal gases. It reads:

$$PV = \text{const}; \quad T = \text{const}. \tag{2.14}$$

Gay-Lussac's and Boyle-Mariotte's law describe partial volume changes with temperature and pressure, respectively. To obtain the total state function of volume, $V(P,T)$, the two laws must be combined. Hence, an isobaric and an isothermal change are performed subsequently. We start with V_0, P_0, and T_0, where P_0 = 101325 Pa and T_0 = 273.15 K and V_0 is the volume of the gas. The final state will be characterized by V, P, and T. Following Gay-Lussac's law, the increase of the temperature form T_0 to T at constant pressure P_0 yields:

$$V_{P_0, T} = \frac{V_0 T}{T_0}. \tag{2.15}$$

According to the Boyle-Mariotte law the subsequent change in pressure from P_0 to P at constant temperature T gives:

$$P_0 V_{P_0, T} = P V_{P, T}. \tag{2.16}$$

Substituting for $V_{P_o, T}$ by the expression given in Eq. (2.15) we obtain:

$$PV = \frac{P_o V_o}{T_o} T = \text{const} \times T. \tag{2.17}$$

Using the volume of one mole of a perfect gas, $V_o = 22.414 \times 10^{-3}$ m^3 at ambient pressure conditions, $P_o = 101325$ Pa and $T_o = 273.15$ K, the constant assumes a value that is referred to as the *gas constant*, R, namely:

$$R = \frac{101325 \text{ Pa} \times 22.414 \times 10^{-3} \text{ m}^3}{273.15 \text{ K}} = \underline{8.3144 \text{ Jmol}^{-1}\text{K}^{-1}}.$$

Thus, for 1 mole of a perfect gas the *volume as a function of state* reads:

$$V = \frac{RT}{P}. \tag{2.18}$$

For an arbitrary amount of gas Eq. (2.18) has to be multiplied by the number of moles, n. Since $nV = V$ (total volume,) the state function then takes the following form:

$$V = n\frac{RT}{P}. \tag{2.19}$$

If Eq. (2.18) is differentiated with respect to the variables T and P,

$$dV = \frac{R}{P}dT - \frac{RT}{P^2}dP \tag{2.20}$$

is obtained.

In order to prove whether or not Eq. (2.20) is a total differential the partial derivatives are to be cross-differentiated. This means, the first derivative must be differentiated with respect to pressure and the second one with respect to temperature. Doing this, we obtain:

$$\frac{\partial}{\partial P}\left(\frac{R}{P}\right)_T = \frac{\partial}{\partial T}\left[\frac{-(RT)}{P^2}\right]_P = -\frac{R}{P^2}. \tag{2.21}$$

Eq. (2.21) is referred to as the *reciprocity relation*. It shows that Eq. (2.18) is indeed a function of state.

2.1.3 Volume of real gases

An ideal or perfect gas is a purely hypothetical construct and can never exist in reality. According to the equation of state, the volume of a perfect gas must become zero at absolute zero, which means that the molecules do not have finite volume. This is, of course, impossible, but at low pressures and high temperatures, where the distances separating molecules are large compared to their size, the behavior of a gas can approach ideality. However, as pressure increases, the volume of the molecules increasingly becomes a significant fraction of the total volume. Similarly, the electrostatic interaction between molecules is negligibly small at low pressures and high temperatures and it only becomes significant with decreasing temperatures and increasing pressures. Hence, another equation of state must be constructed to describe properly the behavior of gases at higher pressures and low temperatures.

In order to account for the finite volume of the molecules, van der Waals proposed an additional term, *b*, that is introduced into the equation of state:

$$V = \frac{RT}{P} + b. \tag{2.22}$$

The term *b* represents the volume of a gas at absolute zero, that is, the volume of the molecules.

Because of the finite molecular volume, the free space, that is the space where molecules can move around, is reduced. It equals $V - b$ and is referred to as the 'free volume'. The fact that molecules posses a finite volume determines also how closely the molecules can approach one other before repulsive forces become significant. An rearrangement of Eq. (2.22) yields:

$$P = \frac{RT}{V-b}. \tag{2.23}$$

Eq. (2.23) gives the pressure of a gas corrected for the molecular repulsion. resulting from the closest approach.

On the other hand, the mutual attraction between molecules causes a reduction of pressure that a gas exerts on the walls of a container. The attractive forces are proportional to the number of molecules, which can be expressed as concentration *c*. If we consider that the attractive force acts between two molecules, the concentration, *c*, has to be squared. Because the concentration is inversely proportional to the volume ($c \propto 1/V$), follows farther that the attractive force is proportional to $1/V^2$. Attraction between molecules causes a decrease in pressure, and this means that the term a/V^2 has to be subtracted from the term in Eq (2.23) to obtain the pressure of a real gas, that is

$$P = \frac{RT}{V-b} - \frac{a}{V^2}, \tag{2.24}$$

where a is an empirical parameter.

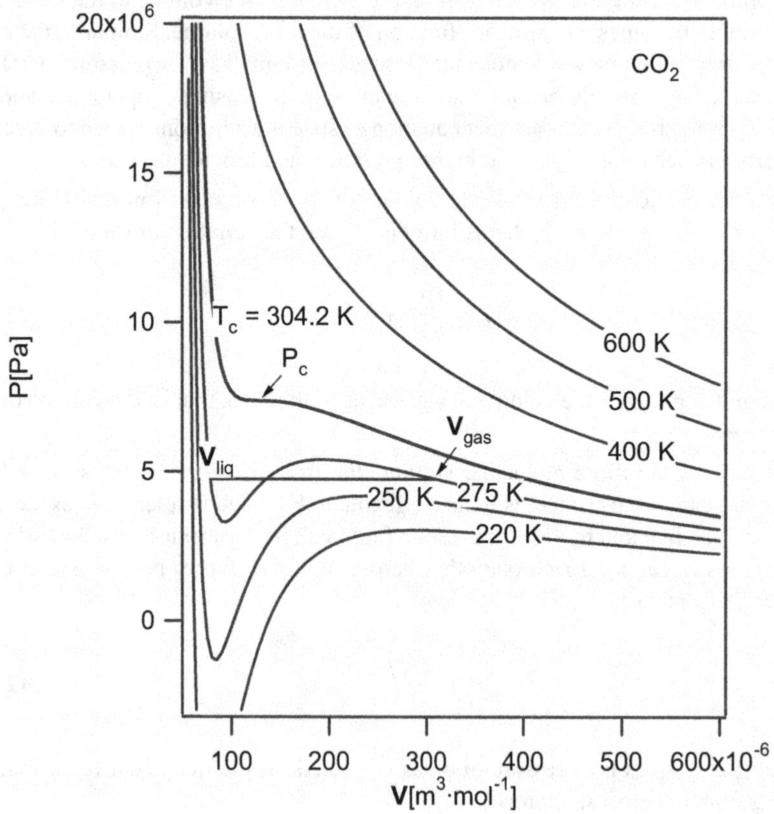

Fig. 2.7 Calculated isotherms for CO_2. V_{liq} and V_{gas} are the volumes of the coexisting liquid and gas, respectively at 275 K. P_c = critical pressure, T_c = critical temperature.

Eq. (2.24) is referred to as the *van der Waals* equation. Its physical meaning can be interpreted as follows: External pressure, P, and the attraction between the molecules, a/V^2, act in the same direction. They push the molecules together. At equilibrium, this pressure is balanced by the thermal pressure, $RT/(V - b)$, holding the molecules apart. Therefore,

$$\frac{RT}{V-b} = \frac{a}{V^2} + P. \tag{2.25}$$

Hence, the constants a and b account for the attractive and repulsive forces between the molecules, respectively. A more precise meaning of these constants and the way how they can be evaluated, will be discussed later on.

The van der Waals equation describes the P-V-T relations of real gases only qualitatively and over a limited range of pressures and temperatures. This fact makes it less important in respect of predicting gas behavior at some arbitrary pressure and temperature conditions. Important, however, is that the equation can be used to describe both gas and liquid states. This is demonstrated in Fig. 2.7.

An rearrangement of Eq. (2.24) shows that it is cubic in volume:

$$V^3 - \left(b + \frac{RT}{P}\right)V^2 + \left(\frac{a}{P}\right)V - \frac{ab}{P} = 0. \tag{2.26}$$

Depending on the temperature, Eq. (2.26) has one or three solutions for V. For temperatures $T \geq T_c$ (see Fig. 2.7) only one root is real. The remaining two are complex. At these temperatures only one phase, a so called *supercritical fluid*, exists. For $T < T_c$ all three roots are real. Two of them, the smallest and the largest one, represent the volumes of liquid and gas, respectively. Hence, at temperatures $T < T_c$ a two-phase field with coexisting gas and liquid exists. Its boundaries are given by the inflection points on the isotherms. The two-phase field ends at the critical point, which is characterized by the *critical temperature*, T_c, *critical pressure*, P_c, and *critical volume* V_c. At the critical point, both inflections of the isotherm coincide with the maximum of the curve limiting the two-phase field. Hence, the following two condition must be fulfilled:

$$\left(\frac{\partial P}{\partial V}\right)_T = 0 \tag{2.27}$$

and

$$\left(\frac{\partial^2 P}{\partial V^2}\right)_T = 0. \tag{2.28}$$

Applying the relations (2.27) and (2.28) to Eq. (2.24) yields:

$$\left(\frac{\partial P}{\partial V}\right)_{T_c} = -\frac{RT_c}{(V_c - b)^2} + \frac{2a}{V_c^3} = 0 \qquad (2.29)$$

and

$$\left(\frac{\partial^2 P}{\partial V^2}\right)_{T_c} = \frac{2RT_c}{(V_c - b)^3} - \frac{6a}{V_c^4} = 0, \qquad (2.30)$$

respectively.

Eqs. (2.29) and (2.30) can be solved to obtain the parameters of the van der Waals equation in terms of critical volume, critical temperature and gas constant, R:

$$a = \frac{9}{8}RT_cV_c \qquad (2.31)$$

and

$$b = \frac{V_c}{3}. \qquad (2.32)$$

Substituting expressions (2.31) and (2.31) into Eq. (2.24) yields the critical pressure, P_c:

$$P_c = \frac{3RT_c}{8V_c}. \qquad (2.33)$$

From Eq. (2.33) it follows:

$$V_c = \frac{3RT_c}{8P_c} \qquad (2.34)$$

and

$$T_c = \frac{8P_cV_c}{3R}. \qquad (2.35)$$

Replacing V_c in Eqs. (2.31) and (2.32) by the expressions given in Eq. (2.34), the parameters a and b are obtained as functions of T_c, P_c, and R, namely:

$$a = \frac{27R^2T_c^2}{64P_c} \tag{2.36}$$

and

$$b = \frac{RT_c}{8P_c}\,. \tag{2.37}$$

From any of Eqs. (2.33) through (2.35) it follows that

$$Z_c = \frac{P_cV_c}{RT_c} = \frac{3}{8} = 0.375, \tag{2.38}$$

where Z_c is the so-called *critical compressibility factor* and it should be the same for all non-ideal gases. However, using the experimentally determined critical data of various non-ideal gases values ranging between 0.25 and 0.30 are obtained. These deviations from the value of 0.375 are the reason why different pairs of critical data yield different values of the parameters a and b.

Example: Consider CO_2 gas. Its critical data are: T_c = 304.14 K, P_c = $73.75\text{x}10^5$ Pa and $V_c = 94.0\text{x}10^{-6}\ \text{m}^3\text{mol}^{-1}$ (Ambrose 1994).

The critical compressibility factor of CO_2 is given by:

$$Z_c = \frac{73.75\times10^5\text{Pa} \times 94.0\times10^{-6}\text{m}^3\text{mol}^{-1}}{8.3144\text{Jmol}^{-1}\text{K}^{-1} \times 304.14\text{K}} = \underline{0.274},$$

which deviates from the theoretical value by +0.101.

Using the critical volume, V_c, and critical temperature, T_c, Eq. (2.31) yields:

$$a = \frac{9}{8} \times 8.3144\text{Jmol}^{-1} \times 304.14\text{K} \times 94.0\times10^{-6}\text{m}^3\text{mol}^{-1} = \underline{0.2674\ \text{Jm}^3\text{mol}^{-2}}.$$

In order to calculate the value of parameter a according to Eq. (2.36), T_c, and P_c, are required. The calculation reads:

$$a = \frac{27 \times 8.3144^2 \mathrm{J}^2 \mathrm{mol}^{-2} \mathrm{K}^{-2} \times 304.14^2 \mathrm{K}}{64 \times 73.75 \times 10^5 \mathrm{Pa}} = \underline{0.3658\ \mathrm{Jm}^3 \mathrm{mol}^{-2}}.$$

Parameter b can be calculated using either Eq. (2.32) or Eq. (2.37). If Eq. (2.32) is used

$$b = \frac{94.0 \times 10^{-6} \mathrm{m}^3 \mathrm{mol}^{-1}}{3} = \underline{31.333 \times 10^{-6} \mathrm{m}^3 \mathrm{mol}^{-1}},$$

is obtained.

Using Eq. (2.37) gives:

$$b = \frac{8.3144 \mathrm{Jmol}^{-1} \mathrm{K}^{-1} \times 304.2 \mathrm{K}}{8 \times 73.825 \times 10^5 \mathrm{Pa}} = \underline{42.825 \times 10^{-6} \mathrm{m}^3 \mathrm{mol}^{-1}}.$$

Another equation of state, also based on both repulsive and attractive forces between gas molecules, was proposed by Redlich and Kwong (1949). It differs from the van der Waals equation in that it has a more complicated expression for the attractive forces as a function of temperature and volume:

$$P = \frac{RT}{(V-b)} - \frac{a}{T^{1/2} V(V+b)}. \tag{2.39}$$

As in the van der Waals equation, in the Redlich-Kwong equation a and b are also considered constant for each gas. Its applicability is, therefore, restricted to a limited range of temperature and pressure. In an attempt to extend it to geologically relevant high pressure and temperature conditions, several workers developed *modified Redlich-Kwong* equations. All of them exhibit a corrected attractive term or repulsive term or both.

De Santis et al. (1974) treated a as a linear function of temperature and b as a constant. Holloway (1977) did the same using the volume data for H_2O of Burnham et al. (1969a,b). Kerrick and Jacobs (1981) developed an equation adopting Carnahan and Starling's (1969) version of a modified Redlich-Kwong equation where

$$P = \frac{RT(1+y+y^2-y^3)}{V(1-y)^3} - \frac{a(T)}{V(V+b)T^{1/2}}, \tag{2.40}$$

with $y = b/4V$. a is a linear function of temperature and b is a constant. Kerrick and Jacobs' (1981) modification assumes that a is not only a function of temperature but also of pressure. Their equation of state reads:

$$P = \frac{RT(1+y+y^2-y^3)}{V(1-y)^3} - \frac{a(T,P)}{V(V+b)T^{1/2}}. \tag{2.41}$$

Halbach and Chatterjee (1982) developed yet another modified Redlich-Kwong equation based on a careful analysis of *P*-*V*-*T* data for H_2O. They determined that *a* is not so much a function of pressure but primarily of temperature. The constant *b*, on the other hand, is considered a monotonous function of *P*. Their equation has the following form:

$$P = \frac{RT}{V-b(P)} - \frac{a(T)}{VT^{1/2}[V+b(P)]}, \tag{2.42}$$

with $a(T) = A_1 + A_2T + A_3/T$ and $b(P) = (1 + B_1P + B_2P^2 + B_3P^3)/(B_4 + B_5P + B_6P^2)$. According to the authors, Eq. (2.42) allows calculations of *P*-*V*-*T* for temperatures between 100-1000°C and pressures up to 200 kbar. Less satisfactory results are obtained only in the vicinity of the saturation curve, particularly at temperatures between 300 and 500°C.

Virial equation offers another approach to quantitative description of the gas behavior. It expresses the compressibility factor,

$$Z = \frac{PV}{RT} \tag{2.43}$$

as a power series in terms of pressure (or density):

$$Z = 1 + BP + CP^2 + DP^3 + \ldots \tag{2.44}$$

Coefficients *B*, *C*, *D*... are termed the second, third, and fourth *virial coefficient.* They are functions of temperature, but not of pressure, and they must be determined experimentally. The virial equation is the only equation of state that has a solid theoretical basis. Coefficients reflect the nature of molecular interactions. The equation preforms remarkably well for many fluids at low pressures and temperatures, but it fails at conditions generally encountered by metamorphic rocks, for example.

2.1.4 Volume of solid phases

For solids, no equations comparable to those for gases exist. It is possible, though, to derive an algebraic expression that describes the volume behavior of solids over limited ranges of temperature and pressure. Let us assume that temperature increases from T_0 to T at the constant pressure of P_0. In order to obtain the volume change caused by this finite temperature change, the expression for the thermal expansion has to be integrated between the limits of T and T_0. Before the integration, the variables must be separated, that is, the increment dT must be brought to the left side of the equation and we have:

$$\int_{T_0}^{T} \alpha dT = \int_{V(T_0 P_0)}^{V(T,P)} \frac{1}{V} dV. \tag{2.45}$$

If thermal expansion can be considered constant within a given temperature interval, the integration of Eq. (2.45) yields:

$$\alpha(T - T_0) = \ln[V(T, P_0)] - \ln[V(T_0, P_0)] \tag{2.46}$$

or

$$V(T, P_0) = V(T_0, P_0)\exp[\alpha(T - T_0)]. \tag{2.47}$$

Because thermal expansion is normally small compared to molar volume, the assumption of constant expansion does not introduce significant error, as long as the temperature interval remains moderate (a few hundreds of degrees).

In the next step, the pressure is changed from P_0 to P, while the temperature is kept constant at T. Now, the expression giving the compressibility of a solid has to be integrated between the limits of P and P_0, that is

$$-\int_{P_0}^{P} \beta dP = \int_{V(T, P_0)}^{V(T,P)} \frac{1}{V} dV. \tag{2.48}$$

Assuming a constant compressibility, integration of Eq. (2.48) gives:

$$-\beta(P - P_0) = \ln[V(T, P)] - \ln[V(T, P_0)] \tag{2.49}$$

or

$$V(T, P) = V(T, P_o)\exp[-\beta(P - P_o)]. \tag{2.50}$$

Replacing $V(T,P_o)$ in Eq. (2.50) by the expression given in Eq. (2.47) yields:

$$,V(p,T) = V(T_o, P_o)\exp[\alpha(T-T_o)-\beta(P-P_o)] \quad (2.51)$$

If the starting temperature, T_o, and the starting pressure, P_o, are set to the standard conditions that is 298 K[1] and 0.1 MPa, respectively, Eq. (2.51) reads:

$$V(P,T) = V_o\exp[\alpha(T-298)-\beta(P-10^5)], \quad (2.52)$$

where V_o is the volume at 298 K and 0.1 MPa. It is generally called the *standard volume*. Compressibilities of solids are generally small compared to molar volumes, therefore, pressures on the order of GPa are required to obtain a marked changes in volume. Hence, 10^5 Pa is much smaller than P and can, therefore, be neglected. Eq. (2.52) thus simplifies to:

$$V(P,T) = V_o\exp[\alpha(T-298)-\beta P]. \quad (2.53)$$

Eq. (2.53) is a transcendent function, that can be approximated by the Mac Laurin's series according to

$$e^x = 1+x+\frac{x^2}{2!}+\ldots \quad (2.54)$$

In our case x is very small, and hence the series can be truncated after the second term. Eq. (2.53) then becomes:

$$V(P,T) = V_o[1+\alpha(T-298)-\beta P]. \quad (2.55)$$

Calculated example: According to the experimental data of Pavese et al. (2001), the thermal expansion, α, of grossular is 27.7 x 10^{-6} K^{-1}. The compressibility coefficient, β, is calculated to be 5.9 x 10^{-3} GPa^{-1}, and the standard volume, V_o = 125.424 $cm^3 mole^{-1}$. With these data, we can calculate the molar volume of grossular at any arbitrary conditions occurring within the pressure and temperature range in which Pavese et al. (2001) performed their experiments. This constraint is a consequence of the simplification made in connection with the derivation of Eq. (2.55). It does not yield reliable results for P-T conditions lying appreciably outside the pressure and temperature range, in which the thermal expansion and compressibility

1. For the sake of convenience, the decimals are neglected

were determined. In order to compare directly the result of our calculation with an experimentally measured value, we choose $P = 2.1$ GPa and $T = 800$ K.

Inserting the values for volume, compressibility, and thermal expansion into Eq. (2.55), the calculation reads:

$$\begin{aligned} V^{grt}_{Ca_3Al_2Si_3O_{12}}(800, 2.1) &= 125.424\text{cm}^3\text{mol}^{-1}[1 + 27.7\times10^{-6}\text{K}^{-1}(800\text{K} \\ &\quad - 298\text{K}) - 5.9\times10^{-3}\text{GPa}^{-1} \times 2.1\text{GPa}] \\ &= \underline{125.614\text{cm}^3\text{mol}^{-1}}. \end{aligned}$$

Pavese et al. (2001) determined for the same *P-T* conditions a value of 125.677 $cm^3 mole^{-1}$. Although Eq. (2.55) gives a rough approximation, the difference between the calculated and measured values is small (0.05%). This is due to the fact that thermal expansion and compressibility have opposite signs.

Calculation of molar volume using x-ray diffraction data

Molar volumes of crystalline materials can be calculated using lattice constants as determined by x-ray diffraction experiments. The following relationship holds:

$$V = \frac{N_A}{z}\{\vec{a} \cdot [\vec{b} \times \vec{c}]\}, \tag{2.56}$$

where N_A designates Avogadro's constant (6.022×10^{23} $mole^{-1}$). z gives the number of formula units per unit cell and $\vec{a}$, $\vec{b}$, and $\vec{c}$ are the lattice vectors. The scalar triple product in braces represents the volume of the crystallographic unit cell. Hence, the molar volume of a crystalline substance is defined as the volume that is occupied by N_A formula units in the spatial arrangement as defined by the crystal structure. In the case of a triclinic substance Eq. (2.56) reads:

$$V = \frac{N_A}{z}\left\{abc[1 - (\cos\alpha)^2 - (\cos\beta)^2 - (\cos\gamma)^2 + 2\cos\alpha\cos\beta\cos\gamma]^{1/2}\right\}, \tag{2.57}$$

where α, β, and γ are the angles between the vectors $\vec{b}$ and $\vec{c}$, $\vec{a}$ and $\vec{c}$, $\vec{a}$ and $\vec{b}$, respectively.

Example: Yang et al. (1997) determined the following lattice constants for kyanite, Al_2SiO_5, at standard conditions: $a_o = 7.1200$ Å, $b_o = 7.8479$ Å, $c_o = 5.5738$ Å, $\alpha = 89.974°$, $\beta = 101.117°$, and $\gamma = 106.000°$. The unit cell of kyanite contains 4 formula units. Using these data, the molar volume of kyanite is calculated as follows:

$$V_{Al_2SiO_5}^{ky} = \frac{6.022\times10^{23}\,\text{mol}^{-1}}{4}\{7.1200\times10^{-8}\,\text{cm}$$

$$\times 7.8479\times10^{-8}\,\text{cm}\times 5.5738\times10^{-8}\,\text{cm}$$

$$\times sqrt[1-\cos^2(89.974°)-\cos^2(101.117°)-\cos^2(106.0°)$$

$$+2\cos(89.974°)\cos(101.117°)\cos(106.0°)]\}$$

$$= \underline{44.157\ \text{cm}^3\text{mol}^{-1}}.$$

For a monoclinic system, where only one angle (β) differs from 90°, Eq. (2.57) simplifies to:

$$V = \frac{N_A}{z}abc\sqrt{1-(\cos\beta)^2} = \frac{N_A}{z}abc\sin\beta \tag{2.58}$$

Example: Diopside, $CaMgSi_2O_6$, possesses a monoclinic structure, C2/c. Its lattice constants are: a = 9.7485 Å, b = 8.9252 Å, c = 5.2518 Å, and β = 105.899° (Tribaudino et al. 2000). Each unit cell contains 4 formula units of diopside. Its molar volume is calculated according to Eq. (2.58):

$$V_{CaMgSi_2O_6}^{cpx} = \frac{6.022\times10^{23}\,\text{mol}^{-1}}{4}[9.7485\times10^{-8}\,\text{cm}$$

$$\times 8.9252\times10^{-8}\,\text{cm}\times 5.2518\times10^{-8}\,\text{cm}\times\sin(105.899°)]$$

$$= \underline{66.161\ \text{cm}^3\text{mol}^{-1}}.$$

2.2 Volume of solutions

While the volume of a pure phase is completely defined by two variables P and T, additional variables are required to describe the volume of a solution. These additional variables are the numbers of moles of the components making up the solution. Accordingly, the equation of state for a volume of a solution reads:

$$V = f(P, T, n_1, n_2, \ldots), \tag{2.59}$$

where n_i is the number of moles of the components i.

At constant temperature and pressure the total volume of a solution, V, represents, in the simplest case, the sum of the molar volumes of all the components multiplied by their respective number of moles. For k components the equation reads:

$$V = \sum_1^k n_i V_i. \tag{2.60}$$

Eq. (2.60) holds for mechanical mixtures and ideal solutions. The precondition for its validity is that by the process of mixing no or only very little strain is introduced into the solution and that any chemical interaction effects between the components plays a minor role.

Example: Consider an olivine single crystal consisting of 92 mole percent forsterite and 8 mole percent fayalite. Its mass is 5 g. What is the volume of the crystal if an ideal mixing between forsterite and fayalite is assumed? To answer this question, the numbers of moles and the molar volumes of the two mixing components must be known. The molar masses of forsterite and fayalite are 140.6936 g and 203.7776 g, respectively. The molar volume of forsterite is 43.79 cm^3 mol^{-1} and that for fayalite 46.26 cm^{3}mole^{-1}. Using the molar masses, the number of moles of forsterite and fayalite in the olivine crystal can be calculated as follows:

$$n^{ol}_{Mg_2SiO_4} = \frac{5 \times 140.6936 \times 92}{100 \times 140.6936^2} = \frac{5 \times 92}{100 \times 140.6936} = \underline{0.0327 \text{ mol}}$$

$$n^{ol}_{Fa_2SiO_4} = \frac{5 \times 203.7776 \times 8}{100 \times 203.7776^2} = \frac{5 \times 8}{100 \times 203.7776} = \underline{0.0020 \text{ mol.}}$$

Inserting the calculated numbers of moles together with the molar volumes of forsterite and fayalite into Eq. (2.60) yields:

$$V = 0.0327\text{mol} \times 43.79\text{cm}^3\text{mol}^{-1} + 0.0020\text{mol} \times 46.26\text{cm}^3\text{mol}^{-1} = \underline{1.52 \text{ cm}^3}.$$

2.2.1 Partial molar volume

If the mixing of components introduces strain into a solid solution, for example, the total volume can not be represented simply by the weighted sum of the molar volumes of the end-member phases. In such cases, the molar volumes must be replaced by the *partial molar volumes*. The partial molar volume of the component i, V_i, is defined as the change in the volume of solution by adding or removing one mole of component i at constant temperature, pressure, and number of all other components j. Mathematically this definition is expressed as follows:

$$V_i = \left(\frac{\partial V}{\partial n_i}\right)_{P,\,T,\,n_{j\neq i}}. \tag{2.61}$$

The partial molar volume depends on the composition *and* on the nature of the solution.

Using the above definition (Eq. (2.61)), the total volume of a non-ideal solution containing k components reads:

$$V = \sum_1^k n_i V_i. \tag{2.62}$$

Example: The total volume of a mechanical mixture consisting of 0.02 mole of sphalerite, ZnS, and 0.01 mole of stoichiometric pyrrhotite, FeS, can be calculated as follows:

$$V = 0.02\,V_{ZnS}^{sph} + 0.01\,V_{FeS}^{po}.$$

V_{ZnS}^{sph} and V_{FeS}^{po} are the molar volumes of pure sphalerite and pure pyrrhotite, respectively. Sphalerite is cubic with the sphalerite-type structure, while pyrrhotite is hexagonal and posses the NiAs-type structure. With V_{ZnS}^{sph}= 23.830 cm^{3}mole^{-1} and V_{FeS}^{po}= 18.187 cm^{3}mole^{-1}, the total volume

$$V = 0.02\text{mol} \times 23.830\text{cm}^3\text{mol}^{-1} + 0.01\text{mol} \times 18.187\text{cm}^3\text{mol}^{-1} = \underline{0.658\text{ cm}^3}$$

is obtained.

If this mixture is annealed at 700°C for 2 days, a homogeneous (Zn,Fe)S single phase solid solution forms. In the solution zinc is partially replaced by ferrous iron, but the crystal structure of sphalerite persists. In this case, the total volume of the solution can not be calculated simply by multiplying the molar volumes of pure sphalerite and pyrrhotite with their respective numbers of moles and summing up the products. The partial molar volumes of the components must be taken instead and we write:

$$V = 0.02 \times V_{ZnS}^{sph} + 0.01 \times V_{FeS}^{sph}.$$

Entering the solid solution FeS formally changes its crystal structure from the NiAs type to the sphalerite type. This means its partial molar volume is based on a non-existing hypothetical cubic modification. Hence, in the case of FeS the difference between the partial molar volume and the molar volume is particularly manifest.

For the given composition Cemič (1983) determined the following partial molar volumes:

$$V_{ZnS}^{sph} = 24.094\ \text{cm}^3\text{mole}^{-1}$$

and

$$V_{FeS}^{sph} = 24.305\ \text{cm}^3\text{mole}^{-1}.$$

Using these values, the total volume of the solid solution is calculated as:

$$V = 0.02\text{mol} \times 24.094\text{cm}^3\text{mol}^{-1} + 0.01\text{mol} \times 24.305\text{cm}^3\text{mol}^{-1} = \underline{0.725\ \text{cm}^3}.$$

A comparison of the total volume of the solid solution with that one of the mechanical mixture gives a difference, ΔV, of 0.067 cm^3. Hence, the same mass of a homogeneous solution takes up a larger volume than does the corresponding mechanical mixture. The thermodynamic consequences of this fact will be discussed later.

The total volume is an extensive property that depends on the amount of material under consideration. The results of the above calculations, therefore, only make sense if the amount of the mechanical mixture or solution is known. If we want to make volume an intensive property, we must divide it by the total number of moles. The resulting property is referred to as the *molar volume of a solution,* $\overline{V}$, and in the case of ideal mixing it holds:

$$\overline{V}^{id} = \frac{V}{\sum_1^C n_i} = \sum_1^C \frac{n_i}{\sum_1^C n_i} V_i, \tag{2.63}$$

where n_i and C give the number of moles and components, respectively. V_i is the molar volume of the end-member phase i.

According to Eq. (1.25) the term $n_i / \left(\sum_1^C n_i\right)$ corresponds to the mole fraction of the component i, and so Eq. (2.63) can be rewritten as follows:

$$\bar{V}^{id} = \sum_1^C x_i V_i . \quad (2.64)$$

Similarly, the molar volume of a non-ideal solution reads:

$$\bar{V} = \sum_1^C x_i V_i , \quad (2.65)$$

with V_i being the partial molar volume of the *i*th-component.

Example: Using the data from the preceding example, the molar volume of a (Zn,Fe)S solid solution can be calculated according to Eq. (2.65) as follows:

$$\bar{V} = x_{ZnS}^{sph} V_{ZnS}^{\text{sph}} + x_{FeS}^{sph} V_{FeS}^{sph} .$$

The mole fraction of ZnS in the sphalerite solid solution is given by:

$$x_{ZnS}^{sph} = \frac{n_{ZnS}^{sph}}{n_{ZnS}^{sph} + n_{FeS}^{sph}} = \frac{0.02}{0.02 + 0.01} = 0.667$$

and that of FeS by:

$$x_{FeS}^{sph} = \frac{n_{FeS}^{sph}}{n_{ZnS}^{sph} + n_{FeS}^{sph}} = \frac{0.01}{0.02 + 0.01} = 0.333.$$

Using these mole fractions and the partial molar volumes of the components the molar volume of the sphalerite solid solution

$$\bar{V} = \frac{V}{0.3} = 0.667\text{mol} \times 24.094\text{cm}^3\text{mol}^{-1} + 0.333\text{mol} \times 24.305\text{cm}^3\text{mol}^{-1}$$

$$= \underline{24.164\ \text{cm}^3\text{mol}^{-1}}$$

is obtained.

2.2.2 Volume relationships in binary solutions

Because the sum of all mole fractions present in every phase equal unity, the molar volume of a binary solution *(A,B)* can be expressed as follows:

$$\bar{V} = (1 - x_B)V_A + x_B V_B \tag{2.66}$$

in the general case, and

$$\bar{V}^{id} = (1 - x_B)V_A + x_B V_B \tag{2.67}$$

in the case that the solution behaves ideal.

In Eqs. (2.66) and (2.67) x_B gives the mole fraction of the component B in the solution. V_i and V_i designate the molar volume and the partial molar volume of the component, respectively.

According to Eq. (2.61), the partial molar volumes of the components A and B in the solution *(A,B)* read:

$$V_A = \left(\frac{\partial V}{\partial n_A}\right)_{P,\,T,\,n_B} \text{ and } V_B = \left(\frac{\partial V}{\partial n_B}\right)_{P,\,T,\,n_A}. \tag{2.68}$$

Substituting the molar volume into Eq. (2.68),

$$V_A = \left\{\frac{\partial}{\partial n_A}[\bar{V}(n_A + n_B)]\right\}_{P,\,T,\,n_B} \text{ and } V_B = \left\{\frac{\partial}{\partial n_B}[\bar{V}(n_A + n_B)]\right\}_{P,\,T,\,n_A} \tag{2.69}$$

is obtained.

From Eq. (2.69) it follows further that:

$$V_A = \bar{V} + n_A\left(\frac{\partial \bar{V}}{\partial n_A}\right)_{P,\,T,\,n_B} + n_B\left(\frac{\partial \bar{V}}{\partial n_A}\right)_{P,\,T,\,n_B} \tag{2.70}$$

and

$$V_B = \bar{V} + n_A\left(\frac{\partial \bar{V}}{\partial n_B}\right)_{P,\,T,\,n_A} + n_B\left(\frac{\partial \bar{V}}{\partial n_B}\right)_{P,\,T,\,n_A}. \tag{2.71}$$

Considering the chain rule of differentiation, one can write

$$\left(\frac{\partial \bar{V}}{\partial n_A}\right)_{P,\,T,\,n_B} = \left(\frac{\partial \bar{V}}{\partial x_B}\right)\left(\frac{\partial x_B}{\partial n_A}\right)_{P,\,T,\,n_B}. \tag{2.72}$$

Using the relationship:

$$x_B = \frac{n_B}{n_A + n_B}, \tag{2.73}$$

one obtains:

$$\left(\frac{\partial x_B}{\partial n_A}\right)_{P,\,T,\,n_B} = -\frac{n_B}{(n_A + n_B)^2} \tag{2.74}$$

and

$$\left(\frac{\partial \bar{V}}{\partial n_A}\right)_{P,\,T,\,n_B} = -\frac{n_B}{(n_A + n_B)^2}\left(\frac{\partial \bar{V}}{\partial x_B}\right). \tag{2.75}$$

Replacing $\left(\frac{\partial \bar{V}}{\partial n_A}\right)_{P,\,T,\,n_B}$ in Eq. (2.70) by the expression (2.75) yields:

$$V_A = \bar{V} - x_B\left(\frac{\partial \bar{V}}{\partial x_B}\right)_{P,\,T}. \tag{2.76}$$

An analogous procedure can be applied to derive the partial molar volume of the component B. The result is given in Eq. (2.77).

$$V_B = \bar{V} - x_A\left(\frac{\partial \bar{V}}{\partial x_A}\right)_{P,\,T}. \tag{2.77}$$

If the mole fraction x_A in Eq. (2.77) is replaced by (1- x_B), the commonly used expression

$$V_B = \bar{V} + (1 - x_B)\left(\frac{\partial \bar{V}}{\partial x_B}\right)_{P,\,T} \tag{2.78}$$

is obtained.

The volume relationships for a binary solution are illustrated graphically in Fig. 2.8. The curved line represents the molar volume of the solution as a function of the composition. The points where this curve meets the ordinate at $x_B = 0$ and $x_B = 1$

correspond to the molar volumes of the end-member phases A and B, respectively.

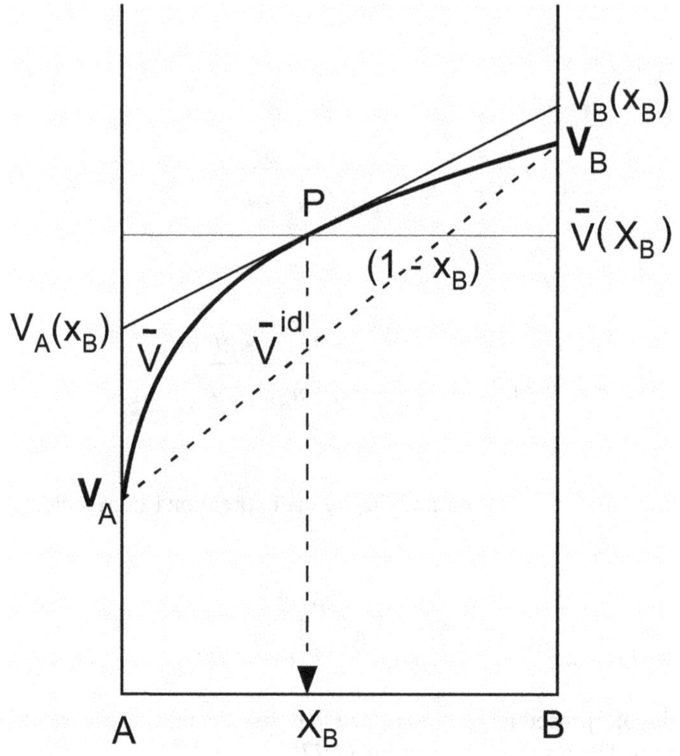

Fig. 2.8 Graphic representation of the volume relationships for a hypothetical binary system *A-B*: The solid and the dashed curve represent the molar volume of the solution *(A,B)* as a function of the mole fraction x_B in the case of non-ideal and ideal mixing of components A and B, respectively. V_A is the molar volume of the end-member phase A and V_B that of the end-member phase B. $V_A(x_B)$ and $V_B(x_B)$ are the partial molar volumes of the components and $\overline{V}(x_B)$ the molar volume of the solution for the composition x_B. For further explanation see text.

The slope of the curve at P can be expressed mathematically as follows:

$$\left(\frac{\partial \overline{V}}{\partial x_B}\right)_{P,T} = \frac{V_B - \overline{V}(x_B)}{(1 - x_B)} \tag{2.79}$$

or

$$\left(\frac{\partial \bar{V}}{\partial x_B}\right)_{P,T} = \frac{\bar{V} - V_A}{x_B}. \tag{2.80}$$

A simple rearrangement of Eq. (2.79) and Eq. (2.80) yields:

$$V_B = \bar{V} + (1 - x_B)\left(\frac{\partial \bar{V}}{\partial x_B}\right)_{P,T} \tag{2.81}$$

and

$$V_A = \bar{V} - x_B\left(\frac{\partial \bar{V}}{\partial x_B}\right)_{P,T}. \tag{2.82}$$

Equations (2.81) and (2.82) are identical to Eqs. (2.78) and (2.76), respectively.

As is apparent from Fig. 2.8, the partial molar volumes of the components A and B in the solution are given by the intersection points of the tangent to the $\bar{V}$ vs. x_B curve with the ordinate at $x_B = 0$ and $x_B = 1$, respectively. In other words, the partial molar volume of a component i can be defined as the volume that it would have if it would occur as an end-member phase, with the same interatomic distances as in the mixture. This means the volume state possessed by the component in the solution is extrapolated to the pure phase. The fact that the tangent defining the partial molar volumes of the components touches the $\bar{V}$ vs. x_B curve at x_B, demonstrates the validity of Eq. (2.65). It is obvious from Fig. 2.8 that the partial molar volume of a component depends on its mole fraction and the bulk composition of the system.

Special cases of partial molar volume arise at both ends of the concentration line, where x_B approaches 0 and 1, respectively. As $x_B \to 0$ the component B is infinitely diluted in A. Thus, its partial molar volume is referred to as the partial molar volume of component B at infinite dilution, V_B^{∞}. At the opposite end of the concentration line, $x_B \to 1$, the concentration of component A in the solution approaches zero. Correspondingly, its partial molar volume is designated as the partial molar volume of the component A at infinite dilution, V_A^{∞}.

2.2.3 Volume changes on mixing

In pure mechanical mixtures all components preserve their molar volumes. This means that the molar volume of the mixture is the sum of the molar volumes of the components. Thus, Eq. (2.64) holds. In solid solutions, however, the mixing of components occurs on the molecular or atomic scale and, thereby, the components lose partly their chemical and physical identities. The molar volume of a solid solution, therefore, often deviates from the linear sum of the molar volumes of the pure components. Hence, if a difference between the molar volume of the solid solution and the linear sum of the molar volumes of the pure components is observed, it is the result of the mixing process. This difference is referred to as the *volume of mixing* or *excess volume*, $\Delta_m \overline{V}^{ex}$. Using the same notification as in Fig. 2.8, the volume of mixing reads:

$$\Delta_m \overline{V}^{ex} = \overline{V} - \overline{V}^{id}. \tag{2.83}$$

The volume of mixing can be positive as well as negative. The behavior depends primarily on the size difference between the substituting atoms and on their bonding character.

The difference between the partial molar volume of a component i in a mixture and its molar volume as a pure phase is defined as the *partial molar volume of mixing* or *partial molar excess volume*, V_i^{ex}. Using this definition, the partial excess volume of component A, can be expressed as follows:

$$V_A^{ex} = V_A - V_A. \tag{2.84}$$

In the case of infinite dilution of A, where x_B approaches unity, Eq. (2.84) reads:

$$V_A^{ex\infty} = V_A^{\infty} - V_A. \tag{2.85}$$

$V_A^{ex\infty}$ is called the *partial molar volume of mixing* or *the partial molar excess volume* of component A *at infinite dilution.*

Analogous expressions can be written to describe the volume relationships for the component B as:

$$V_B^{ex} = V_B - V_B \tag{2.86}$$

and

$$V_B^{ex\infty} = V_B^{\infty} - V_B \tag{2.87}$$

giving the partial volume of mixing for component *B* in general and at infinite dilution of *B*, respectively.

As shown in Fig. 2.9, the volume of mixing represents a weighted sum of the partial molar excess volumes of the components constituting the solution. Thus, it reads:

$$\Delta_m \overline{V}^{ex} = (1 - x_B) V_A^{ex} + x_B V_B^{ex}. \tag{2.88}$$

By definition, the molar volume of mixing at both ends of the compositional line is zero. Partial molar volumes of mixing can be calculated using expressions analogous to those given in Eqs. (2.81) and (2.82). Hence, for the partial molar volume of mixing of component *A* holds:

$$V_A^{ex} = \Delta_m \overline{V}^{ex} - x_B \left[\frac{\partial \Delta_m \overline{V}^{ex}}{\partial x_B} \right]_{P,T}. \tag{2.89}$$

Correspondingly, the partial molar volume of mixing of component *B* can be written:

$$V_B^{ex} = \Delta_m \overline{V}^{ex} + (1 - x_B) \left[\frac{\partial \Delta_m \overline{V}^{ex}}{\partial x_B} \right]_{P,T}. \tag{2.90}$$

Fig. 2.9 shows volumes of mixing and their relationships to one another in a hypothetical binary system *A-B*.

For practical reasons, the molar volume of a binary mixture is often presented in a form of the polynomial with the mole fraction x_B as the independent variable:

$$\overline{V} = a + bx_B + cx_B^2 + dx_B^3 + \ldots \tag{2.91}$$

The coefficients *a, b, c, d,...* are purely empirical and have no physical meaning. They are obtained by fitting an appropriate curve to the measured data set. Thompson (1967) showed that a link between the coefficients and the volumes of mixing exists.

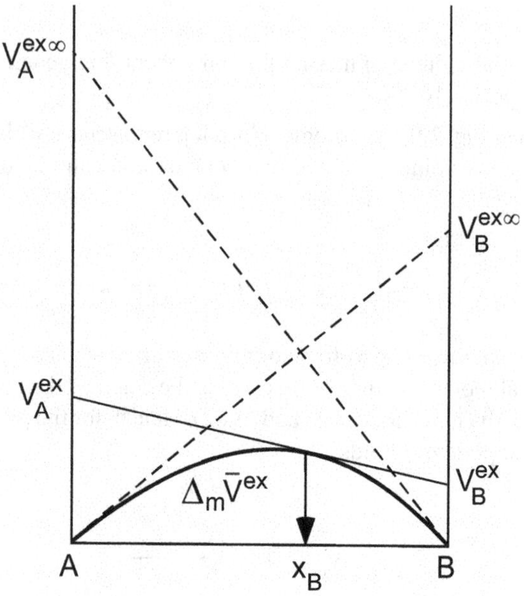

Fig. 2.9 Volumes of mixing. $\Delta_m \bar{V}^{ex}$ = volume of mixing, V_A^{ex} = partial molar volume of mixing of the component A, V_B^{ex} = partial molar volume of mixing of the component B, $V_A^{ex\infty}$ = partial molar volume of mixing of the component A at infinite dilution, $V_B^{ex\infty}$ = partial molar volume of mixing of component B at infinite dilution.

Assume that a second-order polynomial of the form

$$\bar{V} = a + bx_B + cx_B^2 \tag{2.92}$$

fits the experimentally determined volume data.

At the point where x_B equals zero it holds that:

$$\bar{V} = a = V_A. \tag{2.93}$$

At the point where x_B equals one, the molar volume takes the following value:

$$\bar{V} = a + b + c = V_B . \tag{2.94}$$

Following Eq. (2.77), the expression:

$$V_A = a - cx_B^2 \tag{2.95}$$

is obtained for the partial molar volume of the component A.

In the case of infinite dilution of the component A, where x_B is 1, V_A equals V_A^{∞}, and Eq. (2.95) yields:

$$V_A^{\infty} = a - c. \tag{2.96}$$

Substituting Eq. (2.93) into Eq. (2.96) one obtains:

$$c = -(V_A^{\infty} - V_A) = -V_A^{ex\infty}. \tag{2.97}$$

The coefficient c in Eq. (2.91) is therefore, equal to the negative partial molar volume of mixing of the component A at infinite dilution.

The partial molar volume of component B can be calculated analogously using Eq. (2.78), as:

$$V_B = a + b + 2cx_B - cx_B^2. \tag{2.98}$$

If B is infinitely diluted in A, x_B approaches zero and Eq. (2.98) reads:

$$V_B^{\infty} = a + b. \tag{2.99}$$

According to Eq. (2.93), the coefficient a in Eq. (2.92) is equivalent to the molar volume of the pure component A, such that:

$$b = V_B^{\infty} - V_A. \tag{2.100}$$

On the other hand, if Eq. (2.99) is substituted into Eq. (2.94), a value for the coefficient c is obtained again, namely:

$$c = -(V_B^{\infty} - V_B) = -V_B^{\mathrm{ex}\infty}. \tag{2.101}$$

A comparison of Eqs. (2.97) and (2.101) shows that the coefficient c in Eq. (2.91) is equivalent to the partial molar volumes of mixing of both components, A and B, at infinite dilution. In order to understand the implications of this result on the mixing behavior of the components, the volume of mixing has to be expressed in terms of the above discussed volume relationships, namely:

$$\Delta_m \bar{V}^{ex} = [V_A + (V_B^{\infty} - V_A)x_B - V_B^{ex\infty} x_B^2] - [V_A + (V_B - V_A)x_B]. \qquad (2.102)$$

The expression in the first square bracket represents the molar volume of the solution. The second square bracket describes the molar volume of the solution in the case of ideal mixing. Simple rearrangement and factoring out of the term $V_B^{ex\infty}$ yields:

$$\Delta_m \bar{V}^{ex} = x_B(1 - x_B)V_B^{ex\infty}. \qquad (2.103)$$

In Eq. (2.103) $V_B^{ex\infty}$ can as well be replaced by $V_A^{ex\infty}$, because both have the same value.

According to Eq. (2.103) the curve representing the volume of mixing is symmetrical around $x_B = 0.5$. Mixtures that exhibit this kind of behavior are called *symmetric*.

In the case that a cubic term is added to the polynomial in order to describe the behavior of the molar volume as a function of composition, the expression for the volume of mixing reads:

$$\Delta_m \bar{V}^{ex} = x_B^2(1 - x_B)V_A^{ex\infty} + x_B(1 - x_B)^2 V_B^{ex\infty}. \qquad (2.104)$$

Here $V_A^{ex\infty}$ differs from $V_B^{ex\infty}$. Hence, the volume vs. composition function is asymmetric. Mixtures of this type are called *asymmetric*. It can easily be shown that Eq. (2.104) equals Eq. (2.103), if the partial molar volumes of mixing at infinite dilution of both components are the same. In the mineralogical literature $V_i^{ex\infty}$ is often replaced by the symbol W_{ij}^V. It is then referred to as the *volume interaction parameter* between the components i and j.

2.2.4 Examples for the volumes of binary solutions

Example 1: Geiger and Feenstra (1997) measured the unit-cell constants of almandine-pyrope solid solutions and determined their molar volumes as a function of composition. Their results are summarized in Tab. 2.3.

Table 2.3 Compositions, unit-cell constants and the molar volume of almandine-pyrope garnets (Geiger and Feenstra 1997)

$x^{grt}_{Fe_3Al_2Si_3O_{12}}$	a_0[Å]	$\bar{V}$[cm^{3}mol^{-1}]
0.000*	11.5291*	115.358*
0.071	11.5227	115.166
0.154	11.5170	114.995
0.259	11.5105	114.800
0.393	11.4995	114.472
0.493	11.4925	114.263
0.614	11.4830	113.979
0.752	11.4737	113.703
0.910	11.4612	113.332
1.000	11.4555	113.163

* uncertainties given by the authors are neglected.

Fig. 2.10 shows the molar volumes as a function of composition. The solid line is calculated according to Eq. (2.67) as follows:

$$\bar{V}[\mathrm{cm^3mol^{-1}}] = 115.358(1 - x^{grt}_{Mg_3Al_2Si_3O_{12}}) + 113.163x^{grt}_{Mg_3Al_2Si_3O_{12}}.$$

No systematic deviation from linearity is observed. Hence, the mixing of Mg and Fe is ideal, at least, within the limits of the measuring accuracy.

In this example the molar volumes of the pure components are taken to calculate the line. One could, of course, just as well fit a line to the experimental data and extract the molar volumes of the two components from the resulting polynomial. Generally, the second approach is adopted because it is unreasonable to assume that the molar volumes of the pure components can be determined better than the molar volumes of the mixture. In our example, a linear regression yields:

$$\bar{V}[\text{cm}^3\text{mol}^{-1}] = 115.340 - 2.193 x^{grt}_{Mg_3Al_2Si_3O_{12}}. \tag{2.105}$$

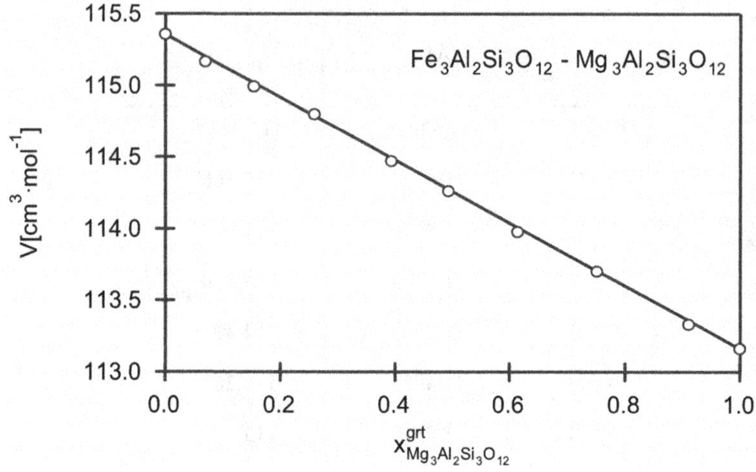

Fig. 2.10 Molar volumes for the solid solution $Fe_3Al_2Si_3O_{12}$ - $Mg_3Al_2Si_3O_{12}$ as a function of the mole fraction $x^{grt}_{Mg_3Al_2Si_3O_{12}}$. The line is calculated using Eq. (2.67) which is valid in case of ideal mixing behavior of the components.

The first term Eq. (2.105) corresponds to the molar volume of pure almandine and the second one gives the difference between the molar volumes of almandine and pyrope. This can be shown by a rearrangement of Eq. (2.67), namely:

$$\bar{V} = V_A + (V_B - V_A) x_B. \tag{2.106}$$

In our case, the following values are obtained:

$$V^{grt}_{Fe_3Al_2Si_3O_{12}} = 115.340 \text{ cm}^3\text{mol}^{-1}$$

and

$$V^{grt}_{Mg_3Al_2Si_3O_{12}} = 113.147 \text{ cm}^3\text{mol}^{-1}.$$

It should be emphasized that the experimental errors were disregarded in these calculations. If they were taken into account, the linear regression would have yielded slightly different coefficients and as a result, slightly different molar volumes for pyrope and almandine would have been obtained.

Example 2: Tab. 2.4 contains the molar volumes of alkali feldspar solid solution series as determined by Hovis (1988).

Table 2.4 Unit-cell parameters of alkali feldspar solid solution series (Hovis 1988)

$x^{fsp}_{KAlSi_3O_8}$	$\overline{V}^{fsp}$[cm^3 mol^{-1}]
0.0029	100.488
0.0099	100.526
0.1437	102.091
0.1510	102.200
0.1553	102.217
0.2753	103.602
0.2898	103.89
0.3508	104.464
0.4346	105.238
0.4406	105.259
0.4911	105.710
0.5557	106.196
0.5993	106.476
0.7083	107.271
0.7333	107.384
0.8074	107.974
0.8359	108.121
0.9600	108.895
0.9917	108.953

If the data of Tab. 2.4 are fitted with a second-order polynomial of the form $\bar{V} = a + bx_B + cx_B^2$, the following coefficients are obtained: a = 100.42, b = 12.747, and c = - 4.1868. Hence, the molar volume of high albite-sanidine solid solutions can be given as follows:

$$\bar{V}^{fsp}[\text{cm}^3\text{mol}^{-1}] = 100.42 + 12.747 x_{KAlSi_3O_8}^{fsp} - 4.1868 (x_{KAlSi_3O_8}^{fsp})^2 .$$

This polynomial can be used to calculate the molar volume for any high albite-sanidine solid solution. For example, at $x_{KAlSi_3O_8}^{fsp} = 0.4$ the calculation reads:

$$\bar{V}^{fsp} = 100.42 + 12.747 \times 0.4 - 4.1868 \times 0.4^2 = \underline{104.85 \text{ cm}^3\text{mol}^{-1}} .$$

According to Eq. (2.93), the coefficient a in the volume polynomial corresponds to the molar volume of the component A. In our case this is the molar volume of high albite and we can write:

$$V_{NaAlSi_3O_8}^{fsp} = 100.42 \text{ cm}^3\text{mol}^{-1} .$$

Correspondingly, the sum $a + b + c$ yields the molar volume of component B. Here it is the molar volume of pure sanidine, namely

$$V_{KAlSi_3O_8}^{fsp} = 100.42 + 12.747 - 4.1868 = \underline{108.98 \text{ cm}^3\text{mol}^{-1}} .$$

These two values can be used to calculate the molar volume of any high albite - sanidine solid solution in the case that the solution behaves ideally. According to Eq. (2.67), the molar volume of an ideal solution is given by a linear sum of the molar volumes of the components. At $x_{KAlSi_3O_8}^{fsp} = 0.4$, for example, one obtains:

$$\bar{V}^{fsp,\, id} = (1 - 0.4) \times 100.42 + 0.4 \times 108.98 = \underline{103.84 \text{ cm}^3\text{mol}^{-1}} .$$

The partial molar volume of high albite is calculated using Eq. (2.76) as:

$$V_{NaAlSi_3O_8}^{fsp} = \bar{V}^{fsp} - x_{KAlSi_3O_8}^{fsp} \left(\frac{\partial \bar{V}^{fsp}}{\partial x_{KAlSi_3O_8}^{\text{fsp}}} \right)_{p,\, T} .$$

If the volume polynomial, derived by least square fit, is introduced into the equation, given above, the following expression is obtained:

$$\begin{aligned} V^{fsp}_{NaAlSi_3O_8}[\text{cm}^3\text{mol}^{-1}] &= 100.42 + 12.747x^{fsp}_{KAlSi_3O_8} - 4.1868(x^{fsp}_{KAlSi_3O_8})^2 \\ &\quad - x^{fsp}_{KAlSi_3O_8}[12.747 - 8.3736(x^{fsp}_{KAlSi_3O_8})] \\ &= 100.42 + 4.1868(x^{\text{fsp}}_{KAlSi_3O_8})^2 . \end{aligned}$$

Hence, at $x^{fsp}_{KAlSi_3O_8} = 0.4$ the partial molar volume of high albite is:

$$V^{fsp}_{NaAlSi_3O_8} = 100.42 + 4.1868 \times 0.4^2 = \underline{101.09 \text{ cm}^3\text{mol}^{-1}}.$$

The partial molar volume of sanidine can be calculated according to Eq. (2.78), that is

$$V^{fsp}_{KAlSi_3O_8} = \bar{V}^{fsp} + (1 - x^{fsp}_{KAlSi_3O_8})\left(\frac{\partial \bar{V}^{fsp}}{\partial x^{fsp}_{KAlSi_3O_8}}\right)_{p,T} .$$

Substituting the volume polynomial for $\bar{V}^{\text{fsp}}$, one obtains:

$$\begin{aligned} V^{fsp}_{KAlSi_3O_8}[\text{cm}^3\text{mol}^{-1}] &= 100.42 + 12.747x^{fsp}_{KAlSi_3O_8} - 4.1868(x^{fsp}_{KAlSi_3O_8})^2 \\ &\quad + (1 - x^{fsp}_{KAlSi_3O_8})[12.747 - 8.3736(x^{fsp}_{KAlSi_3O_8})] \\ &= 100.42 + 12.747 - 8.3736x^{fsp}_{KAlSi_3O_8} + 4.1868(x^{fsp}_{KAlSi_3O_8})^2 . \end{aligned}$$

With $x^{fsp}_{KAlSi_3O_8} = 0.4$, the above expression assumes the following form:

$$\begin{aligned} V^{fsp}_{KAlSi_3O_8} &= 100.42 + 12.747 - 8.3736 \times 0.4 + 4.1868 \times 0.4^2 \\ &= \underline{110.49 \text{ cm}^3\text{mol}^{-1}}. \end{aligned}$$

The molar volume of the high albite-sanidine solid solution at $x^{fsp}_{KAlSi_3O_8} = 0.4$ can now be calculated directly, using the partial molar volumes of the components. According to Eq. (2.67), the molar volume of the solution at any arbitrary composition is given by the linear sum of the partial molar volumes of the components, calculated for the given composition. In our case, we have:

$$\overline{V}^{fsp} = 0.6 \times 101.09 + 0.4 \times 110.49 = \underline{104.85 \text{ cm}^3\text{mol}^{-1}}.$$

The result is, of course, identical with the one, that was obtained previously using the volume polynomial.

The volumes calculated in the above example and the experimental data of Orville (1967) are shown graphically in Fig. 2.11.

The difference between the molar volume of the solution, determined in the experiment, and its molar volume calculated in the case of ideal mixing, yields the molar volume of mixing. For $x^{fsp}_{KAlSi_3O_8} = 0.4$, the molar volume is calculated as follows:

$$\begin{aligned}\Delta_m \overline{V}^{fsp,\, ex} = \overline{V}^{fsp} - \overline{V}^{fsp,\, id} &= 104.85\text{cm}^3\text{mol}^{-1} - 103.84\text{cm}^3\text{mol}^{-1} \\ &= \underline{1.01 \text{ cm}^3\text{mol}^{-1}}.\end{aligned}$$

Following Eq. (2.84), the partial molar volume of mixing or partial molar excess volume of high albite equals the difference between its molar volume and its partial molar volume in the mixture. At $x^{fsp}_{KAlSi_3O_8} = 0.4$ we have:

$$V^{fsp,\, ex}_{NaAlSi_3O_8} = 101.09\text{cm}^3\text{mol}^{-1} - 100.42\text{cm}^3\text{mol}^{-1} = \underline{0.67 \text{ cm}^3\text{mol}^{-1}}.$$

An analogous relationship holds for the partial volume of mixing of sanidine (see Eq. (2.86)). At $x^{fsp}_{KAlSi_3O_8} = 0.4$ one obtains:

$$V^{fsp,\, ex}_{KAlSi_3O_8} = 110.49\text{cm}^3\text{mol}^{-1} - 108.98\text{cm}^3\text{mol}^{-1} = \underline{1.51 \text{ cm}^3\text{mol}^{-1}}.$$

The molar volume of mixing at $x^{fsp}_{KAlSi_3O_8} = 0.4$ can now be calculated using the partial molar volumes of mixing of the two components. It is

$$\Delta_m \overline{V}^{fsp,\, ex} = 0.6 \times 0.67\text{cm}^3\text{mol}^{-1} + 0.4 \times 1.51\text{cm}^3\text{mol}^{-1} = \underline{1.01 \text{ cm}^3\text{mol}^{-1}}.$$

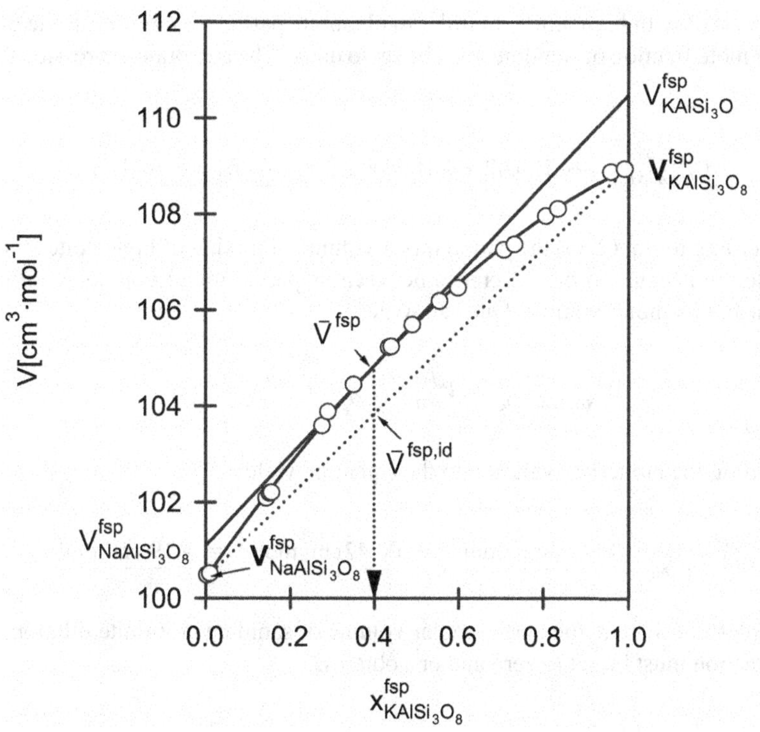

Fig. 2.11 Volume relationships in the binary system high albite - sanidine. Circles represent the experimental data given by Hovis (1988). $\overline{V}^{fsp}$ and $\overline{V}^{fsp,\,id}$ are the molar volumes of sanidine - high albite solid solution at $x^{fsp}_{KAlSi_3O_8} = 0.4$ in the case of non-ideal and ideal mixing, respectively. $V^{fsp}_{NaAlSi_3O_8}$ and $V^{fsp}_{KAlSi_3O_8}$ designate the molar volumes of the end-member phases and $V^{fsp}_{NaAlSi_3O_8}$ and $V^{fsp}_{KAlSi_3O_8}$ are the partial molar volumes of the components at $x^{fsp}_{KAlSi_3O_8} = 0.4$.

If we want to use Eq. (2.103) to calculate the molar volume of mixing, the partial molar volumes of mixing of the components at infinite dilution must be known. These values correspond to the volume interaction parameter, W^V, in Eq. (2.103). As a matter of fact, the partial molar volume of mixing of only one component is actually required in our case, because a second-order polynomial was fit to the experimental data. As demonstrated earlier in this chapter, the partial molar volumes of mixing of the two components are equal in such case. Nontheless, we will calculate the partial molar volumes of mixing of both components.

Let us start with high albite. In order to obtain its partial volume at infinite dilution the mole fraction of sanidine must be set to unity. The algebraic expression then reads:

$$V_{NaAlSi_3O_8}^{fsp,\,\infty} = 100.42 + 4.1868 \times 1.0 = \underline{104.61\ \mathrm{cm}^3\mathrm{mol}^{-1}}.$$

According to Eq. (2.85) the partial molar volume of mixing of high albite at infinite dilution is equal to the difference between its partial molar volume at infinite dilution and its molar volume. One can write:

$$V_{NaAlSi_3O_8}^{fsp,\,ex\infty} = V_{NaAlSi_3O_8}^{fsp,\,\infty} - V_{NaAlSi_3O_8}^{fsp}.$$

Inserting the numerical values into the equation, yields:

$$V_{NaAlSi_3O_8}^{fsp,\,ex\infty} = 104.61\mathrm{cm}^3\mathrm{mol}^{-1} - 100.42\mathrm{cm}^3\mathrm{mol}^{-1} = \underline{4.19\ \mathrm{cm}^3\mathrm{mol}^{-1}}.$$

In order to calculate the partial molar volume of sanidine at infinite dilution, its mole fraction must be set to zero and one obtains:

$$V_{KAlSi_3O_8}^{fsp,\,\infty} = 100.42 + 12.75 = \underline{113.17\ \mathrm{cm}^3\mathrm{mol}^{-1}}.$$

Following Eq. (2.87), the partial molar volume of mixing of sanidine at infinite dilution is defined as follows:

$$V_{KAlSi_3O_8}^{fsp,\,ex\infty} = V_{KAlSi_3O_8}^{fsp,\,\infty} - V_{KAlSi_3O_8}^{fsp}.$$

Substituting the numerical values for $V_{KAlSi_3O_8}^{fsp,\,\infty}$ and $V_{KAlSi_3O_8}^{fsp}$, the partial molar volume of mixing of sanidine at infinite dilution is:

$$V_{KAlSi_3O_8}^{fsp,\,ex\infty} = 113.17\mathrm{cm}^3\mathrm{mol}^{-1} - 108.98\mathrm{cm}^3\mathrm{mol}^{-1} = \underline{4.19\ \mathrm{cm}^3\mathrm{mol}^{-1}}.$$

A comparison of the result with the one obtained for high albite shows that they are equivalent. One can now use this value to calculate the volume of mixing as a function of composition. The mathematical expression is that from Eq. (2.103) and it reads:

$$\Delta_m \overline{V}^{fsp,\,ex}[\text{cm}^3\text{mol}^{-1}] = (1 - x^{fsp}_{KAlSi_3O_8})(x^{fsp}_{KAlSi_3O_8}) \times 4.19.$$

This expression can be used to derive the partial molar volumes of mixing of the components. The partial molar volume of mixing of high albite is obtained using Eq. (2.89), namely:

$$\begin{aligned} V^{fsp,\,ex}_{NaAlSi_3O_8}[\text{cm}^3\text{mol}^{-1}] &= (1 - x^{fsp}_{KAlSi_3O_8})(x^{fsp}_{KAlSi_3O_8}) \times 4.19 \\ &+ \left[(x^{fsp}_{KAlSi_3O_8})^2 - (1 - x^{fsp}_{KAlSi_3O_8})(x^{fsp}_{KAlSi_3O_8})\right] \times 4.19 \\ &= (x^{fsp}_{KAlSi_3O_8})^2 \times 4.19. \end{aligned}$$

The partial molar volume of mixing of sanidine can be calculated according to Eq. (2.90).

$$\begin{aligned} V^{fsp,\,ex}_{KAlSi_3O_8}[\text{cm}^3\text{mol}^{-1}] &= (1 - x^{fsp}_{KAlSi_3O_8})(x^{fsp}_{KAlSi_3O_8}) \times 4.19 \\ &+ \left[(1 - x^{fsp}_{KAlSi_3O_8})^2 - (1 - x^{fsp}_{KAlSi_3O_8})(x^{fsp}_{KAlSi_3O_8})\right] \times 4.19 \\ &= (1 - x^{fsp}_{KAlSi_3O_8})^2 \times 4.19 \end{aligned}$$

One can test the validity for the two expressions by inserting 0.4 for the mole fraction of sanidine and calculate the partial molar volumes of mixing. For high albite one obtains:

$$V^{fsp,\,ex}_{NaAlSi_3O_8} = 0.4^2 \times 4.19 = \underline{0.67\ \text{cm}^3\text{mol}^{-1}}$$

and for sanidine

$$V^{fsp,\,ex}_{KAlSi_3O_8} = (1 - 0.4)^2 \times 4.19 = 1.51\ \text{cm}^3\text{mol}^{-1}.$$

The volumes of mixing, as calculated above, are shown graphically in Fig. 2.12.

Example 3: Bosenick and Geiger (1997) studied the molar volume of pyrope-grossular solid solutions at 295 K. According to them, the molar volume of mixing can be represented by an asymmetric mixture model (see Eq. (2.104)). Their partial molar volumes of mixing at infinite dilution, termed as volume interaction parameters, W_i^V are 1.84 cm^{3}mole^{-1} and 0.12 cm^{3}mole^{-1} for pyrope and grossular, respectively. Using these values, the mathematical expression for the integral molar volume of mixing reads:

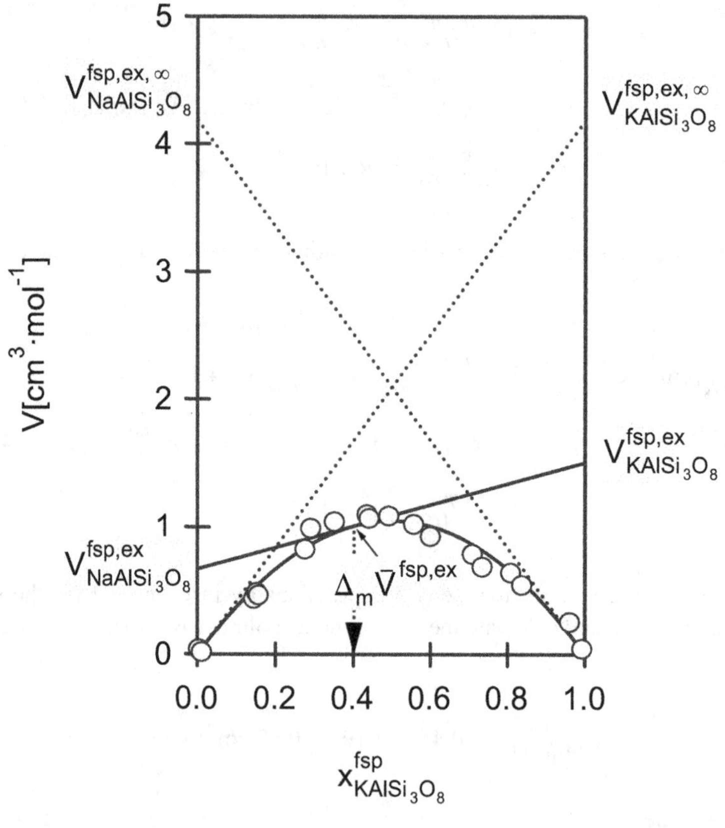

Fig. 2.12 Volumes of mixing in the system high albite-sanidine. $\Delta_m \overline{V}^{fsp,\, ex}$ = molar volume of mixing, $V_{NaAlSi_3O_8}^{fsp,\, ex}$ = partial molar volume of mixing of high albite, $V_{NaAlSi_3O_8}^{fsp,\, \infty}$ = partial molar volume of mixing of sanidine. $V_{NaAlSi_3O_8}^{fsp,\, ex\infty}$ = partial molar volume of mixing of high albite at infinite dilution, $V_{KAlSi_3O_8}^{fsp,\, ex\infty}$ = partial molar volume of mixing of sanidine at infinite dilution. The arrow indicates the mole fraction of sanidine in the solution. (Data from Hovis 1988).

$$\Delta_m \overline{V}^{grt,\,ex}[\mathrm{cm}^3\mathrm{mol}^{-1}] = (1 - x^{grt}_{Ca_3Al_2Si_3O_{12}})(x^{grt}_{Ca_3Al_2Si_3O_{12}})^2 \times 1.84 + (1 - x^{grt}_{Ca_3Al_2Si_3O_{12}})^2 (x^{grt}_{Ca_3Al_2Si_3O_{12}}) \times 0.12. \quad (2.107)$$

Eq. (2.107) is used to calculate the molar volume of mixing as a function of the mole fraction of grossular. The result of the calculation is presented graphically in Fig. 2.13.

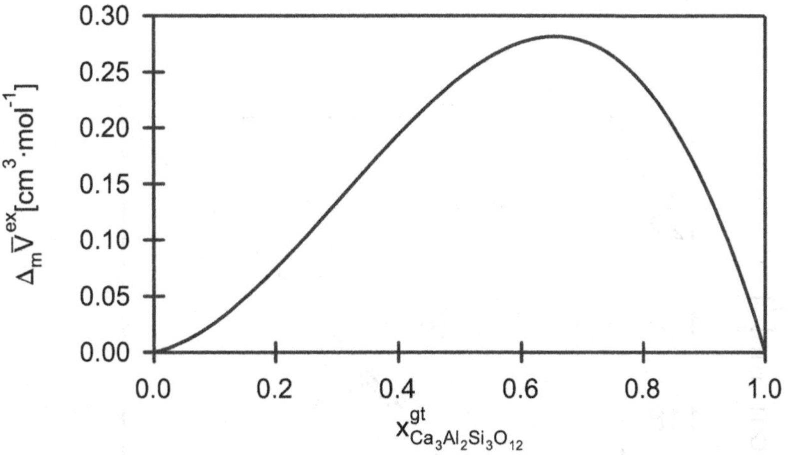

Fig. 2.13 Molar volume of mixing along the join pyrope-grossular calculated using the experimental data of Bosenick and Geiger (1997).

As can be seen from Fig. 2.13, the molar volume of mixing deviates positively from the ideality and the volume vs. composition curve is asymmetric. The greatest volumes of mixing are exhibited by grossular-rich compositions.

The volume of mixing function can also be used to derive a polynomial for the molar volume of the solutions, provided that the molar volumes of the end-member phases are known. This is, of course, the case. The molar volumes of pyrope and grossular have been measured a number of times. We will take the values given by Bosenick and Geiger (1997), where the molar volumes are 113.157 cm^{3}mole^{-1} and 125.293 cm^{3}mole^{-1} for pyrope and grossular, respectively. The uncertainties are not considered.

Following Eq. (2.83), the molar volume for the case of ideal mixing, $\overline{V}^{id}$, has to be added to the molar volume of mixing, $\Delta_m \overline{V}^{ex}$, in order to obtain the molar volume, $\overline{V}$. In our example, the calculation is as follows:

$$\begin{aligned}\overline{V}^{grt}[\text{cm}^3\text{mol}^{-1}] = {} & (1 - x^{grt}_{Ca_3Al_2Si_3O_{12}})(x^{grt}_{Ca_3Al_2Si_3O_{12}})^2 \times 1.84 \\ & + (1 - x^{grt}_{Ca_3Al_2Si_3O_{12}})^2(x^{grt}_{Ca_3Al_2Si_3O_{12}}) \times 0.12 \\ & + 113.157(1 - x^{grt}_{Ca_3Al_2Si_3O_{12}}) + 125.293(x^{grt}_{Ca_3Al_2Si_3O_{12}}).\end{aligned}$$

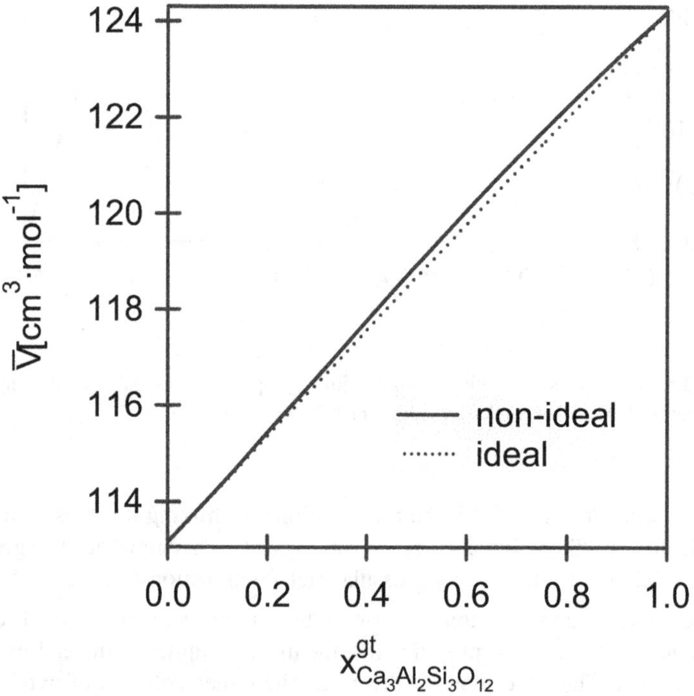

Fig. 2.14 Molar volume of pyrope-grossular solid solutions as a function of mole fraction of grossular. The curves were calculated using the experimental data of Bosenick and Geiger (1997).

Multiplying, factoring out, and rearranging yields:

$$\bar{V}^{grt}[\mathrm{cm^3\,mol^{-1}}] = 113.157 + 11.158(x^{grt}_{Ca_3Al_2Si_3O_{12}}) + 1.6(x^{grt}_{Ca_3Al_2Si_3O_{12}})^2 - 1.72(x^{grt}_{Ca_3Al_2Si_3O_{12}})^3 .$$

Fig. 2.14 shows the molar volumes along the pyrope-grossular join. The solid curve was calculated using the third-order polynomial derived above and it represents the molar volume as a function of the composition. The dotted line gives the molar volume for the case of ideal mixing.

2.2.5 Volume of reaction

Consider the general reaction:

$$\nu_a A + \nu_b B = \nu_c C + \nu_d D \tag{2.108}$$

occurring at constant temperature T and constant pressure P. A, B, C and D represent the chemical symbols of reaction constituents and ν_a, ν_b, ν_c, and ν_d are whole-number stoichiometric coefficients. They are negative for components on the left of the equation and positive for those on the right. A stoichiometric process can thus be represented by the equation

$$\sum \nu_i M_i = 0, \tag{2.109}$$

where M_i designates the chemical formula of the i-th reaction constituent.

The change in total volume of a reacting system is given by the following expression:

$$dV = -\left(\frac{\partial V}{\partial n_A}\right)_{P,\,T,\,n_{j\neq A}} dn_A - \left(\frac{\partial V}{\partial n_B}\right)_{P,\,T,\,n_{j\neq B}} dn_B + \left(\frac{\partial V}{\partial n_C}\right)_{P,\,T,\,n_{j\neq C}} dn_C + \left(\frac{\partial V}{\partial n_D}\right)_{P,\,T,\,n_{j\neq D}} \tag{2.110}$$

If the reactants and the products, both occur as mechanical mixture of pure phases, the partial derivatives of the total volume in Eq. (2.110) represent the molar volumes of the reaction constituents, that is

$$\left(\frac{\partial V}{\partial n_i}\right)_{P,\,T,\,n_{j\neq i}} = V_i . \tag{2.111}$$

Thus, Eq. (2.110) can be rewritten as:

$$dV = \sum_A^D V_i dn_i. \qquad (2.112)$$

In Eqs. (2.110) and (2.112) dn_i are related to each other by the stoichiometry of the reaction. It is convenient, therefore, to introduce a new variable termed the *extent of reaction*, ξ. This variable indicates the degree of advancement of the system from an initial unstable or metastable state towards a stable state. The changes in the variable ξ and in the individual number of moles, dn_i, are related as follows:

$$dn_i = \nu_i d\xi. \qquad (2.113)$$

Using this relationship Eq. (2.112) reads:

$$dV = \sum_A^D \nu_i V_i d\xi \qquad (2.114)$$

or

$$\left(\frac{\partial V}{\partial \xi}\right)_{P,T} = \Delta_r V = \sum_A^D \nu_i V_i. \qquad (2.115)$$

$\Delta_r V$ is termed the *volume of reaction*. It indicates the change in the volume of the reacting system per unit reaction progress variable, ξ.

Example: Consider the solid state reaction

$$Ca_3Al_2Si_3O_{12} + SiO_2 = CaAlSi_3O_8 + 2CaSiO_3$$

taking place at 298 K and 0.1 MPa. The stoichiometric coefficients, ν_i, are: -1, -1, 1 and 2. According to Eq. (2.115) the volume of reaction is calculated as follows:

$$\Delta_r V = -V^{grt}_{Ca_3Al_2Si_3O_{12}} - V^{qtz}_{SiO_2} + V^{fsp}_{CaAl_2Si_2O_8} + 2V^{wo}_{CaSiO_3}.$$

Using the molar volumes given by Robie and Hemingway (1995), the volume of reaction reads:

$$\Delta_r V = -1\,\text{mol} \times 125.28\,\text{cm}^3\text{mol}^{-1} - 1\,\text{mol} \times 22.69\,\text{cm}^3\text{mol}^{-1}$$
$$+ 1\,\text{mol} \times 100.79\,\text{cm}^3\text{mol}^{-1} + 2\,\text{mol} \times 39.90\,\text{cm}^3\text{mol}^{-1} = \underline{32.62\ \text{cm}^3}$$

2.3 Problems

1. At 800°C and 0.5 MPa, 1 mol of an ideal gas has a volume of 17.826 dm^3.
- Calculate the change in the volume of the gas when it is compressed isothermally to 1 MPa.

2. The critical data of a gas are: P_c = 22.06 MPa, T_c = 647.14 K and V_c = 56.0×10^{-6} m^{3}mol^{-1} (Ambrose, 1994).
- Calculate the values of the parameters a and b of the gas assuming van der Waals behavior of the gas. Why do the results depend upon the choice of the formulas adopted?

3. Calculate the volume of a non-ideal gas at 400°C and 40 MPa using the van der Waals equation. The values of the parameters a and b are 0.1765 J m^{3}mol^{-2} and 33×10^{-6} m^{3}mol^{-1}, respectively.
   In order to solve the cubic equation, use the Newton's approximation method according to the formula:

$$x_1 = x_o - \frac{f(x)}{f'(x)},$$

   where x_0 and x_1 are the initial and the succeeding values, respectively. Repeat the calculation until the difference between the two subsequent values becomes negligibly small. As an initial value, use the ideal volume of the gas at the P, T conditions given above.

4. Rhodonite, $MnSiO_3$, has a triclinic structure and has the following lattice constants:

$$a_o = 7.616\ \text{Å},$$
$$b_o = 11.851\ \text{Å},$$
$$c_o = 6.707\ \text{Å},$$
$$\alpha = 92°33',$$
$$\beta = 94°21',$$
$$\gamma = 105°40.2'.$$

Each unit cell contains 10 formula units of rhodonite. $N_A = 6.022 \times 10^{23}$; 1Å = 10^{-10} m.

- Calculate the volume that is occupied by 1 formula unit of rhodonite. Express the volume in [Å^3].
- Calculate the molar volume of rhodonite and give it in units of [$m^3 mol^{-1}$] and [$cm^3 mol^{-1}$].

5. A pyrope single crystal has a mass of 4 g. At standard P,T conditions its density, ρ, is 3.582 $g cm^{-3}$. The value of its bulk modulus, B, is 1.73×10^5 MPa, and that of its thermal expansion, α, is 2.47×10^{-5} K^{-1}.

- Calculate the volume of the crystal at 1000°C and 0.8 GPa, assuming a constant thermal expansion and compressibility.

6. At standard P-T conditions, andalusite, Al_2SiO_5, has a molar volume of 51.52 $cm^3 mol^{-1}$. Its thermal expansion, α, is 2.47×10^{-5} K^{-1} and the compressibility coefficient, β, is 5.43×10^{-12} Pa^{-1}. Under the same P,T conditions, sillimanite has a molar volume of 49.86 $cm^3 mol^{-1}$, a thermal expansion coefficient of 1.44×10^{-5} K^{-1} and the compressibility of 6.22×10^{-12} Pa^{-1} (Holland and Powell, 1990).

- Calculate the change in volume associated with the phase transition andalusite → sillimanite at 985 K and 0.2 GPa. Assume that the thermal expansion and the compressibility are pressure and temperature independent.

7. Newton et al. (1977) measured the lattice constants for a series of Diopside-Ca-Tschermak solid solutions. Using their data, a polynomial describing the molar volume as a function of composition can be derived. It reads:

$$V[\mathrm{cm^3 mol^{-1}}] = 66.043 - 2.9603 x_{CaTs}^{cpx} + 0.52016 (x_{CaTs}^{cpx})^2, \qquad (2.116)$$

where x_{CaTs}^{cpx} is the mole fraction of $CaAlAlSiO_2$ in clinopyroxene.

- Calculate the molar volumes of the pure end-member components.
- Calculate the partial molar volumes of diopside and Ca-Tschermak at x_{CaTs}^{cpx} = 0.4 and at infinite dilution of the components.
- Calculate the partial molar excess volumes of the components at $x_{CaTs}^{cpx} = 0.4$.
- Calculate the partial molar excess volumes of the components at infinite dilution.
- Express the molar excess volume of the solution as a function of composition

using the volume interaction parameter, W^V.

8. Calculate the change in the volume associated with the following chemical reaction:

$$5MgSiO_3 + 5Al_2O_3 \rightarrow Mg_2Al_3AlSi_5O_{18} + 3MgAl_2O_4.$$

$$V^{en}_{MgSiO_3} = 31.31 \text{ cm}^3\text{mol}^{-1},$$

$$V^{cor}_{Al_2O_3} = 25.58 \text{ cm}^3\text{mol}^{-1},$$

$$V^{cord}_{Mg_2Al_3AlSi_5O_{18}} = 233.22 \text{ cm}^3\text{mol}^{-1},$$

$$V^{sp}_{MgAl_2O_4} = 39.71 \text{ cm}^3\text{mol}^{-1},$$

(Robie and Hemingway 1995).

Chapter 3 The first law of thermodynamics

According to the law on the conservation of energy the sum of kinetic and potential energy of a system is constant. This, however, is true only for frictionless systems. If friction is present in the system kinetic energy decreases and heat is produced. There exists a relationship between the dynamic energy dissipated and the heat produced. This relationship serves as the basis for the development of thermodynamics.

3.1 The relationship between heat and work

The basic statement concerning the relationship between heat and work reads: *heat, Q, and work, W, are equivalent*:

$$Q \leftrightarrow W. \tag{3.1}$$

That is, heat and work are just two different forms of energy. For example, the temperature of a crystal can be increased by rubbing it adiabatically, which means without any heat exchange with the surroundings. The same temperature increase can be achieved if the crystal is brought into contact with a "hot" body, that is, by providing the necessary heat directly. The proportionality factor that links work and the resultant heat is termed the mechanical *equivalent of heat*. Its present value is 0.2389 calories. A *calorie* is defined as the quantity of heat that is required to increase the temperature of 1 gram water from 14.5 to 15.5°C. This so-called *thermodynamical calorie* was widely used in thermochemistry until 1960, when S-I units were introduced. Its value now corresponds to 4.184 Joule.

3.2 Internal Energy

Based on the equivalence of heat and work, the statement for the first law of thermodynamics reads:

If a closed system, in the course of a thermodynamic process, moves from state A into state B, the sum of the absorbed heat and performed work equals the change in a function of state, termed the internal energy, U.

Thus the total change in internal energy, ΔU, for the system is:

$$\Delta U = U_B - U_A = Q + W. \tag{3.2}$$

For an infinitesimal change of state Eq. (3.2) can be written as a differential

$$dU = \delta Q + \delta W. \tag{3.3}$$

Because the internal energy, U, is a state function, dU is a total differential. This means, the integration between two states does not depend on the path taken by the system. The quantities δQ and δW are not total differentials and their integrals are path dependent. The symbol 'δ' indicates the differential increments of the quantities Q and W, which are not properties of state. Therefore, δQ and δW cannot be integrated without a knowledge of the path taken by the system during the course of a thermodynamic process.

In a cyclic processes, where the system returns to its initial state, the change in the internal energy is zero:

$$\oint dU = 0. \tag{3.4}$$

The relationship given in Eq. (3.4) is representative for all functions of state.

3.2.1 Work

The term δW in Eq. (3.3) indicates a differential element of work, but it does not say anything about the type. It can be mechanical work, electrical work, or work performed in a gravitational field, etc. Normally, work due to changes in volume and electrical work play the most important role in most thermodynamic processes.

Consider a gas-filled cylinder containing a frictionless moveable piston (Fig. 3.1) with weights sitting on top of it. At equilibrium, the external pressure, P_{ext}, exerted by the piston plus the mass of the weights is balanced by the internal pressure, P_{int}, of the gas i.e. $P_{ext} = P_{int}$. If the external pressure is reduced, for example, by removing one weight, the gas expands until the external pressure and the internal gas pressure are once again equal. If the difference between the gas pressure and external pressure exerted by the piston and the weights is infinitesimally small, the distance the piston travels after the weight reduction will be infinitesimally small too. In the case that the temperature is held constant and friction is negligible, the work done by the gas during this expansion can be expressed as:

$$dW = -(P_{ext} \times A)dx. \tag{3.5}$$

In Eq. (3.5) the letter A defines the area of the piston. Hence, the product within the parenthesis gives the force and dx the distance the piston travels during the expansion. The negative sign is required, because the force exerted by the gas is directed against that exerted by the piston and weights. Taking into account that

$$A dx = dV, \tag{3.6}$$

Eq. (3.5) reads:

$$dW = -P_{ext} dV. \tag{3.7}$$

dW is the infinitesimal increment of work.

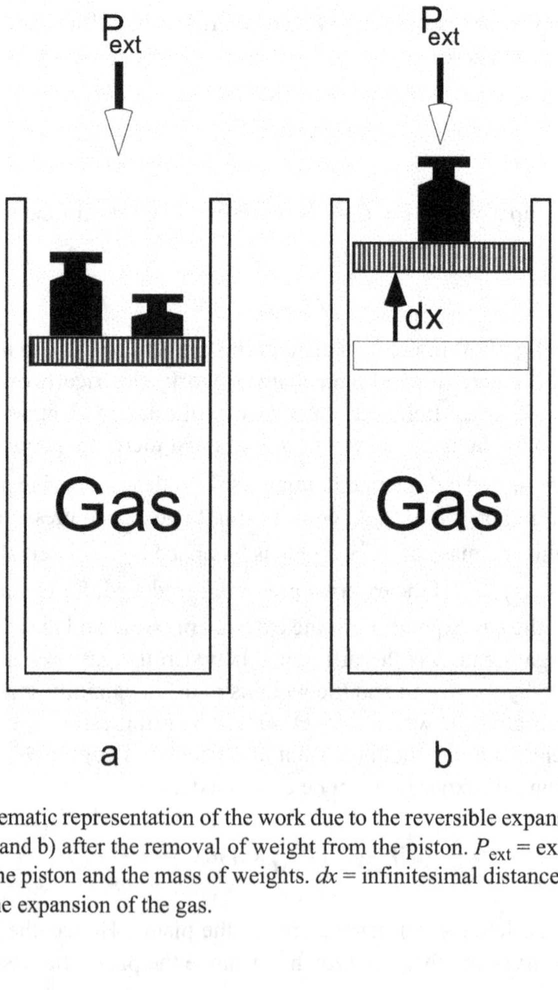

Fig. 3.1 Schematic representation of the work due to the reversible expansion of a gas. a) state before and b) after the removal of weight from the piston. P_{ext} = external pressure exerted by the piston and the mass of weights. dx = infinitesimal distance the piston travels during the expansion of the gas.

For an infinite number of infinitesimally small incremental steps from an initial volume, V_i, to an end volume, V_e, the work done by the gas is

$$W = -\int_{V_i}^{V_e} P dV. \tag{3.8}$$

In the case of an *isothermal* change in volume of an *ideal gas,* one can substitute RT/V into in Eq. (3.8) and obtains:

$$W = -RT\int_{V_i}^{V_e} \frac{dV}{V} = -RT\int_{V_i}^{V_e} d\ln V = -RT\ln\frac{V_e}{V_i}. \tag{3.9}$$

Note that the sign convention is obeyed in Eq. (3.9) because the gas expands, and the final volume, V_e, is larger than the starting volume, V_i. Thus, the logarithm is positive and the expression remains negative. This complies with the standard convention that the work and the energy released by a system are negative.

In order to calculate the work done on a solid through isothermal compression, the pressure, P, in Eq. (3.8) has to eliminated. This can be done using Eq. (2.4). A small rearrangement of this equation yields:

$$dP = -\frac{dV}{\beta V}, \tag{3.10}$$

where V in the denominator designates the reference volume. As stated earlier (see section 2.1.1, page 25), for solids the volume at ambient pressure and temperature, V_0, can be taken as the reference without introducing significant error into the calculation. If, in addition, the compressibility coefficient, β, is assumed to be constant, Eq. (3.10) can be integrated from P_0 to P and from V_0 to V_P:

$$P - P_o = -\frac{V_P - V_o}{\beta V_o}. \tag{3.11}$$

P_0 is the ambient pressure and is small compared to the pressure P. It can, therefore, be neglected. Substituting P in Eq. (3.8) by the expression given in Eq. (3.11) yields:

$$W = \int_{V_o}^{V_P} \left(\frac{V_P - V_o}{\beta V_o} \right) dV = \frac{(V_P - V_o)^2}{2\beta V_o}. \tag{3.12}$$

Another approach for calculating the volume work done on or by a solid is to replace dV in Eq. (3.8) using again the definition of the compressibility coefficient. The result reads:

$$dV = -\beta V_0 dP. \tag{3.13}$$

After substitution of Eq. (3.13) into Eq. (3.8), P is the only remaining variable. In order to obtain the work done by the solid, the integration has to be carried out between the limits P_0 and P:

$$W = \int_{P_o}^{P} \beta V_0 P dP. \tag{3.14}$$

If P_0 is again neglected and the compressibility coefficient, β, is considered to be constant, the integration of Eq. (3.14) yields:

$$W = \frac{V_0 P^2 \beta}{2}. \tag{3.15}$$

Calculated example: Consider a pyrope crystal with a mass of 40.32 g. Its molar volume at room temperature is 113.176 $cm^3 mole^{-1}$ (Robie and Hemingway 1995). The average compressibility coefficient, β, in the pressure range between 0.1 MPa and 5.0 GPa is 6.19 x 10^{-3} GPa^{-1}(Hazen and Finger 1978). What is the work performed on the crystal due to an isothermal (T = 298 K) pressure increase from 0.1 MPa to 5.0 GPa?

The molar volume of pyrope at 298 K and 5.0 GPa can be calculated using Eq. (2.52). It is

$$V_P = V_o[1 - \beta P] = 113.176\ \text{cm}^3\text{mol}^{-1}[1 - 6.19\times10^{-12}\text{Pa}^{-1} \times 5.0\times10^9\text{Pa}] = \underline{109.673\ \text{cm}^3\text{mol}^{-1}}.$$

The molar mass, M, of pyrope is 403.1508 g. For a mass of 40.32 g, this corresponds to

$$n = \frac{m}{M} = \frac{40.32 \text{ g}}{403.1508 \text{ gmol}^{-1}} = \underline{0.1 \text{mol}}.$$

In order to obtain the volumes required for the calculation of the work performed during the course of compression, the molar volume of pyrope at 0.1 MPa and 5.0 GPa, respectively, are to be multiplied by the number of moles, $n = 0.1$. The following values are obtained:

$$V_o = 11.3176 \text{ cm}^3$$

and

$$V_P = 10.9727 \text{ cm}^3.$$

Inserting these values into Eq. (3.12), the calculation reads:

$$W = \frac{[(11.3176 - 10.9727)\times 10^{-6} \text{m}^3]^2}{2 \times 6.19\times 10^{-12} \text{Pa}^{-1} \times 10.9727\times 10^{-6} \text{m}^3} = \underline{876 \text{ J}}.$$

Using Eq. (3.15) one obtains:

$$W = \frac{10.9727\times 10^{-6} \text{m}^3 \times (5.0\times 10^9 \text{Pa})^2 \times 6.19\times 10^{-12} \text{Pa}^{-1}}{2} = \underline{849 \text{ J}}.$$

The two results differ by 27 J. This difference is due to the fact that the formulas used in the calculations represent only approximations, since the compressibility coefficients are assumed to be pressure independent.

3.2.2 Enthalpy

If the incremental volume work, pdV, is substituted for δA in Eq. (3.3), the total differential of the internal energy reads:

$$dU = \delta Q - PdV. \tag{3.16}$$

In the case of an isobaric volume change, the integration of Eq. (3.16) between states I and II yields:

$$U_{II} - U_I = \Delta Q - P(V_{II} - V_I) \tag{3.17}$$

or rearranged:

$$\Delta Q = (U_{II} + PV_{II}) - (U_I + PV_I) \equiv H_{II} - H_I. \tag{3.18}$$

The term H designates a state function called the *enthalpy*. This function was derived assuming that the work exchange between the system and the surroundings occurs at constant pressure. Under this assumption, the change in the enthalpy, ΔH, equals the heat, ΔQ, absorbed or released by the system. Thus, it holds that

$$(\Delta H)_P = \Delta Q \tag{3.19}$$

or in differential form

$$(dH)_P = dQ. \tag{3.20}$$

The property, given in Eq. (3.20), makes the enthalpy particularly useful for the description of isobaric processes.

According to Eq. (3.18), the definition of enthalpy reads:

$$H = U + PV. \tag{3.21}$$

Its total differential is:

$$dH = dU + PdV + VdP. \tag{3.22}$$

Substituting the expression (3.16) for dU in Eq. (3.22), yields:

$$dH = \delta Q + VdP. \tag{3.23}$$

3.3 Application of the first law of thermodynamics to pure phases

As stated in Chap. 1, functions of state of pure phases are determined completely by two intensive variables. In the case of the internal energy, temperature and volume are normally chosen. The internal energy, U, is thus given by:

$$U = f(T, V). \tag{3.24}$$

Temperature and pressure are the standard variables that are chosen to define the enthalpy, H, of pure phases. Using them one obtains:

$$H = f(T, P). \tag{3.25}$$

The total differentials of the internal energy and enthalpy read:

$$dU = \left(\frac{\partial U}{\partial T}\right)_V dT + \left(\frac{\partial U}{\partial V}\right)_T dV \tag{3.26}$$

and

$$dH = \left(\frac{\partial H}{\partial T}\right)_P dT + \left(\frac{\partial H}{\partial P}\right)_T dP. \tag{3.27}$$

The combination of Eq (3.26) with Eq. (3.16) yields:

$$\delta Q - PdV = \left(\frac{\partial U}{\partial T}\right)_V dT + \left(\frac{\partial U}{\partial V}\right)_T dV \tag{3.28}$$

or rearranged

$$dQ = \left(\frac{\partial U}{\partial T}\right)_V dT + \left[\left(\frac{\partial U}{\partial V}\right)_T + P\right] dV. \tag{3.29}$$

Note, that under the specific set of variables given in Eq. (3.29), dQ is an exact differential.

For isochoric processes, where dV equals 0, one obtains:

$$dQ = \left(\frac{\partial U}{\partial T}\right)_V dT. \tag{3.30}$$

Dividing of Eq. (3.30) by dT yields:

$$\left(\frac{dQ}{dT}\right)_V = \left(\frac{\partial U}{\partial T}\right)_V = c_V. \tag{3.31}$$

In Eq. (3.31) c_V designates the *heat capacity* of a single phase or of a given system *at constant volume*. It represents the heat required to increase the temperature of the system by one degree, if the volume is held constant.

An expression similar to that in Eq. (3.28) is obtained, if Eq. (3.23) is combined with Eq. (3.27), namely:

$$\delta Q + VdP = \left(\frac{\partial H}{\partial T}\right)_P dT + \left(\frac{\partial H}{\partial P}\right)_T dP. \tag{3.32}$$

A rearrangement of Eq. (3.32) gives:

$$dQ = \left(\frac{\partial H}{\partial T}\right)_P dT + \left[\left(\frac{\partial H}{\partial P}\right)_T - V\right] dP. \tag{3.33}$$

For isobaric processes, where *dP* equals 0, Eq. (3.33) simplifies to

$$dQ = \left(\frac{\partial H}{\partial T}\right)_P dT. \tag{3.34}$$

After dividing of Eq. (3.34) by *dT*, the *heat capacity* of a single phase or of a system *at constant pressure* is obtained, that is

$$\left(\frac{dQ}{dT}\right)_P = \left(\frac{\partial H}{\partial T}\right)_P = c_p. \tag{3.35}$$

According to Eq. (3.31), c_v corresponds to the change in the internal energy per temperature unit in the case that the volume is held constant. Correspondingly, c_p represents the change in the enthalpy per temperature unit at constant pressure. Both c_v and c_p are extensive properties. Dividing either quantity by the number of moles yields the molar heat capacity which is an intensive property. It is given as:

$$\frac{c_v}{n} = C_v \quad \text{and} \quad \frac{c_p}{n} = C_p. \tag{3.36}$$

C_v and C_p are referred to as the *molar heat capacity at constant volume* and the *molar heat capacity at constant pressure*, respectively.

3.3.1 Heat capacities C_v and C_p

For practical reasons, heat capacity is measured almost exclusively at constant pressure. However, the heat capacity at constant volume is occasionally required, as for example for lattice dynamics calculations. The difference between their values is typically small, and is given by:

$$C_p - C_v = \frac{TV\alpha^2}{\beta}. \tag{3.37}$$

The symbols in Eq. (3.37) have their usual meaning. T designates the temperature in K, V gives the molar volume, α is the thermal expansion coefficient, and β the compressibility coefficient.

Example 1: If the heat capacity at constant volume of forsterite is to be calculated using the heat capacity at constant pressure, the following data are required according to Eq. (3.37): the molar volume of forsterite, its heat capacity at constant pressure, its thermal expansion coefficient, α, and its compressibility coefficient, β. All these quantities are temperature and pressure dependent. Their values must, therefore, correspond to the conditions at which the heat capacity at constant volume is to be calculated. For a pressure and temperature of 0.1 MPa and 298 K, respectively, Holland and Powell (1990) give the following values:

$$V^{ol}_{Mg_2SiO_4} = 43.66\ \mathrm{cm^3 mol^{-1}},$$

$$C^{ol}_{p,\,Mg_2SiO_4} = 118.70\ \mathrm{Jmol^{-1}K^{-1}},$$

$$\alpha = 3.66\times10^{-5}\ \mathrm{K^{-1}} \quad \text{and}$$

$$\beta = 7.33\times10^{-12}\ \mathrm{Pa^{-1}}.$$

Inserting these values into Eq. (3.37) yields:

$$(C_p - C_v)^{ol}_{Mg_2SiO_4} = \frac{298\mathrm{K} \times 43.66\times10^{-6}\mathrm{m^3 mol^{-1}} \times (3.66\times10^{-5}\mathrm{K^{-1}})^2}{7.33\times10^{-12}\mathrm{Pa^{-1}}}$$

$$= \underline{2.38\ \mathrm{Jmol^{-1}K^{-1}}}.$$

Hence, the heat capacity at constant volume for ambient pressure and temperature conditions equals:

$$C^{ol}_{v,\,Mg_2SiO_4} = 118.70\ \mathrm{Jmol^{-1}K^{-1}} - 2.38\ \mathrm{Jmol^{-1}K^{-1}} = \underline{116.32\ \mathrm{J{\cdot}mol^{-1}K^{-1}}}.$$

Example 2: Following Holland and Powell (1990), the values required to calculate the heat capacity of diopside, $CaMgSi_2O_6$, under the condition of constant volume using Eq. (3.37), are:

$$V^{cpx}_{CaMgSi_2O_6} = 66.19\ \mathrm{cm^3 mole^{-1}},$$

$$C^{cpx}_{p,\,CaMgSi_2O_6} = 166.63\ \mathrm{Jmole^{-1}K^{-1}},$$

$$\alpha = 3.32\times10^{-5}\mathrm{K^{-1}} \quad \text{and}$$

$$\beta = 8.31 \times 10^{-12} \text{Pa}^{-1}.$$

With these values the difference

$$(C_p - C_v)^{cpx}_{CaMgSi_2O_6} = \frac{298\text{K} \times 66.19 \times 10^{-6} \text{m}^3\text{mol} \times (3.32 \times 10^{-5} \text{K}^{-1})^2}{8.31 \times 10^{-12} \text{Pa}^{-1}}$$

$$= \underline{2.62 \text{ Jmol}^{-1}\text{K}^{-1}}.$$

is obtained.

Thus, C_v, for diopside is 1.6 % smaller than C_p. In the case of forsterite, the corresponding values differ by 2 %. These differences are typical for all solids at ambient conditions. Hence, the assumption that both capacities are equal, does not effect most thermodynamic calculations significantly. At high temperatures, however, the differences become larger. The calculation using Eq. (3.37) yields acceptable results only if the temperature dependencies of the molar volume, thermal expansion and compressibility are known. This is not always the case. Particularly, the temperature dependence of the compressibility is rarely available.

Comparably good results are obtained if the *Grüneisen constant*, γ, is introduced into the calculations because it is almost temperature independent. It relates the volume, compressibility, thermal expansion coefficient and heat capacity, C_v, as follows:

$$\gamma = \frac{\alpha V}{\beta C_v}. \tag{3.38}$$

A combination of Eqs. (3.37) and (3.38) yields:

$$C_v = \frac{C_p}{(1 + \alpha\gamma T)}. \tag{3.39}$$

Example: Consider forsterite, Mg_2SiO_4, once more. At 1000 K, its heat capacity, C_p, is 175.14 $\text{Jmole}^{-1}\text{K}^{-1}$ (Holland and Powell 1990). Using the elastic properties determined for forsterite by Suzuki et al. (1983), the Grüneisen constant is calculated to be 1.34.

In order to calculate the heat capacity at constant volume for 1000 K, the data, given above, are inserted into Eq. (3.39) and one obtains:

$$C^{ol}_{v,\,Mg_2SiO_4,1000} = \frac{175.14\ \mathrm{Jmol^{-1}K^{-1}}}{(1 + 3.66\times10^{-5}\mathrm{K^{-1}} \times 1.34 \times 1000\ \mathrm{K})} = \underline{166.95\ \mathrm{Jmol^{-1}K^{-1}}}.$$

Thus, at 1000 K, the values for the heat capacity of forsterite at constant pressure and at constant volume differ by ca. 5%.

If the heat capacity at constant volume is to be calculated using Eq. (3.37), all values should hold for 1000 K. The fulfillment of this requirement is less critical in the case of the thermal expansion and compressibility, because their values are very small. Disregarding their temperature dependence normally does not influence the result significantly. The molar volume for forsterite at 1000 K can, however, be calculated using Eq. (2.52). It is

$$\begin{aligned} V^{ol}_{Mg_2SiO_4,1000} &= V^{ol}_{Mg_2SiO_4,\,298}[1 + \alpha(T - 298)] \\ &= 43.66\mathrm{cm^3mol^{-1}}[1 + 3.66\times10^{-5}\mathrm{K^{-1}}(1000\mathrm{K} - 298\mathrm{K})] \\ &= \underline{44.78\ \mathrm{cm^3mol^{-1}}}. \end{aligned}$$

Inserting this value, together with the molar volume, thermal expansion, compressibility coefficient and the heat capacity, C_p, at 1000 K, into Eq. (3.37) the calculation reads:

$$\begin{aligned} (C_p - C_v)^{ol}_{Mg_2SiO_4,\,1000} &= \frac{1000\mathrm{K} \times 44.78\times10^{-6}\mathrm{cm^3mol^{-1}} \times (3.66\times10^{-5}\mathrm{K^{-1}})^2}{8.31\times10^{-12}\mathrm{Pa^{-1}}} \\ &= \underline{7.22\ \mathrm{Jmol^{-1}K^{-1}}}. \end{aligned}$$

Using this result, the heat capacity, $C^{ol}_{v,\,Mg_2SiO_4}$, is calculated as:

$$C^{ol}_{v,\,Mg_2SiO_4} = 175.14\ \mathrm{Jmol^{-1}K^{-1}} - 7.22\ \mathrm{Jmol^{-1}K} = \underline{167.92\ \mathrm{Jmol^{-1}K^{-1}}}.$$

This value, if compared with the one obtained using the Grüneisen constant, gives a difference of only 0.58%.

Using the elastic properties of forsterite, determined by Suzuki et al. (1983), a C_v of 166.61 $\mathrm{Jmol^{-1}K^{-1}}$ is obtained for 1000 K. Thus, the two values calculated applying Eqs. (3.37) and (3.39), are only ca. 1% too large. This is a very small difference especially if one considers that all values applied in the calculations have their own uncertainties.

Based on the experimental results, Dulong and Petit stated that, at moderate temperatures, the molar heat capacity of most solid elements has a value of 6.4 cal g-atom^{-1}K^{-1} (= 26.78 Jg-atom^{-1}K^{-1}). This statement, that is known as the *rule of Dulong-Petit,* corresponds to the prediction of the quantum theory of heat capacity at constant volume, which states that, at elevated temperatures, the heat capacity of the elements approaches the value of $3R$ (= 24.93 Jg-atom^{-1}K^{-1}). Comparing the two values one has to consider that $\boldsymbol{C_p}$ is slightly lager than $\boldsymbol{C_v}$. As an expansion of the Dulong-Petite rule, Neumann and Kopp stated that the molar heat of a solid compound is approximately the sum of the molar heat capacities of its constituents. Both, Dulong-Petite and Neumann-Kopp rule, have only restricted availability. Nonetheless, the heat capacity of a complex silicate can be calculated using the heat capacities of its constituent oxides.

Example 1: Consider beryl, $Be_3Al_2Si_6O_{18}$. Its oxide formula reads: $3BeO{\cdot}Al_2O_3{\cdot}6SiO_2$. Applying the Neumann-Kopp rule, its heat capacity for 298 K is calculated as follows:

$$C^{crd,\,calc}_{p,\,Mg_2Al_3[AlSi_5O_{18}]} = 3C^{ber}_{p,\,BeO} + C^{cor}_{p,\,Al_2O_3} + 6C^{qtz}_{p,\,SiO_2}.$$

Robie and Hemingway (1995) give 25.56 Jmole^{-1}K^{-1}, 79.10 Jmole^{-1}K^{-1} and 44.59 Jmole^{-1}K^{-1} for $C^{ber}_{p,\,BeO}$, $C^{cor}_{p,\,Al_2O_3}$ and $C^{qtz}_{p,\,SiO_2}$, respectively. Using these values the heat capacity of cordierite is calculated as follows:

$$C^{crd,\,calc}_{p,\,Mg_2Al_3[AlSi_5O_{18}]} = 3 \times 25.56\ \mathrm{Jmol^{-1}K^{-1}} + 79.10\ \mathrm{Jmol^{-1}K^{-1}} + 6 \times 44.59\ \mathrm{Jmol^{-1}K^{-1}} = \underline{423.32\ \mathrm{Jmol^{-1}K^{-1}}}.$$

The experimentally determined value for beryl is 417.00 Jmol^{-1}K^{-1} (Robie and Hemingway 1995). It is, thus, only 1.5% smaller than the value calculated above.

Example 2: In order to calculate the heat capacity of pyrope, $Mg_3Al_2Si_3O_{12}$, the value for periklas, MgO, is required in addition to the heat capacities of corundum and quartz. According to Robie and Hemingway (1995) its value is 37.26 Jmole^{-1}K^{-1}. Consequently, the calculation reads:

$$C^{grt,\,calc}_{p,\,Mg_3Al_2Si_3O_{12}} = 3 \times 37.26\ \mathrm{Jmol^{-1}K^{-1}} + 79.10\ \mathrm{Jmol^{-1}K^{-1}} + 3 \times 44.59\ \mathrm{Jmol^{-1}K^{-1}} = \underline{324.65\ \mathrm{Jmol^{-1}K^{-1}}}.$$

The obtained result is very close to the experimental value for pyrope (325.76 Jmol^{-1}K^{-1}, as published by Robie and Hemingway (1995)). The difference is smaller than 0.5%.

In Tab. 3.1 calculated heat capacities of a some of minerals are presented together with the corresponding experimental values. The last column contains the differences between the two values, given in percent.

Table 3.1 Molar heat capacities calculated using the Neumann-Kopp rule compared with the experimental values (Robie and Hemingway 1995).

Mineral	C_p^{calc} [$\mathrm{Jmol^{-1}K^{-1}}$]	C_p^{obs} [$\mathrm{Jmol^{-1}K^{-1}}$]	Δ%
Forsterite	118.61	119.11	-0.42
Enstatite	166.18	163.70	1.49
Diopside	166.78	168.51	-1.04
Anorthite	211.34	210.35	0.47
Grossular	330.22	339.08	-2.90
Spinel	115.94	116.36	-0.36
Muscovite	327.85	325.99	0.57
Tremolite	660.75	655.44	0.81
Anthophyllit	651.13	664.02	-1.94
Kaolinite	235.46	243.37	-3.25

As apparent from Tab. 3.1, the agreement between calculated and measured heat capacities is considerably good in the case of waterfree phases. The largest deviation is found for grossular. But the result can be improved using C_p of wollastonite, $CaSiO_3$, instead of lime, CaO, and low quartz, SiO_2 ($C_{p,calc} = 337.67\ \mathrm{Jmole^{-1}K^{-1}}$).

Considerably poorer results are obtained for phases with water as a constitutional part of the structure such as kaolinite or anthophyllite. In the past, attempts were made to improve the results by accounting for the hydrogen bonding in these phases. Significant improvements were achieved by the separation of the of H_2O contributions to C_p into 'structural' and 'zeolithic' (Berman and Brown 1985).

The temperature dependence of heat capacity

In the vicinity of 0 K, the internal energy, U, and enthalpy, H, of solids tend towards constant values. Because the heat capacities C_v and C_p represent the derivatives of the internal energy and enthalpy, respectively, their values approach 0. The mathematical expressions for this relationship read:

$$\lim_{T\to 0}\left(\frac{\partial U}{\partial T}\right)_V = \lim_{T\to 0} C_V(T) = 0 \tag{3.40}$$

and

$$\lim_{T\to 0}\left(\frac{\partial H}{\partial T}\right)_p = \lim_{T\to 0} C_p(T) = 0. \tag{3.41}$$

In addition, the difference between C_p and C_V decreases with falling temperatures and vanishes at 0 K. Different solids approach the limiting value in different ways. For example, the heat capacity of diamond is immeasurably small at temperatures below 50 K and remains far below the value of $3R$ at room temperature. On the other hand, lead and copper obey the Dulong-Petit's rule quite well, which means that their heat capacity values are close to $3R$ at room temperature.

The increase in the heat capacity of a solid with increasing temperature is due to the excitation of atomic vibrations. According to kinetic theory of classical mechanics, the average energy per vibrating atom is twice $1/2kT$ per degree of freedom, where k is Boltzmann's constant which is obtained by dividing the universal gas constant R, by the Avogadro number, N_A. A vibrating atom has both potential and kinetic energy. Their average contributions to the total energy are equal, namely $1/2kT$ for each. Thus, the value of $1/2kT$ has to be doubled. The potential energy is associated with stretching of electrostatic bonds between the vibrating atoms. Its maximum is reached at the extremes of the vibratory path, when the atom is momentarily at rest. In contrast, the maximum kinetic energy is reached, when the atom passes the vibrational midpoint. At this point it travels with a maximum velocity. The sum of kinetic and potential energy, however, is constant, regardless of the position of the atom on its vibratory path. The interaction between the constituent atoms of a solid is such that the number of vibrational modes per atom is always three. In total, the energy of a mole of an one-atomic solid is therefore $3 \times 2 \times 1/2kN_AT = 3RT$. Thus, the heat capacity at constant volume, C_V, should have a value of $3R$ multiplied with the number of atoms per formula unit and it should not change with temperature. However, this is not the case. As shown in Fig. 3.2, heat capacities are strongly temperature dependent and the value predicted by the theory from classical mechanics is approached only at high temperatures. Their true behavior can only be explained by quantum mechanical theory.

Einstein considered a crystal as an array of q atoms, each of which behaves as a *harmonic oscillator* vibrating independently about its defined lattice site. The behavior of the individual oscillators is not influenced by that of its neighbors and all oscillators vibrate with a single fixed frequency given as ν. An oscillator has three degrees of freedom, that is, it can vibrate in three directions in space and the total number of vibrational modes per mole of a solid is thus 3 qN_A.

According to *Planck*, the energy of an atomic of molecular system is quantized. The energy quantum, ε_i, that can be absorbed or released by the *i*-th harmonic oscillator is

$$\varepsilon_i = \left(i + \frac{1}{2}\right)h\nu, \tag{3.42}$$

where *i* is an integer designating the energy level, ranging from zero to infinity, and *h* is Planck's constant of action. ν is the frequency of the vibrating oscillator.

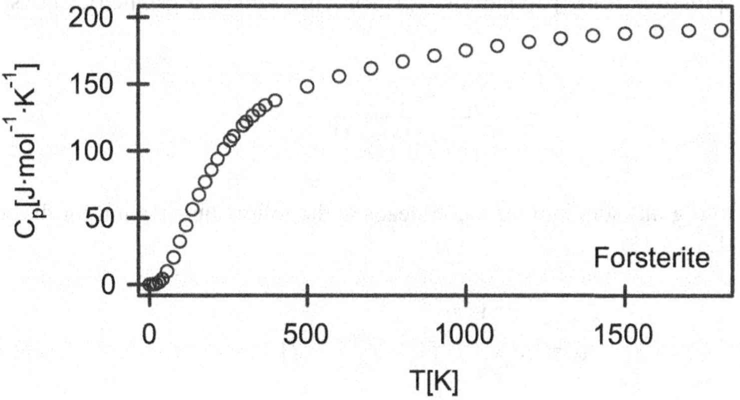

Fig. 3.2 Heat capacity of forsterite, Mg_2SiO_4, as a function of temperature (Robie et al. 1982).

In one mole of a *monatomic* crystal there are a total of N_A atoms and, hence, $3N_A$ oscillators. At any fixed temperature, the energies of N_A atoms are distributed over different levels, whereby n_o of them are in the zeroth energy level (ground state) having an energy of $3\varepsilon_o = 3 \times 1/2h\nu$, n_1 are in the first energy level and have an energy of $3\varepsilon_1 = 3(1 + 1/2)h\nu$, n_2 are in the second energy level and possess an energy of $3\varepsilon_2 = 3(2 + 1/2)h\nu$ and so on. In the *i*-th energy level, there are n_i atoms and they have an energy of $3\varepsilon_i = 3(i + 1/2)h\nu$.

The populations of different energy levels is governed by Boltzmann's statistics. Thus, the number of atoms occupying the *i*-th energy level equals

$$n_i = ae^{-\varepsilon_i/kT}, \tag{3.43}$$

where T is the temperature, a a proportionality factor and k Boltzmann's constant.

Because there are N_A atoms in one mole of a monatomic crystal, it holds that

$$N_A = (n_o + n_1 + n_2 + \ldots n_i). \tag{3.44}$$

Applying Eq. (3.43) to Eq. (3.44) yields:

$$N_A = ae^{-\frac{1}{2}\cdot\frac{h\nu}{kT}}\left(1 + e^{-\frac{h\nu}{kT}} + e^{-\frac{2h\nu}{kT}} + e^{-\frac{3h\nu}{kT}} + \ldots e^{-\frac{ih\nu}{kT}}\right). \tag{3.45}$$

The expression in the parenthesis represents the sum of a geometric series. Its value is

$$\frac{1}{1 - e^{-(h\nu/kT)}}.$$

Substituting this sum into Eq. (3.45) leads to the following equation for the constant a in Eq. (3.43):

$$a = N_A e^{\frac{1}{2}\cdot\frac{h\nu}{kT}}[1 - e^{-(h\nu/kT)}]. \tag{3.46}$$

Inserting Eq. (3.46) in Eq. (3.43) yields

$$n_i = N_A[1 - e^{-(h\nu/kT)}]\cdot e^{-i(h\nu/kT)}. \tag{3.47}$$

The total vibrational energy, U, of a monatomic crystal consisting of $3N_A$ harmonic oscillators at temperature T equals

$$U = 3(n_o\varepsilon_o + n_1\varepsilon_1 + n_2\varepsilon_2 \ldots n_1\varepsilon_i) = 3\sum n_i\varepsilon_i. \tag{3.48}$$

Considering the relationship given in Eq. (3.42), Eq. (3.48) can be rewritten in terms of the vibrational frequency, ν:

$$U = 3\left[n_o \cdot \frac{1}{2}h\nu + n_1 \cdot \frac{3}{2}h\nu + n_2 \cdot \frac{5}{2}h\nu + \ldots n_1\left(i + \frac{1}{2}\right)h\nu\right] \tag{3.49}$$
$$= 3\sum n_i\left[\left(i + \frac{1}{2}\right)h\nu\right].$$

Substituting the expression in Eq. (3.47) for n_0, n_1, $n_2 \ldots n_i$ into Eq. (3.49), the total vibrational energy for one mole of monatomic crystal is obtained. It reads:

$$U = 3N_A\left[\frac{1}{2}h\nu + \frac{h\nu}{e^{(h\nu/kT)} - 1}\right]. \tag{3.50}$$

In the case of a polyatomic crystal, the right side of Eq. (3.50) has to be multiplied by the number of atoms, q, in the chemical formula and one obtains:

$$U = 3qN_A\left[\frac{1}{2}h\nu + \frac{h\nu}{e^{(h\nu/kT)} - 1}\right]. \tag{3.51}$$

The temperature independent term in Eqs. (3.50) and (3.51) gives the energy at 0 K. Hence, an oscillator never loses its energy completely. At absolute zero, it equals $1/2h\nu$ and this is called the *zero point vibrational energy*. It derives from the vibrational motion of the atom in its ground state.

In order to obtain the heat capacity, C_v, the energy of vibration, U, has to be differentiated with respect to temperature, that is

$$\left(\frac{\partial U}{\partial T}\right)_V = C_v = 3qN_A k\left(\frac{h\nu}{kT}\right)^2 \frac{e^{(h\nu/kT)}}{(e^{(h\nu/kT)} - 1)^2}. \tag{3.52}$$

The multiplication of the frequency, ν, by h/k yields the so-called characteristic temperature, Θ. In this particular case, the characteristic temperature is referred to as the *Einstein temperature*, Θ_E.

Substituting the Einstein temperature, Θ_E, into Eq. (3.52) the heat capacity equations reads:

$$C_V = 3qR\left(\frac{\Theta_E}{T}\right)^2 \frac{e^{(\Theta_E/T)}}{(e^{(\Theta_E/T)} - 1)^2} = 3qR \cdot f(\Theta_E). \tag{3.53}$$

The term $f(\Theta_E)$ is called *Einstein heat capacity function*. It ranges from 0 to 1.

Einstein's model equation gives the temperature dependence of the heat capacity of the correct general form. It can be shown that its value approaches $3R$ at temperatures $T \gg \Theta_E$ in agreement with Dulong-Petit's law. At low temperatures ($T \ll \Theta_E$), it converges towards 0. However, calculated C_v values approach zero more rapidly than do the experimental C_v values. This discrepancy is mainly due to the fact that atoms do not vibrate with a single frequency. A better agreement between measured and calculated values is obtained for temperatures $T \geq \Theta_E$.

Debye considered a crystal as a continuum and assumed that the atoms do not vibrate independently of each other. Atomic vibrational behavior is influenced by the behavior of their neighbors. Instead of a single frequency, there is a frequency spectrum ranging from $\nu = 0$ to $\nu = \nu_{max}$. Debye assumed further that within this frequency range the number of vibrations increases parabolically with increasing frequency. Thus, he integrated Einstein's equation over the range from 0 to ν_{max} to obtain the heat capacity through

$$C_v = \frac{9N_A h^3}{k^2 \Theta_D} \int_0^{\nu_{max}} \left(\frac{h\nu}{kT}\right)^2 \frac{e^{h\nu/kT}}{(e^{h\nu/kT} - 1)^2} \nu^2 d\nu, \tag{3.54}$$

and by substituting z and Θ_D for $h\nu/kT$ and $h\nu_{max}/k$, respectively, as

$$C_v = 9R\left(\frac{T}{\Theta_D}\right)^3 \int_0^{\Theta_D/T} \frac{z^4 e^z}{(e^z - 1)^2} dz = 3R \cdot f\left(\frac{\Theta_D}{T}\right). \tag{3.55}$$

The term $f(\Theta_D/T)$ is called the *Debye heat capacity function* and it equals

$$3\left(\frac{T}{\Theta_D}\right)^3 \int_0^{\Theta_D/T} \frac{z^4 e^z}{(e^z - 1)^2} dz. \tag{3.56}$$

Eq. (3.55) holds for one mole of a monatomic solid. For a polyatomic crystal, the right side of the equation has to be multiplied by the number of atoms per chemical formula. Θ_D is referred to as the *Debye temperature.*

At high temperatures where $z \ll 1$, C_v approaches the classical value of $3R$. At temperatures in the vicinity of 0 K, Eq. (3.55) simplifies to

$$C_v \approx \frac{36RT^3\pi^4}{15\Theta_D^3} = 1943.7\left(\frac{T}{\Theta_D}\right)^3 = aT^3. \tag{3.57}$$

The relation, given in Eq. (3.57), is called the *Debye* T^3 law.

If Debye's model was completely correct, the Debye temperature, extracted from experimentally determined heat capacities, would be a constant. This is generally, however, not the case. It rather varies somewhat with the temperature of the experiment.

Heat capacity values calculated using the Debye function (Eq. (3.55)) agree reasonably with experimental values for monatomic solids having cubic symmetry (e.g. diamond, silicon, copper, lead, etc.) For silicates the agreement is rather poor. The main reason for the discrepancy lies in the simple parabolic form used to describe the vibrational spectrum. Debye's theory also does not account for anharmonicity in the oscillators.

Vibrational spectra can be measured using infrared spectroscopy, Raman spectroscopy and inelastic neutron scattering.

Fig. 3.3 shows the temperature dependence of C_V for forsterite calculated using the Einstein and Debye models. Experimental C_V's are given for comparison. They were obtained from measured C_p values (Robie et al. 1982) using Eq. (3.39).

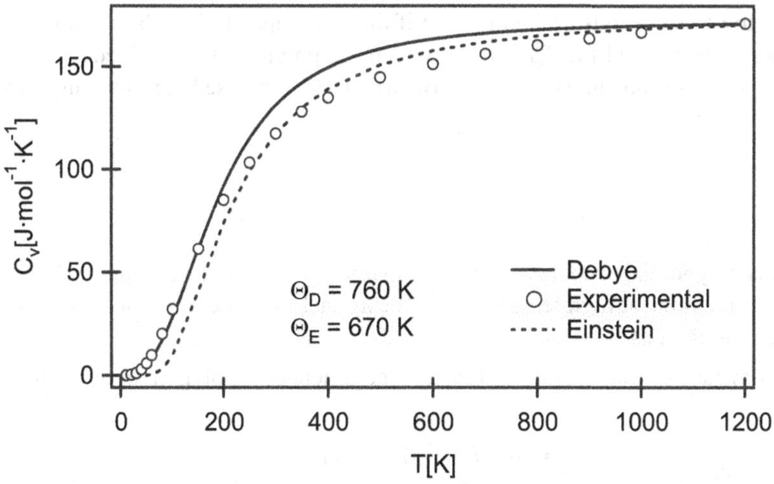

Fig. 3.3 Heat capacity at constant volume for forsterite calculated using the Einstein and Debye models. Θ_D = Debye temperature (Sumino et al. 1983), Θ_E = Einstein temperature (Kieffer 1985). Experimental data are from Robie et al. 1982.

Because of the various uncertainties in the theoretical calculations it is common practice to determine the heat capacities experimentally. The measurements are undertaken at constant pressure.

The empirical representation of heat capacities

Heat capacities are represented by empirical expressions that have the form of a polynomial for use, for example, in thermodynamical calculations. Several polynomial were developed over time. The oldest and the most simple expression is referred to as Maier-Kelly polynomial. It has the following form:

$$C_p = a + bT + cT^{-2}. \tag{3.58}$$

The coefficients *a*, *b*, and *c* are obtained by fitting the polynomial to the experimental data. Although they are empirically determined, their values can be linked to the Dulong-Petit law, to the Güneisen constant and to the Debye temperature.

Haas and Fisher (1976) extended the Maier-Kelly polynomial by adding two more terms, namely $dT^{-1/2}$ and eT^{-2}. Their expression reads:

$$C_p = a + bT + cT^{-2} + dT^{-1/2} + eT^{2}. \tag{3.59}$$

The Haas-Fisher polynomial reproduces experimental C_p data with a high precision. However, serious problems arise if the heat capacity is to be extrapolated at temperatures beyond the T region in which the polynomial was fitted. In order to overcome this problem, Berman and Brown (1985) proposed the following expression:

$$C_p = a + bT^{-2} + cT^{-1/2} + dT^{-3}. \tag{3.60}$$

This polynomial accounts for the fact that heat capacity should approach the Dulong-Petite value at high temperatures. It can, therefore, be used for high temperature extrapolations.

For the same reason, Holland (1981) recommended a polynomial of the form:

$$C_p = a + bT + cT^{-2} + dT^{-1/2}. \tag{3.61}$$

Robie et al. (1995), adopted in their thermodynamic data base the five term expression given in Eq. (3.59). Because of the quadratic term, it yields inappropriately high C_p values at high temperature. And, therefore, the authors warn of extrapolations beyond a given temperature.

Tab. 3.2 gives the coefficients for the five term C_p polynomial for a few selected silicates.

Table 3.2 Coefficients for the heat capacity polynomial of the form: $C_p[\text{Jmol}^{-1}\text{K}^{-1}] = a + bT + cT^{-2} + dT^{-1/2} + eT^2$ (Robie and Hemingway 1995).

Mineral	a x 10^{-2}	b x 10^3	c x 10^{-6}	d x 10^{-3}	e x 10^5
Kyanite	2.794	-7.124	-2.056	-2.289	
Andalusite	2.773	-6.588	-1.9141	-2.2656	
Sillimanite	2.8019	-6.900	-1.376	-2.399	
Forsterite	0.8736	87.17	-3.699	0.8436	-2.237
Pyrope	8.730	-137.4	0.0045	-8.794	3.3415
Cordierite	8.123	43.34	-8.211	-5.000	
Wollastonite	2.0078	-25.89	-0.1579	-1.826	0.74.34
Diopside	4.7025	-98.64	0.2454	-4.823	2.813
Tremolite	61.31	-4189.0	51.39	-85.66	175.7
Muscovite	9.177	-81.11	2.834	-10.35	
Sanidine	6.934	-171.7	3.462	-8.305	4.919
Albite	5.839	-92.85	1.678	-6.424	2.272

Fig. 3.4 shows the temperature dependence of heat capacity for different oxides and silicates standardized to one atom. The heat capacity values for corundum, andalusite, grossular, wollastonite and low quartz are very similar, whereas those of high quartz and lime are quite different. This is due to the difference in lattice dynamic behavior of these oxides in comparison to that of complex silicates. Thus, the underlying physical assumption inherent to the Neumann-Kopp approximation is not valid in this case and the simple oxide summation fails. Robinson and Haas (1983) attempted to circumvent this problem by introducing an empirically-based model. They constructed empirical heat capacity polynomials for fictive oxide components with cations in a given polyhedral coordination, such as $Mg^{[4]}O$, $Mg^{[6]}O$, $Al_2{}^{[4]}O_3$, $Al_2{}^{[6]}O_3$, etc. These polynomial were obtained by a mathematical least-squares procedure that used as input data the experimentally determined heat capacity of a number of different silicates.

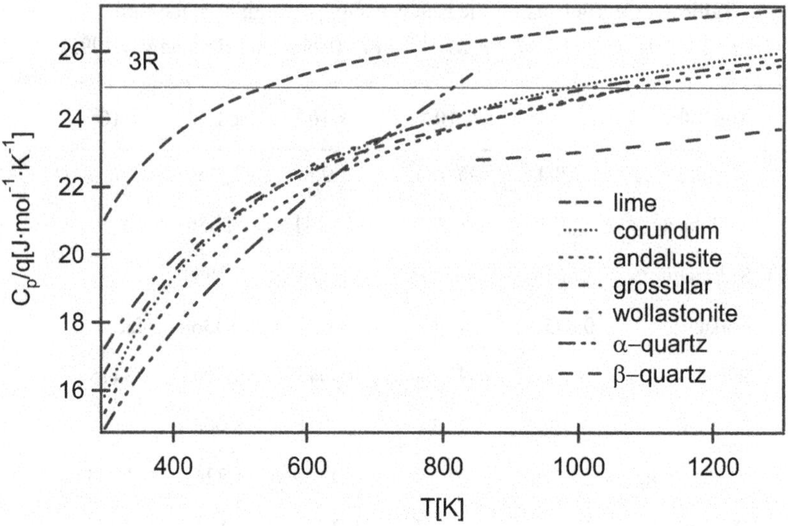

Fig. 3.4 Heat capacities, C_p, as a function of temperature, standardized on a one atom basis. The horizontal line corresponds to Dulong-Petit's value of $3R$.

Example 1: Consider grossular, $Ca_3Al_2Si_3O_{12}$. Here, calcium is 8-fold coordinated, aluminium has six oxygens as neighbors and silicon's coordination number is four. In order to evaluate the C_p polynomial for grossular, the polynomials for $Ca^{[8]}O$, $Al_2{}^{[6]}O_3$ and $Si^{[4]}O_2$ are required. In Robinson and Haas (1983) one finds:

$$C_p^{Ca[8]O} = 83.6079 - 5.95782\times10^{-3}T + 19661.5T^{-2} - 716.401T^{-1/2},$$

$$C_p^{Al_2[6]O_3} = 222.74 - 16.409\times10^{-3}T - 2464.56T^{-1/2} \text{ and}$$

$$C_p^{Si[4]O_2} = 109.383 - 5.5518\times10^{-3}T - 1083.05T^{-1/2}, \text{ where}$$

C_p[Jmol^{-1}K^{-1}].

In order to obtain the C_p polynomial for grossular, the C_p polynomials for the oxides are to be summed according to their stoichiometric proportions in garnet:

$$C_{p,\,Ca_3Al_2Si_3O_{12}}^{calc} = 3C_p^{Ca[8]O} + C_p^{Al_2[6]O_3} + 3C_p^{Si[4]O_2}.$$

The summation yields:

$$C^{calc}_{p,\,Ca_3Al_2Si_3O_{12}}[\mathrm{Jmol^{-1}K^{-1}}] = 801.713 - 50.938\times10^{-3}T + 58984.5T^{-2} - 7862.91T^{-1/2}.$$

For $T = 800$ K,

$$C^{calc}_{p,\,Ca_3Al_2Si_3O_{12}} = \underline{483.06\ \mathrm{Jmol^{-1}K^{-1}}}$$

is obtained. This value may be compared with a value of 477.67 $\mathrm{Jmole^{-1}K^{-1}}$ that is obtained using the experimental data given by Bosenick et al. (1996).

If the $\boldsymbol{C}_p$ polynomials for the oxides (lime, quartz and corundum) are used, according to the Neumann-Kopp's rule, larger discrepancies between calculated and measured values result. For a temperature of 800 K, for example, a $\boldsymbol{C}_p$ value of 500 $\mathrm{Jmole^{-1}K^{-1}}$ is obtained. This large discrepancy is a consequence of the fact that the normalized $\boldsymbol{C}_p$ vs. temperature curve for lime is much different than those for corundum, quartz and grossular (see Fig. 3.4).

At high temperatures, $\boldsymbol{C}_p$ values exceed Dulong-Petits limit of $3R$. One of the reasons for this behavior is the formation of large concentrations of point defects as the melting temperature is approached.

3.3.2 Enthalpy, H, of pure phases as a function of temperature

Using the definition of the heat capacity, c_p, as given in Eq. (3.35), the change in enthalpy, dH, with change in temperature, dT, at constant pressure is given by:

$$dH = c_p dT. \tag{3.62}$$

For one mole of a pure phase, the enthalpy, H, and heat capacity, c_p, in Eq. (3.62) must be replaced by the molar quantities $\boldsymbol{H}$ and $\boldsymbol{C}_p$, respectively.

In order to obtain the molar enthalpy of a pure phase at an arbitrary temperature, T, Eq. (3.62) has to be integrated over the temperature range from 0K to T, that is

$$\boldsymbol{H}_T = \boldsymbol{H}_{T=0} + \int_0^T \boldsymbol{C}_p dT. \tag{3.63}$$

The integration constant $\boldsymbol{H}_{T=0}$ corresponds to the enthalpy of the pure phase at

0 K. Its value is unknown, because it cannot be determined experimentally. Only changes, i. e. $\Delta \boldsymbol{H}$, are amenable to measurement. Therefore, a standard state to which the changes refer has to be chosen. It is taken as the enthalpy change associated with the formation of a phase from the elements at 0.1 MPa and 298 K. (More details about this will be given later).

In thermodynamic tables one finds the so-called *heat content functions* $[H_T - H_{298}]/T$. The term in the numerator is referred to as a *standard enthalpy content* and is calculated by integrating Eq. (3.63) between 298 (more precisely 298.15) and T:

$$[\boldsymbol{H}_T - \boldsymbol{H}_{298}] = \int_{298}^{T} \boldsymbol{C}_p dT. \tag{3.64}$$

Inserting Eq. (3.59) into Eq. (3.64) yields

$$\begin{aligned}[\boldsymbol{H}_T - \boldsymbol{H}_{298}] &= \int_{298}^{T} (a + bT + cT^{-2} + dT^{-1/2} + eT^2)dT \\ &= a(T-298) + \frac{b}{2}(T^2 - 298^2) - c\left(\frac{1}{T} - \frac{1}{298}\right) \\ &\quad + 2d(\sqrt{T} - \sqrt{298}) + \frac{e}{3}(T^3 - 298^3).\end{aligned} \tag{3.65}$$

Eq. (3.65) gives the heat amount that is required to increase the temperature of one mole of a pure phase from 289.15 K (room temperature) to T.

Example 1: The temperature of one mole of diopside, $CaMgSi_2O_6$, is increased from 298 to 1000 K. How much heat is absorbed by the mineral?

According to Eq. (3.64), the heat that is transferred to diopside equals its molar heat content, that is

$$[\boldsymbol{H}_{1000} - \boldsymbol{H}_{298}]^{Di}_{CaMgSi_2O_6} = \int_{298}^{T} \boldsymbol{C}^{Di}_{p,\,CaMgSi_2O_6} dT.$$

Using the coefficients of the $\boldsymbol{C}_p$-polynomial for diopside given in Tab. 3.2, the calculation reads:

$$[H_{1000} - H_{298}]^{Di}_{CaMgSi_2O_6} = 470.25(1000 - 298) - \frac{9.864\times10^{-2}}{2}$$

$$\times (1000^2 - 298^2) - 2.454\times10^5 \times \left(\frac{1}{1000} - \frac{1}{298}\right)$$

$$- 2 \times 4.823\times10^3 \times (\sqrt{1000} - \sqrt{298})$$

$$+ \frac{2.813\times10^{-5}}{3}(1000^3 - 298^3) = \underline{156.364 \text{ kJmol}^{-1}}.$$

Thus, a heat of 156.364 kJ is required to increase the temperature of one mole of diopside from 298 to 1000 K.

In the reverse process, when the temperature of a mineral is decreased, the heat value has a negative sign.

Example 2: A sanidine crystal, $KAlSi_3O_8$, has a mass of 27.734 gram.

How much heat is released to the surroundings if the crystal is cooled at constant pressure from 800 to 298 K?

We first calculate the molar change in the enthalpy that is associated with the cooling process.

$$[H_{298} - H_{800}]^{san}_{KAlSi_3O_8} = -\int_{298}^{800} C^{san}_{p,\,KAlSi_3O_8}\, dT.$$

Here, we use again the coefficients given for sanidine from Tab. 3.2 and calculate:

$$[H_{298} - H_{800}]^{san}_{KAlSi_3O_8} = -\left[693.4(800 - 298) - \frac{0.1717}{2}(800^2 - 298^2)\right.$$

$$- 3.462\times10^6\left(\frac{1}{800} - \frac{1}{298}\right) - 2 \times 8305$$

$$\left.\times (\sqrt{800} - \sqrt{298}) + \frac{4.919\times10^{-5}}{3}(800^3 - 298^3)\right]$$

$$= \underline{-132.949 \text{ kJmol}^{-1}}.$$

The sanidine mass has to be converted to the number of moles. Using its molar mass of 277.34 gram, one has

$$n^{san}_{KAlSi_3O_8} = \frac{27.734\ \text{g}}{277.34\ \text{gmol}^{-1}} = 0.1\ \text{mol}.$$

In order to obtain the heat that is released to the surroundings due to the cooling process, the calculated molar heat of -132.949 Jmole^{-1} has to be multiplied by the number of moles, as

$$Q = -132.949\ \text{kJmol}^{-1} \times 0.1\ \text{mol} = \underline{-13.295\ \text{kJ}}.$$

Eq. (3.64) gives the heat content for a phase only in the case where no phase transition occurs over the considered temperature range.

If a phase transition takes place between 298.15 K and T, the amount of heat, associated with the transition has to be is taken into account. It is the difference in the enthalpy between the disappearing (educt) and the forming phase (product) and is referred to as the *heat of transition*, $\Delta_{tr}\boldsymbol{H}$:

$$\Delta_{tr}\boldsymbol{H} = \boldsymbol{H}_{prod} - \boldsymbol{H}_{educt}. \tag{3.66}$$

In order to calculate the heat content of a substance undergoing a phase transition, Eq. (3.64) has to be modified to

$$[\boldsymbol{H}_T - \boldsymbol{H}_{298}] = \int_{298}^{T_{tr}} \boldsymbol{C}_{p,1} dT + \Delta\boldsymbol{H}_{tr} + \int_{T_{tr}}^{T} \boldsymbol{C}_{p,2} dT, \tag{3.67}$$

where $\boldsymbol{C}_{p,1}$ and $\boldsymbol{C}_{p,2}$ are the molar heat capacities for the phases in the temperature range between 298.15 and T_{tr} and between T_{tr} and T, respectively.

Example: Fayalite, Fe_2SiO_4, melts incongruently at 1490 K and its enthalpy of melting is 89.3 kJmole^{-1} (Robie and Hemingway 1995). At 1800 K, the heat content of the melt can be calculated according to Eq. (3.67) as follows:

$$[\boldsymbol{H}_{1800} - \boldsymbol{H}_{298}]^{fa}_{Fe_2SiO_4} = \int_{298}^{1490} \boldsymbol{C}^{fa}_{p,\,Fe_2SiO_4} dT + \Delta_{tr}\boldsymbol{H} + \int_{1490}^{1800} \boldsymbol{C}^{melt}_{p,\,Fe_2SiO_4} dT.$$

The expressions for the heat capacity of fayalite and for fayalite melt are taken from Robie and Hemingway (1995). They read:

$$C^{fa}_{p,\,Fe_2SiO_4}[\mathrm{Jmol^{-1}K^{-1}}] = 176.02 - 8.808\times10^{-3}T - 3.889\times10^{6}T^{-2} + 2.471\times10^{-5}T^{2}$$

and

$$C^{melt}_{p,\,Fe_2SiO_4} = 240.60\ \mathrm{Jmole^{-1}K^{-1}}.$$

Using these data

$$\begin{aligned}[H_{1800} - H_{298}]^{fa}_{Fe_2SiO_4} &= 176.02(1490 - 298) - \frac{8.808\times10^{-3}}{2}(1490^2 - 298^2)\\ &+ 3.889\times10^{6}\left(\frac{1}{1490} - \frac{1}{298}\right) + \frac{2.471\times10^{-5}}{3}(1490^3 - 298^3)\\ &+ 89300 + 240.60(1800 - 1490)\\ &= \underline{380.904\ \mathrm{kJmol^{-1}}}\end{aligned}$$

is obtained.

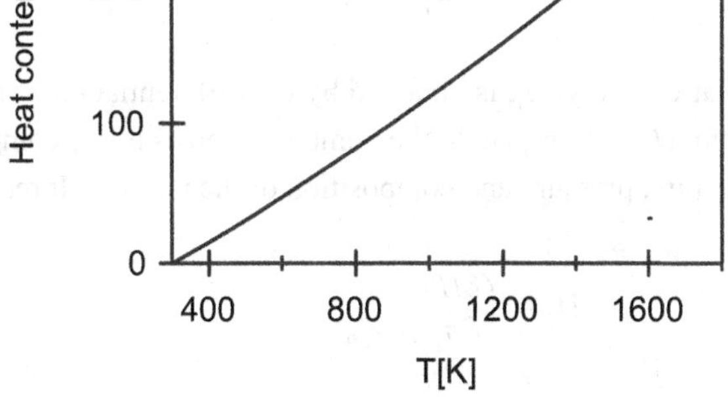

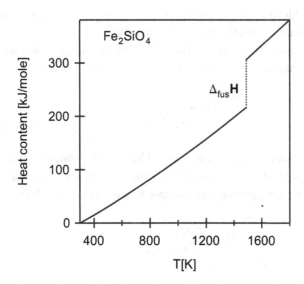

Fig. 3.5 Heat content of Fe_2SiO_4 as a function of temperature. $\Delta_{fus}H$ is the heat of melting.

Fig. 3.5 shows the heat content of Fe_2SiO_4 as a function of temperature up to 1800 K. $\Delta_{tr}\boldsymbol{H}$ is the heat that is required to melt one mole of fayalite at 1490 K and 0.1 MPa.

At this point, it has to be emphasized that Eq. (3.67) gives the heat content quantitatively only if the enthalpy of a material under consideration shows a finite difference at the transition temperature. This, however, is not always the case, as discussed further in section 6.1.2.

3.4 Enthalpy of solutions

Similar to the volume, the enthalpy of a solution is a function of temperature, pressure and composition that is given by the number of moles, n_i

$$H = f(T, P, n_1, n_2 \ldots n_k). \tag{3.68}$$

At constant temperature and pressure the enthalpy of an ideal solution is simply the sum of the molar enthalpies of the components making up the solution multiplied by the corresponding number of moles:

$$H = \sum_1^k n_i \boldsymbol{H}_i. \tag{3.69}$$

Eq. (3.69) holds of course for mechanical mixtures, where the components can be separated from each other using physical methods.

In the case of non-ideal solutions the molar enthalpy in Eq. (3.69) has to be replaced by the *partial molar enthalpies* of the components, where

$$H = \sum_1^k n_i H_i. \tag{3.70}$$

The partial molar enthalpy, H_i, is obtained by the differentiation of the total enthalpy of the system, H, with respect to the number of moles of the component i, n_i, at constant temperature, pressure and composition of the mixture. It reads:

$$H_i = \left(\frac{\partial H}{\partial n_i}\right)_{P, T, n_{j \neq i}} \tag{3.71}$$

and gives the change in the total enthalpy of a solution when 1 mole of component i is added at constant temperature and pressure. A requirement for Eq. (3.71) is that the quantity of the solution must be so large that an addition of n_i causes virtually no change in the composition.

3.4.1 Enthalpy of binary solution

Consider a binary solution consisting of the components *A* and *B*. According to Eq. (3.70) the total enthalpy of the solution is

$$H = n_A H_A + n_B H_B, \tag{3.72}$$

where n_A, n_B, H_A, and H_B are the molar fractions and partial molar enthalpies of the components *A* and *B*, respectively.

Dividing Eq. (3.72) by the sum of the moles in the system, $n_A + n_B$, yields:

$$\frac{H}{n_A + n_B} = \overline{H} = \frac{n_A}{n_A + n_B} H_A + \frac{n_B}{n_A + n_B} H_B = x_A H_A + x_B H_B. \tag{3.73}$$

$\overline{H}$ is referred to as the *molar enthalpy of solution.*

As stated earlier, the absolute enthalpy can not be determined experimentally. Only changes in enthalpy can be measured and, therefore, it is necessary to select some *standard state* and refer the changes to this state. In the case of solutions it is the state of the pure components at the temperature and pressure of mixing. Adopting this standard state, the enthalpy change associated with the process of mixing at constant temperature and pressure is given by the difference between the molar enthalpy of the mechanical mixture and that of the solution. In other words, it is the difference in the enthalpy of a system before and after the mixing process takes place.

The molar enthalpy of a binary system consisting of components *A* and *B*, before the process of mixing takes place, is given as

$$\overline{H}^{mm} = x_A \boldsymbol{H}_A + x_B \boldsymbol{H}_B, \tag{3.74}$$

where $\overline{H}^{mm}$, $\boldsymbol{H}_A$ *and* $\boldsymbol{H}_B$ represent the molar enthalpy of the mechanical mixture and the molar enthalpies of the pure components *A* and *B*, respectively.

After the completion of the mixing process, the molar enthalpy for the system under consideration reads:

$$\overline{H}^{sol} = x_A H_A + x_B H_B. \tag{3.75}$$

Here, $\overline{H}^{sol}$, H_A and H_B represent the molar enthalpy of the solution and the partial molar enthalpies of the components A and B, respectively.

In order to obtain the change in the enthalpy due to the process of mixing Eq. (3.74) has to be subtracted from Eq. (3.75), that is

$$\overline{H}^{sol} - \overline{H}^{mm} = \Delta_m \overline{H}^{ex} \equiv \Delta_m \overline{H} = x_A(H_A - \boldsymbol{H}_A) + x_B(H_B - \boldsymbol{H}_B). \tag{3.76}$$

$\Delta_m \overline{H}$ is the quantity of heat that is exchanged between the system and its surroundings when A and B form a homogeneous solution at constant temperature and pressure. It is referred to as the *enthalpy of mixing*. In the case of an ideal solution the molar enthalpy of solution corresponds to the stoichiometric sum of the molar enthalpies of the pure components. Thus, in Eq. (3.76) the partial molar enthalpies of the components and their molar enthalpies are identical and the terms in the parentheses are zero. Consequently, the enthalpy of mixing for ideal solutions equals 0. In non-ideal solutions, however, the partial molar enthalpies of the components differ from their molar enthalpies in the pure state $\Delta_m \overline{H} \neq 0$ and, the molar enthalpy of mixing is identical with the *molar excess enthalpy* of solution, $\Delta_m \overline{H}^{ex}$. The terms in the parentheses in Eq. (3.76) correspond to the *partial molar excess enthalpies* or *partial molar enthalpies of mixing* of the components. They are:

$$H_A - \boldsymbol{H}_A = H_A^{ex} \tag{3.77}$$

and

$$H_B - \boldsymbol{H}_B = H_B^{ex}. \tag{3.78}$$

Considering the fact that the sum of the molar fractions of all components making up a solution is 1, the molar heat of mixing for a binary solution can be rewritten as follows:

$$\Delta_m \overline{H} = (1 - x_B)H_A^{ex} + x_B H_B^{ex}. \tag{3.79}$$

Similarly as to the case of the partial molar volumes, the partial molar excess enthalpies of the components can be derived from the molar heat of mixing. The condition for this is that the molar heat of mixing as a function of molar fraction is known over the compositional range under consideration. The partial molar excess enthalpies of the components in a binary solution then read:

$$H_A^{ex} = \Delta_m \overline{H} - x_B \left(\frac{\partial \Delta_m \overline{H}}{\partial x_B} \right)_{p,\,T} \tag{3.80}$$

and

$$H_B^{ex} = \Delta_m \overline{H} + (1 - x_B) \left(\frac{\partial \Delta_m \overline{H}}{\partial x_B} \right)_{p,\,T} . \tag{3.81}$$

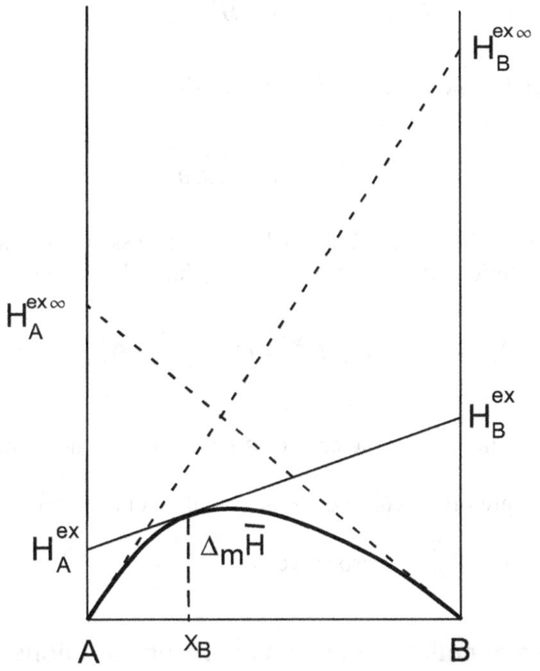

Fig. 3.6 Enthalpy of mixing and partial molar excess enthalpies of the components, *A* and *B* as functions of composition in a binary solution *(A,B)*. $\Delta_m \overline{H}$ designates the molar heat of mixing, H_A^{ex} and H_B^{ex} are the partial molar excess enthalpies of the components *A* and *B*, respectively, at the composition x_B, $H_A^{ex\infty}$ and $H_B^{ex\infty}$ represent the partial molar excess enthalpies of the components for the case where they are infinitely diluted.

The relationship between the molar heat of mixing and the partial molar excess enthalpies of the components in a binary solution *(A,B)* is shown schematically in

Fig. 3.6. Because there is no heat of mixing for pure components, the curve of the molar enthalpy of mixing is zero at $x_B = 0$ and at $x_B = 1$. The intercepts of the tangent to the curve of the molar enthalpy of mixing with the ordinate at $x_B = 0$ and $x_B = 1$ give the partial molar excess enthalpies for the components A and B, respectively. In the special case solutions at infinite dilution, that is when x_B approaches 0 and 1, the partial excess enthalpies are designated as $H_A^{ex\infty}$ and $H_B^{ex\infty}$, respectively. They are used to express the enthalpy of mixing as a function of composition at constant temperature and pressure. In the case of a simple binary mixture where the partial excess enthalpies of the components at infinite dilution are equal, i.e.

$$H_A^{ex\infty} = H_B^{ex\infty} = H_{A,B}^{ex\infty}, \tag{3.82}$$

the expression giving the enthalpy of mixing reads:

$$\Delta_m \bar{H} = (1 - x_B) x_B H_{A,B}^{ex\infty}. \tag{3.83}$$

In the case of a subregular solution where the excess enthalpies of the components at infinite dilution differ one from another, the enthalpy of mixing is given by:

$$\Delta_m \bar{H} = (1 - x_B) x_B^2 H_A^{ex\infty} + (1 - x_B)^2 x_B H_B^{ex\infty}. \tag{3.84}$$

In the literature the partial molar excess enthalpies of the components $H_{A,B}^{ex\infty}$, $H_A^{ex\infty}$ and $H_B^{ex\infty}$ are often replaced by the so-called enthalpic interaction parameters W_{ij}^H, W_{A-B}^H and W_{B-A}^H, respectively.

3.4.2 Examples of enthalpy of mixing in binary solutions

Example 1: Tab. 3.3 gives the enthalpy of mixing for the high albite-high sanidine solid solution series derived from experimental heat of solution measurements made at 50°C by Hovis (1988).

The data in Tab. 3.3 are shown graphically in Fig. 3.7. Their values as a function of concentration suggest that a simple solution best describes the mixing behavior according to Eq. (3.83). A least-squares fit to the data yields the value for the partial molar enthalpy of mixing for the components at infinite dilution $H_{Na,K}^{ex\infty}$, of 19.587 kJmole^{-1}. Thus, the expression for the heat of mixing curve reads:

$$\Delta_m \overline{H}[\text{kJmol}^{-1}] = (1 - x^{san}_{KAlSi_3O_8})x^{san}_{KAlSi_3O_8} \cdot 19.587.$$

Table 3.3 Enthalpies of mixing for the solid solution high albite-high sanidine (Hovis 1988)

$x^{san}_{KAlSi_3O_8}$	$\Delta_m \overline{H}[\text{kJmol}^{-1}]$
0.009	0.415±0.822
0.1437	2.268±0.790
0.3508	4.335±0.549
0.4911	4.855±0.502
0.5993	4.961±0.613
0.8074	3.058±0.597
0.960	0.524±0.927

The above equation can be used to derive expressions for the calculation of the partial molar excess enthalpies of the components at any arbitrary composition. For this purpose, the right hand term of the equation has to be substituted for the heat of mixing in Eqs (3.80) and (3.81), respectively. The partial molar excess enthalpies then read:

$$H^{ex}_{NaAlSi_3O_8} = (x^{san}_{KAlSi_3O_8})^2 \cdot 19.587 \text{ kJmole}^{-1}$$

and

$$H^{ex}_{KAlSi_3O_8} = (1 - x^{san}_{KAlSi_3O_8})^2 \cdot 19.587 \text{ kJmole}^{-1}.$$

In the special case where the mole fraction of high sanidine in the solution equals 0.4, the partial molar excess enthalpy of high albite is

$$H^{ex}_{NaAlSi_3O_8} = (0.4)^2 \times 19.587 = \underline{3.134 \text{ kJmole}^{-1}}.$$

The partial molar excess enthalpy of high sanidine at this concentrations is given by:

$$H^{ex}_{KAlSi_3O_8} = (1.0 - 0.4)^2 \times 19.587 = \underline{7.051 \text{ kJmole}^{-1}}.$$

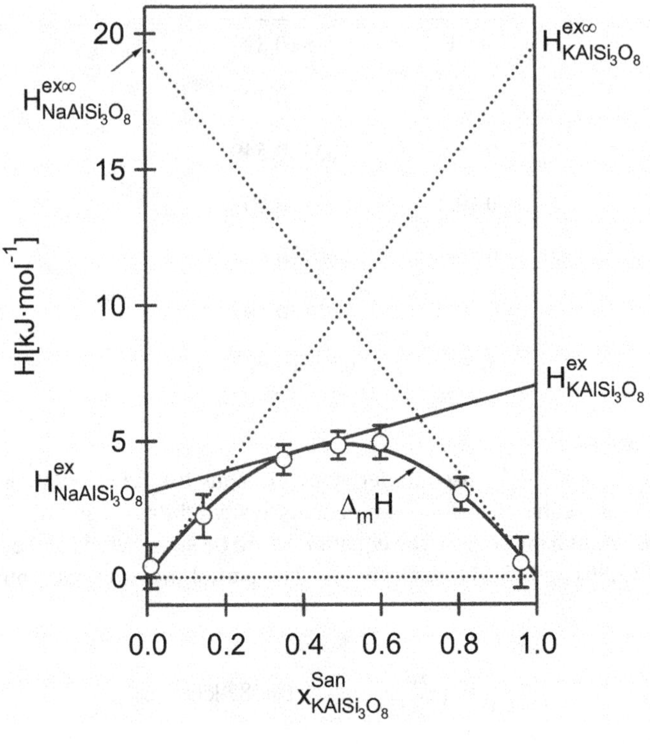

Fig. 3.7 Enthalpy relationships in the system high albite-high sanidine (Hovis 1988). $\Delta_m \bar{H}$ designates the molar enthalpy of mixing, $H^{ex}_{NaAlSi_3O_8}$ and $H^{ex}_{KAlSi_3O_8}$ are the partial molar excess enthalpies of high albite and high sanidine, respectively, at $x^{san}_{KAlSi_3O_8} = 0.4$. $H^{ex\infty}_{NaAlSi_3O_8}$ and $H^{ex\infty}_{KAlSi_3O_8}$ represent the partial molar excess enthalpies of the components at infinite dilution.

The two partial molar excess enthalpies can be used to calculate the enthalpy of mixing at $x^{san}_{KAlSi_3O_8} = 0.4$, following Eq. (3.79). It is

$$\Delta_m \overline{H} = (1.0 - 0.4) \times 3.134 + 0.4 \times 7.051 = \underline{4.701 \text{ kJmole}^{-1}}.$$

Of course, the same result is obtained if the equation developed for simple solutions is used, namely

$$\Delta_m \overline{H} = (1 - 0.4) \times 0.4 \times 19.587 = \underline{4.701 \text{ kJmole}^{-1}}.$$

The partial molar excess enthalpies of the components at infinite dilution are obtained when $x_{KAlSi_3O_8}^{san}$ approaches 1 and 0, respectively. The calculation reads:

$$H_{NaAlSi_3O_8}^{ex\infty} = 1^2 \times 19.587 = \underline{19.587 \text{ kJmole}^{-1}}$$

and

$$H_{NaAlSi_3O_8}^{ex\infty} = (1 - 0)^2 \times 19.587 = \underline{19.587 \text{ kJmole}^{-1}}.$$

The two values are equal, because a simple solution model was used to fit the experimental data.

Example 2: Geiger et al. (1987) determined the enthalpy of mixing for a series of synthetic garnet solid solutions on the join almandine-pyrope. They fitted their data to the subregular solution model and obtained the following enthalpic interaction parameters:

$$W_{Al-Py}^{H} = -15.76 \text{ kJmol}^{-1}$$

and

$$W_{Py-Al}^{H} = 36.17 \text{ kJmol}^{-1}.$$

Thus, the expression for the molar enthalpy of mixing according to Eq. (3.84) reads:

$$\Delta_m \overline{H}[\text{kJmol}^{-1}] = (1 - x_{Mg_3Al_2Si_3O_{12}}^{grt})(x_{Mg_3Al_2Si_3O_{12}}^{grt})^2 \times (-15.76) + (1 - x_{Mg_3Al_2Si_3O_{12}}^{grt})^2 (x_{Mg_3Al_2Si_3O_{12}}^{grt}) \times 36.17.$$

In order to derive equations for the partial molar excess enthalpies of the components, the expression given above has to be substituted for the integral molar enthalpy of mixing in Eqs. (3.80) and (3.81). We start with the partial molar excess enthalpy of almandine. Inserting the expression for the molar enthalpy of mixing into Eq. (3.80) yields:

$$
\begin{aligned}
H^{ex}_{Fe_3Al_2Si_3O_{12}} &= (1 - x^{grt}_{Mg_3Al_2Si_3O_{12}})(x^{grt}_{Mg_3Al_2Si_3O_{12}})^2 \times (\text{- } 15.76) \\
&\quad + (1 - x^{grt}_{Mg_3Al_2Si_3O_{12}})^2 x^{grt}_{Mg_3Al_2Si_3O_{12}} \times 36.17 \\
&\quad - x \left\{ \frac{\partial}{\partial x^{grt}_{Mg_3Al_2Si_3O_{12}}} \left[(x^{grt}_{Mg_3Al_2Si_3O_{12}})^2 (1 - x^{grt}_{Mg_3Al_2Si_3O_{12}}) \right.\right. \\
&\quad \left.\left. \times (\text{- } 15.76) + (1 - x^{grt}_{Mg_3Al_2Si_3O_{12}})^2 x^{grt}_{Mg_3Al_2Si_3O_{12}} \times 36.17 \right] \right\} \\
&= (x^{grt}_{Mg_3Al_2Si_3O_{12}})^2 [(-15.76) + 2 \times (36.17 + 15.76) \\
&\quad \times (1 - x^{grt}_{Mg_3Al_2Si_3O_{12}})].
\end{aligned}
$$

For $x^{grt}_{Mg_3Al_2Si_3O_{12}} = 0.8$ one obtains:

$$
\begin{aligned}
H^{ex}_{Fe_3Al_2Si_3O_{12}} &= 0.8^2 [(-15.76) + 2 \times (36.17 + 15.76)(1 - 0.8)] \\
&= \underline{3.208 \text{ kJmol}^{-1}}.
\end{aligned}
$$

The mathematical expression for the partial molar excess enthalpy of pyrope is obtained by inserting the term for the molar enthalpy of mixing into Eq. (3.81), i.e.

$$H^{ex}_{Mg_3Al_2Si_3O_{12}} = (1 - x^{grt}_{Mg_3Al_2Si_3O_{12}})(x^{grt}_{Mg_3Al_2Si_3O_{12}})^2 \times (-15.76)$$

$$+ (1 - x^{grt}_{Mg_3Al_2Si_3O_{12}})^2 \times (x^{grt}_{Mg_3Al_2Si_3O_{12}}) \times 36.17$$

$$+ (1 - x^{grt}_{Mg_3Al_2Si_3O_{12}}) \times \frac{\partial}{\partial x^{grt}_{Mg_3Al_2Si_3O_{12}}} \Big[(x^{grt}_{Mg_3Al_2Si_3O_{12}})^2$$

$$\times (1 - x^{grt}_{Mg_3Al_2Si_3O_{12}}) \times (-15.76) + (1 - x^{grt}_{Mg_3Al_2Si_3O_{12}})^2$$

$$\times x^{grt}_{Mg_3Al_2Si_3O_{12}} \times 36.17 \Big] = (1 - x^{grt}_{Mg_3Al_2Si_3O_{12}})[36.17$$

$$+ 2 \times (-15.76 - 36.17) x^{grt}_{Mg_3Al_2Si_3O_{12}}].$$

If one chooses 0.8 as the mole fraction of pyrope the calculation reads:

$$H^{ex}_{Mg_3Al_2Si_3O_{12}} = (1 - 0.8)^2 [36.17 + 2(-15.76 - 36.17) \times 0.8]$$
$$= \underline{-1.877 \text{ kJmol}^{-1}}.$$

The two partial molar excess enthalpies can be used to calculate the integral heat of mixing for a an almandine-pyrope solid solution containing 0.8 moles almandine. The calculation has to be carried out according to Eq. (3.79). It is:

$$\Delta_m \overline{H} = (1 - 0.8) \times 3.207 + 0.8 \times (-1.877) = \underline{-0.86 \text{ kJmol}^{-1}}.$$

This result can be verified using the expression for the heat of mixing (Eq. (3.84)), that is

$$\Delta_m \overline{H} = (1 - 0.8) \times 0.8^2 \times (-15.76) + (1 - 0.8)^2 \times 0.8 \times 36.17$$
$$= \underline{-0.86 \text{ kJmol}^{-1}}.$$

The enthalpic interaction parameters given in the expression for the heat of mixing are identical to the partial excess enthalpies of the components at infinite dilution. This can be verified by setting $x^{grt}_{Mg_3Al_2Si_3O_{12}}$ to one and to zero in the equations giving the partial excess enthalpies of mixing of almandine and pyrope.

For almandine holds:

$$H^{ex\infty}_{Fe_3Al_2Si_3O_{12}} = 1^2[(-15.76) + 2 \times (36.17 + 15.76)(1 - 1)] = \underline{-15.76 \text{ kJmol}^{-1}}$$

and for pyrope

$$H^{ex\infty}_{Mg_3Al_2Si_3O_{12}} = (1 - 0)^2[36.17 + 2(-15.76 - 36.17) \times 0] = \underline{36.17 \text{ kJmol}^{-1}}.$$

3.5 Enthalpy of reaction

Consider a reacting system consisting of the components A, B, C and D and suppose a reaction of the type

$$\nu_a A + \nu_b B = \nu_c C + \nu_d D$$

takes place at constant temperature and pressure. What is the change in the enthalpy of the system? To answer this question, the enthalpy of each single reacting component constituting the system must be known. Mathematically, they are given by the partial derivatives of the total enthalpy with respect to the considered component at constant pressure, temperature and composition of the system. Considering the fact that in a closed system the mass remains constant, the change in the total enthalpy due to an infinitely small extent of reaction is given by:

$$dH = -\left(\frac{\partial H}{\partial n_A}\right)_{p,\,T,\,n_{j \neq A}} dn_A - \left(\frac{\partial H}{\partial n_B}\right)_{p,\,T,\,n_{j \neq B}} dn_B + \left(\frac{\partial H}{\partial n_C}\right)_{p,\,T,\,n_{j \neq C}} dn_C + \left(\frac{\partial H}{\partial n_D}\right)_{p,\,T,\,n_{j \neq D}} dn_D. \quad (3.85)$$

If the reactants as well as products are mechanical mixtures of the pure components, the partial derivatives in Eq. (3.85) can be replaced by the molar enthalpies of the pure components. In addition, as shown earlier (see Eq. (2.113)), the molar increments, dn_i, are related through the stoichiometry of the reaction. Thus, Eq. (3.85) can be written as follows:

$$dH = \sum_A^D \nu_i \boldsymbol{H}_i d\xi. \tag{3.86}$$

In Eq. (3.86) ν_i and $\boldsymbol{H}_i$ are the stoichiometric coefficient and the molar enthalpy of the component i, respectively. ξ is the extent of reaction. It indicates the extent to which the reaction has progressed from the initial non equilibrium state towards equilibrium.

Division of Eq. (3.86) by $d\xi$ yields:

$$\left(\frac{\partial H}{\partial \xi}\right)_{P,T} = \Delta_r \boldsymbol{H} = \sum_A^D \nu_i \boldsymbol{H}_i. \tag{3.87}$$

$\Delta_r \boldsymbol{H}$ is referred to as the *enthalpy of reaction*. It gives the amount of heat that is exchanged between the system and its surroundings when a chemical reaction occurs at constant temperature and pressure. Reactions that absorb heat in order to maintain a constant temperature of the system are called *endothermic*. Conversely, when heat is released to the surroundings the reaction is called *exothermic*.

In the case of the generalized reaction given above, the enthalpy of reaction reads:

$$\Delta_r \boldsymbol{H} = -\nu_a \boldsymbol{H}_A - \nu_b \boldsymbol{H}_B + \nu_c \boldsymbol{H}_C + \nu_d \boldsymbol{H}_D. \tag{3.88}$$

Eq. (3.88) has no practical meaning, because the molar enthalpies of the components are unknown. In order to assign an enthalpy value to a substance, the *standard enthalpy of formation* from the elements, $\Delta_f \boldsymbol{H}_{298}$, is defined. It is the enthalpy associated with the formation of substance from the elements in their stable form at 298.15 K and 0.1 MPa. The standard enthalpies of the elements in their stable form are assigned arbitrarily a value of zero.

Example: Consider the formation of pyrite, FeS_2 from the elements. The reaction is:

$$Fe + 2S = FeS_2.$$

In its reference state, iron has a body-centered cubic cell and is termed α-Fe. Sulfur is orthorhombic. At standard conditions, the formation of pyrite is exothermic having an enthalpy value of -171.5 kJmole^{-1}. Formally, the calculation reads:

$$\Delta_r \boldsymbol{H}_{298} \equiv \Delta_f \boldsymbol{H}^{py}_{298,\, FeS_2} = -0 - 0 + (-171.5) = \underline{-171.5\ \text{kJ·mol}^{-1}}.$$

The standard heats of formation for most minerals are tabulated in thermodynamic tables. They can be used to calculate the *enthalpy of reaction* at the standard conditions. The enthalpy of reaction is therefore termed the *standard enthalpy of reaction,* $\Delta_r \boldsymbol{H}_{298}$.

Example: Consider the reaction:

$$3CaAl_2Si_2O_8 = Ca_3Al_2Si_3O_{12} + 2Al_2SiO_5 + SiO_2.$$

In order to calculate the standard enthalpy of this reaction, the standard enthalpies of formation for the reactant and for the products must be known. The calculation then reads:

$$\Delta_r \boldsymbol{H}_{298} = -3\Delta_f \boldsymbol{H}^{an}_{298,\, CaAl_2Si_2O_8} + \Delta_f \boldsymbol{H}^{grt}_{298,\, Ca_3Al_2Si_3O_{12}} + 2\Delta_f \boldsymbol{H}^{sill}_{298,\, Al_2SiO_5} + \Delta_f \boldsymbol{H}^{\alpha-qtz}_{298,\, SiO_2}.$$

The required standard enthalpies of formation are found in Robie and Hemingway (1995). The following values are given:

$$\Delta_f \boldsymbol{H}^{an}_{298,\, CaAl_2Si_2O_8} = -4234.0 \text{ kJmol}^{-1},$$

$$\Delta_f \boldsymbol{H}^{grt}_{298,\, Ca_3Al_2Si_3O_{12}} = -6640.0 \text{ kJmol}^{-1},$$

$$\Delta_f \boldsymbol{H}^{sill}_{298,\, Al_2SiO_5} = -2586.1 \text{ kJmol}^{-1} \text{ and}$$

$$\Delta_f \boldsymbol{H}^{\alpha-qtz}_{298,\, SiO_2} = -910.7 \text{ kJmol}^{-1}.$$

Using them, the standard enthalpy of reaction is calculated as follows:

$$\Delta_r \boldsymbol{H}_{298} = -3(-4234.0) + (-6640.0) + 2(-2586.1) + (-910.7) = \underline{-20.9 \text{ kJ}}.$$

Hence, if the reaction 3 anorthite = grossular + 2 sillimanite + quartz could take place at standard conditions, 20.9 kJ of heat would be released to the surroundings. Although, this is a high-pressure high-temperature reaction, the calculation is reasonable, because the temperature and pressure dependence of the heat of formation of the reactants and products is similar and the heat of reaction does not change strongly with temperature and pressure.

3.5.1 Temperature dependence of the enthalpy of reaction

The temperature dependence of the enthalpy of reaction is given by the Kirchhoff's law that reads:

$$\left(\frac{\partial \Delta_r \boldsymbol{H}}{\partial T}\right)_P = \Delta_r \boldsymbol{C}_p, \tag{3.89}$$

where $\Delta_r \boldsymbol{C}_p$ gives the change in the heat capacity of the reacting system. It represents the stoichiometric sum of the molar heat capacities of the components that participate in the reaction. One can write:

$$\Delta_r \boldsymbol{C}_p = \sum_1^k \nu_i \boldsymbol{C}_{p,i}. \tag{3.90}$$

In Eq. (3.90), $\boldsymbol{C}_{p,i}$ and ν_i are the molar heat capacity of the i-th component and the corresponding stoichiometric coefficient, respectively.

In the case of a general reaction (see Eq. (2.108)), the heat capacity change is given by:

$$\Delta_r \boldsymbol{C}_p = -\nu_a \boldsymbol{C}_{p,A} - \nu_b \boldsymbol{C}_{p,B} + \nu_c \boldsymbol{C}_{p,C} + \nu_d \boldsymbol{C}_{p,D}. \tag{3.91}$$

Using Kirchhoff's law, the enthalpy of reaction can be calculated for any arbitrary temperature, provided that the heat capacities of the components are known for the temperature range considered. For example, the transformation from temperature T_1 to temperature T_2 reads:

$$\Delta_r \boldsymbol{H}_{T_2} = \Delta_r \boldsymbol{H}_{T_1} + \int_{T_1}^{T_2} \Delta_r \boldsymbol{C}_p dT. \tag{3.92}$$

Because the heat capacities are also functions of temperature, their temperature dependence must also be known. If this is not the case, different approximations of the temperature dependence for the heat capacity change associated with the reaction can be made. The simplest case is to assume that the heat capacity change is zero ($\Delta_r \boldsymbol{C}_p = 0$) and the enthalpy of reaction is constant. Another approximation is to assume that the heat capacity change remains constant over the temperature range considered ($\Delta_r \boldsymbol{C}_p = a$, where a is a constant). This means that the heat of reaction is a linear function of temperature. Both approximations yield reasonable results only in the case when no fluid phases are involved in the reaction and when the temperature interval is small.

Example: Consider the reaction: 3 anorthite = grossular + 2 sillimanite + quartz once again. The calculated value for the heat of reaction holds for 298.15 K. If we want to convert its value to 800 K, the molar heat capacities and the temperature dependence of the reactants and the products are required. The following C_p polynomials are given by Robie and Hemingway (1995):

$$C_{p,\,CaAl_2Si_2O_8}^{an} = 516.8 - 9.249\times10^{-2}T - 1.408\times10^{6}T^{-2} - 4.589\times10^{3}T^{-1/2} + 4.188\times10^{-5}T^{2}\,[\mathrm{Jmol^{-1}K^{-1}}],$$

$$C_{p,\,Ca_3Al_2Si_3O_{12}}^{grt} = 1529.3 - 0.699T + 7.443\times10^{6}T^{-2} - 1.894\times10^{4}T^{-1/2} + 2.53\times10^{-4}T^{2}\,[\mathrm{Jmol^{-1}K^{-1}}],$$

$$C_{p,\,Al_2SiO_5}^{sill} = 280.19 - 6.9\times10^{-3}T - 1.376\times10^{6}T^{-2} - 2.399\times10^{3}T^{-1/2}\,[\mathrm{Jmol^{-1}K^{-1}}] \quad \text{and}$$

$$C_{p,\,SiO_2}^{\alpha-qtz} = 81.145 + 1.828\times10^{-2}T - 1.81\times10^{5}T^{-2} - 6.985\times10^{2}T^{-1/2} + 5.406\times10^{-6}T^{2}\,[\mathrm{Jmol^{-1}K^{-1}}].$$

There are different ways to evaluate the integrals in Eq. (3.92). One can first sum the C_p polynomials of the components according to their stoichiometric proportions to obtain $\Delta_r C_p$ as a function of temperature and then integrate it over the temperature range in question. Another way is to calculate the heat contents of the components first and then sum them together by taking into account the stoichiometric coefficients. The latter method is used most frequently, particularly in the case when the system undergoes phase transitions in the temperature range considered.

In our example, we will adopt the second way to calculate the heat contents of the components. One has:

$$[H_{800} - H_{298}]_{CaAl_2Si_2O_8}^{an} = 516.8(800-298) - \frac{9.249\times10^{-2}}{2}(800^2-298^2) + 1.408\times10^{6}\left(\frac{1}{800}-\frac{1}{298}\right) - 2\times4.589\times10^{3}(\sqrt{800} - \sqrt{298}) + \frac{4.188\times10^{-5}}{3}(800^3-298^3) = \underline{136.6\ \mathrm{kJmol^{-1}}},$$

$$[\boldsymbol{H}_{800} - \boldsymbol{H}_{298}]^{grt}_{Ca_3Al_2Si_2O_{12}} = 1529.3(800-298) - \frac{0.699}{2}(800^2 - 298^2)$$
$$- 7.443\times10^6\left(\frac{1}{800} - \frac{1}{298}\right) - 2\times1.894\times10^4(\sqrt{800}$$
$$- \sqrt{298}) + \frac{2.53\times10^{-4}}{3}(800^3 - 298^3) = \underline{214.2\ \text{kJmol}^{-1}},$$

$$[\boldsymbol{H}_{800} - \boldsymbol{H}_{298}]^{sill}_{Al_2SiO_5} = 280.19(800-298) - \frac{6.9\times10^{-3}}{2}$$
$$\times(800^2 - 298^2) + 1.376\times10^6\left(\frac{1}{800} - \frac{1}{298}\right)$$
$$- 2\times2.399\times10^3(\sqrt{800} - \sqrt{298}) = \underline{83.0\ \text{kJmol}^{-1}}$$

and

$$[\boldsymbol{H}_{800} - \boldsymbol{H}_{298}]^{\alpha-qtz}_{SiO_2} = 81.145(800-298) + \frac{1.828\times10^{-2}}{2}$$
$$\times(800^2 - 298^2) + 1.81\times10^5\left(\frac{1}{800} - \frac{1}{298}\right)$$
$$- 2\times6.985\times10^2(\sqrt{800} - \sqrt{298})$$
$$+ \frac{5.406\times10^{-6}}{3}(800^3 - 298^3) = \underline{30.9\ \text{kJmol}^{-1}},$$

giving

$$\Delta\boldsymbol{H}_{r,\,800} = \Delta\boldsymbol{H}_{r,\,298} - 3[\boldsymbol{H}_{800} - \boldsymbol{H}_{298}]^{an}_{CaAl_2Si_2O_8}$$
$$+ [\boldsymbol{H}_{800} - \boldsymbol{H}_{298}]^{grt}_{Ca_3Al_2Si_3O_{12}} + 2[\boldsymbol{H}_{800} - \boldsymbol{H}_{298}]^{sill}_{Al_2SiO_5}$$
$$+ [\boldsymbol{H}_{800} - \boldsymbol{H}_{298}]^{\alpha-qtz}_{SiO_2},$$

and

$$\Delta_r\boldsymbol{H}_{800} = -20.9\ \text{kJmol}^{-1} - 3\times136.6\ \text{kJmol}^{-1} + 214.2\ \text{kJmol}^{-1}$$
$$+ 2\times83.0\ \text{kJmol}^{-1} + 30.9\ \text{kJmol}^{-1} = -\ \underline{19.6\ \text{kJmol}^{-1}}.$$

At 800 K, the enthalpy of reaction is only 1.3 kJ/mol smaller than at 298 K. Thus, the simplifying assumption of $\Delta_r C_p \approx 0$ does not introduce a large error. This is generally the case when only solids participate in the reaction.

3.5.2 Hess's law

A reaction can occur directly or it can proceed over several intermediate steps. Independently of the way in which a reaction takes place, its value of the enthalpy of reaction will always remain the same. In the case that the intermediate steps are involved, the enthalpy value is the sum of the enthalpy values associated with each of the intermediate reactions. This phenomenon is known as *Hess's law*. It is a consequence of the fact that enthalpy is a function of state, and its change in any process depends only on the initial and final states of the system.

Hess's law is frequently used to determine the enthalpy of reaction for those reactions, that occur at *P*-*T*-conditions at which a direct measurement is not possible.

Example: Suppose that the standard enthalpy of reaction, $\Delta_r \boldsymbol{H}_{298}$, for the decomposition reaction:

$$NaAlSi_3O_8 = NaAlSi_2O_6 + SiO_2 \quad \Delta_r \boldsymbol{H}_{298}$$

is not known and we want to calculate it using Hess's law. For this purpose, reactions are required which can be combined to yield the reaction given above. Furthermore, their standard enthalpy of reaction must be known. These two requirements are met by the following reactions:

$$a)\ NaAlSiO_4 + NaAlSi_3O_8 = 2NaAlSi_2O_6$$

and

$$b)\ NaAlSiO_4 + SiO_2 = NaAlSi_2O_6.$$

If reaction *b*) is subtracted from reaction *a*), the decomposition reaction results. The standard enthalpy of reaction is, therefore, calculated as follows:

$$\Delta_r \boldsymbol{H}_{298} = \Delta_r \boldsymbol{H}_{298,\mathrm{a}} - \Delta_r \boldsymbol{H}_{298,\mathrm{b}}.$$

Using $\Delta_r \boldsymbol{H}_{298,\mathrm{a}}$ = -33.2 kJ and $\Delta_r \boldsymbol{H}_{298,\mathrm{b}}$ = -29.6 kJ, the calculation of the enthalpy of reaction reads:

$$\Delta_r \boldsymbol{H}_{298} = -33.2\ \mathrm{kJ} + 29.6\ \mathrm{kJ} = \underline{-3.6\ \mathrm{kJ}}.$$

The Hess's law provides the basis for high-temperature solution calorimetry. This is an important method used to determine the enthalpy of formation of minerals. The method measures the heat associated upon the dissolution of a compound and its constituent components in an appropriate solvent. The components are often oxides in the case of silicate. The measured heats of solution are summed according to the oxide-based reaction of formation. This yields a value of the same magnitude as the heat of formation. Its sign, however, is reversed, because the dissolution process is opposite in nature to the formation reaction. If the molar heat capacities of oxides and of the silicate are known, the heat of formation from the oxide can be converted to the room temperature value. The standard heat of formation is then obtained by adding the standard heats of formation of the oxides to the heat of formation, which is determined in the calorimetric experiment. The procedure can best be demonstrated by a concrete example.

Faßhauer and Cemič (2001) determined the enthalpy of formation of petalite, $LiAlSi_4O_{10}$. The reaction from the oxides to form petalite is:

$$LiAlO_2 + 4SiO_2 = LiAlSi_4O_{10}.$$

The enthalpies of solution of $LiAlO_2$, SiO_2 and $LiAlSi_4O_{10}$ were measured in a $2PbO{\cdot}B_2O_3$ melt at 1001 K. The following values were obtained:

$$\Delta_{sol}\boldsymbol{H}_{LiAlSi_4O_2} = 32.60 \text{ kJmol}^{-1},$$

$$\Delta_{sol}\boldsymbol{H}_{SiO_2}^{\alpha-qtz} = -3.84 \text{ kJmol}^{-1} \text{ and}$$

$$\Delta_{sol}\boldsymbol{H}^{pet} = 63.68 \text{ kJmol}^{-1}.$$

Taking the enthalpies of solution of each of the component oxides, the enthalpy of formation of petalite is given by:

$$\Delta_f\boldsymbol{H}_{1001}^{pet,\,ox} = -(-\Delta_{sol}\boldsymbol{H}_{LiAlO_2}) - 4\times\left(-\Delta_{sol}\boldsymbol{H}_{SiO_2}^{\alpha-qtz}\right) + \left(-\Delta_{sol}\boldsymbol{H}_{LiAlSi_4O_{10}}^{pet}\right).$$

In the above given equation the enthalpies of solution are negative. This is done to preserve the conventional formalism used to calculate the enthalpy of reaction. In the literature, this formalism is avoided and the he equation is written as:

$$\Delta_f\boldsymbol{H}_{1001}^{pet,\,ox} = \Delta_{sol}\boldsymbol{H}_{LiAlO_2} + 4\Delta_{sol}\boldsymbol{H}_{SiO_2}^{\alpha-qtz} - \Delta_{sol}\boldsymbol{H}_{LiAlSi_4O_{10}}^{pet}.$$

Inserting the measured values, the calculation reads:

$$\Delta_f H_{1001}^{pet,\,ox} = 32.60\ \text{kJmol}^{-1} - 4 \times 3.84\ \text{kJmol}^{-1} - 63.68\ \text{kJmol}^{-1}$$
$$= \underline{-46.44\ \text{kJmol}^{-1}}.$$

In order to obtain the standard enthalpy of formation for petalite, the heat of formation at 1001 K has to be transformed to room temperature and the standard enthalpies of formation of the oxides must be added. One has:

$$\Delta_f H_{298}^{pet} = \Delta_f H_{1001}^{pet,\,ox} + \left\{ -[H_{298} - H_{1001}]_{LiAlO_2} - 4[H_{298} - H_{1001}]_{SiO_2} + [H_{298} - H_{1001}]_{LiAlSi_4O_{10}}^{pet} \right\} + \Delta_f H_{298,\,LiAlO_4} + 4\Delta_f H_{298,\,SiO_2}^{\alpha - qtz}.$$

The enthalpy contents for each component over the temperature interval from 1001 to 298 K, $[H_{298} - H_{1001}]_i$, are calculated using the C_p polynomials given by Robie and Hemingway (1995). We have:

$$[H_{298} - H_{1001}]_{LiAlO_2} = -64.51\ \text{kJmol}^{-1}$$
$$[H_{298} - H_{1001}]_{SiO_2}^{\alpha - qtz} = -45.59\ \text{kJmol}^{-1} \text{ and}$$
$$[H_{298} - H_{1001}]_{LiAlSi_4O_{10}}^{pet} = -240.52\ \text{kJmol}^{-1}.$$

According to Robie and Hemingway (1995), the values for standard heat of formation of α-quartz and $LiAlO_2$ are -910.7 and -1188.67 kJmol^{-1}. Using these data the standard heat of formation of petalite is calculated as:

$$\Delta_f H_{298}^{pet} = -46.44\ \text{kJmol}^{-1} + 64.51\ \text{kJmol}^{-1} + 4 \times 45.59\ \text{kJmol}^{-1}$$
$$- 240.52\ \text{kJmol}^{-1} - 1188.67\ \text{kJmol}^{-1} - 4 \times 910.7\ \text{kJmol}^{-1}$$
$$= \underline{-4871.56\ \text{kJmol}^{-1}}.$$

3.6 Problems

1. Consider a fayalite single crystal with a mass of 10.189 g. At standard P, T conditions, its molar volume equals 46.31 $\text{cm}^3\text{mol}^{-1}$.

- Calculate the work that is performed by the mineral when it is heated to 800°C at constant pressure of 0.1 MPa.

- Calculate the work that is associated with the pressure increase to 2.5 GPa at 800°C.
- Why does the result depend on choice of the formula?

$$\alpha = 3.045\ \text{K}^{-1},$$

$$\beta = 8.64 \times 10^{-12}\ \text{Pa}^{-1}.$$

2. The Einstein temperature, θ_E, of tremolite, $Ca_2Mg_5Si_8O_{22}(OH)_2$, equals 763 K (Kieffer, 1985).

- Calculate the Einstein frequency, ν_E.
- Calculate the internal energy, U, of tremolite at 300 K.
- Calculate the heat capacity at constant volume, C_v, for tremolite at 300 and 500 K.

$$R = 8.3144\ \text{Jmol}^{-1}\text{K}^{-1},$$

$$h = 6.626\times10^{-34}\ \text{Js}.$$

3. Calculate the heat that is released to the surroundings when, at constant pressure, the temperature of 5×10^3 kg diopside, $CaMgSi_2O_6$, is decreased from 1000 to 25°C. The molar mass of diopside is 216.550 g

$$C_{p,\,CaMgSi_2O_6}^{cpx} = 470.25 - 9.864\times10^{-2}T + 2.454\times10^{5}T^{-2} - 4823\,T^{-1/2} + 2.813\times10^{-5}T^{2}\,[\text{Jmol}^{-1}\text{K}^{-1}].$$

4. Charlu et al. (1981) measured, using the high-temperature solution calorimetry, the heat of solution in the system gehlenite ($Ca_2Al_2SiO_7$)-åkermanite ($CaMgSi_2O_7$) and calculated the heat of mixing. Their results can be represented by the following equation:

$$\Delta_m\bar{H}[\text{kJmol}^{-1}] = 24.288 \times (1 - x_{CaMgSi_2O_7}) \cdot (x_{Ca_2MgSi_2O_7})^2 + 0.502 \times (1 - x_{Ca_2MgSi_2O_7})^2 \cdot x_{Ca_2MgSi_2O_7}. \tag{3.93}$$

- Which type of mixtures does the melilite solid solution represent?
- Calculate the partial molar enthalpies of mixing of the components $Ca_2Al_2SiO_7$ and $Ca_2MgSi_2O_7$ in the case that the solid solution contains 0.3 mole åkermanite.
- Calculate the partial molar enthalpies of mixing of the components at infinite dilution.

- Calculate the molar enthalpy of mixing at $x_{Ca_2MgSi_2O_7} = 0.3$ using the partial molar excess enthalpies of the components at $x_{Ca_2MgSi_2O_7} = 0.3$.

5. Calculate the change in the enthalpy associated with the following reaction:

forsterite + sillimanite → cordierite + spinel

in the case that it takes place at constant pressure of 0.1 MPa and at temperatures 298 K and 1073 K, respectively.

$$\Delta_f H^{fo}_{298,\, Mg_2SiO_4} = -2171.87\ [\text{kJmol}^{-1}],$$

$$\Delta_f H^{sill}_{298,\, Al_2SiO_5} = -2586.67\ [\text{kJmol}^{-1}],$$

$$\Delta_f H^{crd}_{298,\, Mg_2Al_3AlSi_5O_{18}} = -9166.50\ [\text{kJmol}^{-1}]\ \text{and}$$

$$\Delta_f H^{sp}_{298,\, MgAl_2O_4} = -2303.57\ [\text{kJmol}^{-1}].$$

$$C^{fo}_{p,\, Mg_2SiO_4} = 0.2349 + 0.1069\times10^{-5}T - 542.9T^{-2} - 1.9064T^{-1/2}[\text{kJmol}^{-1}\text{K}^{-1}],$$

$$C^{sill}_{p,\, Al_2SiO_5} = 0.2261 + 1.407\times10^{-5}T - 2440T^{-2} - 1.376T^{-1/2}[\text{kJmol}^{-1}\text{K}^{-1}],$$

$$C^{sp}_{p,\, MgAl_2O_4} = 0.2229 + 0.6127\times10^{-5}T - 1685.7T^{-2} - 1.5512T^{-1/2}[\text{kJmol}^{-1}\text{K}^{-1}]$$

and

$$C_{p,\, Mg_2Al_3AlSi_5O_{18}} = 0.8213 + 4.3339\times10^{-5}T - 8211.2T^{-2} - 5.000T^{-1/2}[\text{kJmol}^{-1}\text{K}^{-1}]$$

(Data: Holland and Powell 1990).

6. If anorthite undergoes a reaction with H_2O to give zoisite , kyanite and quartz at 0.1 MPa and 800°C a heat of 67.432 kJ is released to the surroundings. Under the same conditions the reaction anorthite + grossular + H_2O → zoisite + quartz produces a heat of 113.124 kJ.

- Use the Hess's law and calculate the change in the enthalpy associated with the reaction:

anorhtite → grossular + kyanite + quartz.

- Calculate the enthalpy of reaction at 0.1 MPa and 298 K.

$$C_{p,\,CaAl_2Si_2O_8}^{an} = 297.022 + 43.388\times10^{-3}T - 13.535\times10^{6}T^{-2}[\text{Jmol}^{-1}\text{K}^{-1}],$$

$$C_{p,\,Ca_3Al_2Si_3O_{12}}^{grt} = 456.307 + 49.204\times10^{-3}T - 13.142\times10^{6}T^{-2}[\text{Jmol}^{-1}\text{K}^{-1}],$$

$$C_{p,\,Al_2SiO_5}^{ky} = 183.770 + 17.100\times10^{-3}T - 6.121\times10^{6}T^{-2}[\text{Jmol}^{-1}\text{K}^{-1}],$$

$$C_{p,\,SiO_2}^{\alpha-qtz} = 40.497 + 44.601\times10^{-3}T - 0.833\times10^{6}T^{-2}[\text{Jmol}^{-1}\text{K}^{-1}] \text{ and}$$

$$C_{p,\,SiO_2}^{\beta-qtz} = 67.593 + 2.577\times10^{-3}T - 0.138\times10^{6}T^{-2}[\text{Jmol}^{-1}\text{K}^{-1}]$$

$$\Delta_{tr}H_{SiO_2}^{\alpha/\beta} = 728 \text{ Jmol}^{-1}$$

$$T_{tr}^{\alpha/\beta} = 847\text{K}.$$

(Data: Knacke et al. 1991)

7. Akaogi et al. (1984) dissolved the three Mg_2SiO_4 polymorphs: olivine (α), modified spinel (β) and spinel (γ) in a $2PbO{\cdot}B_2O_3$ melt at 975 K and measured the heat of dissolution. Following results were obtained:

$$\Delta_{sol}H_{Mg_2SiO_4}^{\alpha} = 67128 \text{ Jmol}^{-1},$$

$$\Delta_{sol}H_{Mg_2SiO_4}^{\beta} = 37158 \text{ Jmol}^{-1} \text{ and}$$

$$\Delta_{sol}H_{Mg_2SiO_4}^{\gamma} = 30338 \text{ Jmol}^{-1}.$$

- Calculate the change in the enthalpy associated with the phase transitions $\alpha \rightarrow \beta$ and $\beta \rightarrow \gamma$ at the temperature of the calorimeter (975 K) and also at 298 K.
- Calculate the standard enthalpies of formation of modified spinel (β) and spinel (γ).

$$C_{p,\,Mg_2SiO_4}^{\alpha} = 155.854 + 22.23\times10^{-3}T - 40.945\times10^{5}T^{-2}[\text{Jmol}^{-1}\text{K}^{-1}],$$

$$C_{p,\,Mg_2SiO_4}^{\beta} = 151.837 + 23.23\times10^{-3}T - 43.133\times10^{5}T^{-2}[\text{Jmol}^{-1}\text{K}^{-1}] \text{ and}$$

$$C_{p,\,Mg_2SiO_4}^{\gamma} = 155.226 + 14.46\times10^{-3}T - 47.848\times10^{5}T^{-2}[\text{Jmol}^{-1}\text{K}^{-1}].$$

(C_p reproduced using the data from Watanabe, 1982)

The standard enthalpy of formation of olivine (α) equals -2173.0 kJmol^{-1} (Robie and Hemingway 1995).

Chapter 4 Second law of thermodynamics

It is common experience that all spontaneous processes tend to proceed only in one direction. Water always flows downhill, that is, from a high gravitational potential to a lower potential. Salt always diffuses from a state of high concentration to one of lower concentration. Heat flows from a state of higher to lower temperature etc. Processes of this kind are called *natural*. During the course of a natural process, a system changes from an initial to a final state. The two states are entirely determined by the variables of state. The two previously defined state functions internal energy, U, and enthalpy, H, have some definite values. However, these functions tell nothing about the direction of a spontaneous change. Their values can increase as well as decrease as the two following examples show.

Example 1: Consider the reaction

$$CaCO_3 + SiO_2 = CaSiO_3 + CO_2.$$

At 0.1 MPa and 600 K, the value of the enthalpy of reaction is 86.5 kJ. Thus, heat must be transferred from the surroundings to the system in order to keep the temperature of the reacting system constant. Nonetheless, the products are more stable than reactants and the reaction proceeds from left to the right.

Example 2: At 0.1 MPa and 900 K, the enthalpy of reaction

$$NaAlSi_3O_8 = NaAlSi_2O_6 + SiO_2$$

is - 13.99 kJ. The negative sign means that heat is released to the surroundings under the given conditions. One could think that the reaction occurs spontaneously. However, this is not the case. At 0.1 MPa and 900 K, albite is the stable phase.

The search for a state function that describes the tendency to proceed for all processes, led to the definition of *entropy*, S. This function is central to the *second law of thermodynamics*. The entropy measures the tendency for spontaneous change. Combined with the functions defined by the first law, the entropy provides information about the spontaneous direction of chemical reactions.

4.1 Entropy

4.2 Classical definition of entropy

In thermodynamics a distinction is made between *reversible* and *irreversible* processes. Natural processes are irreversible. They proceed in only one direction, that is from a nonequilibrium to an equilibrium state and they can be reversed only by an external agent that leaves changes in the surroundings. In the case of a chemical reaction, the term irreversible applies only for some set of variables. If one or more variables are changed, the direction of irreversibility can change. For example, the reaction

$$Ca_3Al_2Si_3O_{12} + SiO_2 = CaAl_2Si_2O_8 + 2CaSiO_3$$

proceeds from left to right at 0.2 GPa and 1000 K (Huckenholz et al. 1975) while at 0.2 GPa and 700 K, it proceeds from right to left. In either case, the reaction is irreversible.

A reversible process is one that proceeds in such infinitesimally small steps that the system remains at equilibrium all the time. Reversible processes can be closely approached but never realized completely.

In the case that both heat and work are exchanged between a system and its surroundings, δQ is an inexact differential. However, if δQ is divided by T, a new function of state, called *entropy* S, is obtained. It is

$$\frac{\delta Q_{rev}}{T} = dS. \tag{4.1}$$

The subscript '*rev*' means that the relation in Eq. (4.1) holds only for reversible processes. 1/T is called the *integrating factor*. It transforms an inexact differential into an exact one.

The entropy of an isolated system remains constant only if reversible processes occur, that is

$$dS_{rev} = 0. \tag{4.2}$$

In the case that an irreversible process takes place, the entropy change for an isolated system is always positive,

$$dS_{irrev} > 0. \tag{4.3}$$

According to Eqs. (4.2) and (4.3), the entropy of an isolated system can either increase or remain constant. It can never decrease. This is the most important consequence of the second law of thermodynamics.

In the case of a closed system where the heat exchange with the surroundings takes place, the statements given in Eqs. (4.2) and (4.3) hold for the total entropy change (system and surroundings). It is positive for irreversible (natural) and zero for the reversible processes. Mathematically this statement reads:

$$dS_{tot} = dS_{syst} + dS_{surr} \geq 0. \tag{4.4}$$

Rearranging Eq. (4.1) and substituting it into Eq. (3.16) yields a useful form of the first law of thermodynamics for reversible processes, namely:

$$dU = TdS - PdV. \tag{4.5}$$

If Eq. (4.1) is substituted into Eq. (3.23), another important relationship is obtained:

$$dH = TdS + VdP. \tag{4.6}$$

On the atomistic scale, the entropy can be understood as a measure of the degree of disorder. The greater it is, the higher the entropy and vice versa. For substances in different states of aggregation the entropy increases from solid to gaseous state. It is:

$$S_{solid} < S_{fluid} < S_{gas}.$$

The physical interpretation of entropy is a subject matter of statistical mechanics. and will be presented in a separate paragraph to follow.

4.2.1 Entropy of pure phases

If a closed system consists of a pure single phase, only two intensive properties are required to determine exactly the value of any state function. Entropy is a state function and can thus be defined as a function of temperature and volume,

$$S(V, T) \tag{4.7}$$

or as a function of temperature and pressure,

$$S(T, P). \tag{4.8}$$

Differentiating Eqs (4.7) and (4.8) and combining the result with Eq. (4.1) yields:

$$dS = \frac{\delta Q_{rev}}{T} = \left(\frac{\partial S}{\partial T}\right)_V dT + \left(\frac{\partial S}{\partial V}\right)_T dV \tag{4.9}$$

and

$$dS = \frac{\delta Q_{rev}}{T} = \left(\frac{\partial S}{\partial T}\right)_P dT + \left(\frac{\partial S}{\partial P}\right)_T dP, \tag{4.10}$$

respectively.

Substituting the first law definitions for δQ_{rev} in Eqs. (4.9) and (4.10), a relation between the first and second law is obtained, namely

$$dS = \frac{dU + PdV}{T} = \frac{1}{T}\left\{\left(\frac{\partial U}{\partial T}\right)_V dT + \left[\left(\frac{\partial U}{\partial V}\right)_T + P\right]dV\right\} \tag{4.11}$$

and

$$dS = \frac{dH - VdP}{T} = \frac{1}{T}\left\{\left(\frac{\partial H}{\partial T}\right)_P dT + \left[\left(\frac{\partial H}{\partial P}\right)_T - V\right]dP\right\}. \tag{4.12}$$

According to Eqs. (3.31) and (3.35), the partial differentials $\left(\frac{\partial U}{\partial T}\right)_V$ and $\left(\frac{\partial H}{\partial T}\right)_P$ correspond to c_v and c_p, respectively. Thus, one can write:

$$dS = \frac{c_v}{T}dT + \frac{1}{T}\left[\left(\frac{\partial U}{\partial V}\right)_T + P\right]dV \tag{4.13}$$

and

$$dS = \frac{c_p}{T}dT + \frac{1}{T}\left[\left(\frac{\partial H}{\partial P}\right)_T - V\right]dP. \tag{4.14}$$

For isochoric processes, where $dV = 0$ one has:

$$dS = \frac{c_v}{T}dT. \tag{4.15}$$

That is, the change in entropy with temperature at constant volume equals the heat capacity of the system at constant volume, divided by the temperature.

Analogously, it holds for isobaric processes, where $dP = 0$, that

$$dS = \frac{c_P}{T} dT. \tag{4.16}$$

Here, the change in entropy with temperature at constant pressure corresponds to the heat capacity of a system at constant pressure, divided by the temperature.

In order to express the quantities in square brackets in Eqs. (4.11) through (4.14) by more convenient terms, we use the fact that entropy is a state function. A comparison of the coefficients in Eqs. (4.9) through (4.14) shows that:

$$\left(\frac{\partial S}{\partial T}\right)_V = \frac{1}{T}\left(\frac{\partial U}{\partial T}\right)_V, \tag{4.17}$$

$$\left(\frac{\partial S}{\partial V}\right)_T = \frac{1}{T}\left[\left(\frac{\partial U}{\partial V}\right)_T + P\right], \tag{4.18}$$

$$\left(\frac{\partial S}{\partial T}\right)_P = \frac{1}{T}\left(\frac{\partial H}{\partial T}\right)_P \tag{4.19}$$

and

$$\left(\frac{\partial S}{\partial P}\right)_T = \frac{1}{T}\left[\left(\frac{\partial H}{\partial P}\right)_T - V\right]. \tag{4.20}$$

Because entropy is a function of state, the cross-differentiation identity holds, which means that the order of differentiation is irrelevant. Hence, the second derivative of Eq. (4.17) with respect to volume and the second derivative of Eq. (4.18) with respect to temperature are equal,

$$\frac{\partial}{\partial V}\left(\frac{\partial S}{\partial T}\right)_V = \frac{\partial}{\partial T}\left(\frac{\partial S}{\partial V}\right)_T \tag{4.21}$$

or

$$\frac{1}{T}\left[\frac{\partial}{\partial V}\left(\frac{\partial U}{\partial T}\right)_V\right] = \frac{\partial}{\partial T}\left\{\frac{1}{T}\left[\left(\frac{\partial U}{\partial V}\right)_T + P\right]\right\}$$

and

$$\frac{1}{T}\left[\frac{\partial}{\partial V}\left(\frac{\partial U}{\partial T}\right)_V\right] = \frac{1}{T}\left[\frac{\partial}{\partial T}\left(\frac{\partial U}{\partial V}\right)_T + \left(\frac{\partial P}{\partial T}\right)_V\right] - \frac{1}{T^2}\left[\left(\frac{\partial U}{\partial V}\right)_T + P\right].$$

Canceling the identical terms and rearrangement yields:

$$\left[\left(\frac{\partial U}{\partial V}\right)_T + P\right] = T\left(\frac{\partial P}{\partial T}\right)_V. \tag{4.22}$$

The partial derivative $(\partial P/\partial T)_V$ gives the change in pressure with temperature at constant volume of the system. According to Eq. (2.9), this derivative can be expressed through accessible quantities: the thermal coefficient, α, and compressibility, β, as follows:

$$\left(\frac{\partial P}{\partial T}\right)_V = \frac{\alpha}{\beta}.$$

Substituting the thermal expansion and compressibility into Eq. (4.22) gives:

$$\left[\left(\frac{\partial U}{\partial V}\right)_T + P\right] = T\frac{\alpha}{\beta} \tag{4.23}$$

or

$$\left(\frac{\partial U}{\partial V}\right)_T = T\frac{\alpha}{\beta} - P. \tag{4.24}$$

Hence, the change in the internal energy with volume at constant temperature, $(\partial U/\partial V)_T$, is a function of pressure, thermal expansion and compressibility.

In an analogous way, one can find a more convenient expression for the terms in the square brackets of Eqs. (4.12) and (4.14). The starting point is again the cross-differentiation identity, namely

$$\frac{\partial}{\partial P}\left(\frac{\partial S}{\partial T}\right)_P = \frac{\partial}{\partial T}\left(\frac{\partial S}{\partial P}\right)_T \tag{4.25}$$

and therefore

$$\frac{1}{T}\left[\frac{\partial}{\partial P}\left(\frac{\partial H}{\partial T}\right)_P\right] = \frac{\partial}{\partial T}\left\{\frac{1}{T}\left[\left(\frac{\partial H}{\partial P}\right)_T - V\right]\right\}$$

and

$$\frac{1}{T}\left[\frac{\partial}{\partial P}\left(\frac{\partial H}{\partial T}\right)_P\right] = \frac{1}{T}\left[\frac{\partial}{\partial T}\left(\frac{\partial H}{\partial P}\right)_T - \left(\frac{\partial V}{\partial T}\right)_P\right] - \frac{1}{T^2}\left[\left(\frac{\partial H}{\partial P}\right)_T - V\right].$$

Thus

$$\left[\left(\frac{\partial H}{\partial P}\right)_T - V\right] = -T\left(\frac{\partial V}{\partial T}\right)_P. \tag{4.26}$$

According to Eq. (2.3) the partial derivative $(\partial V/\partial T)_P$, can be replaced by αV and one obtains:

$$\left[\left(\frac{\partial H}{\partial P}\right)_T - V\right] = -T\alpha V. \tag{4.27}$$

A rearrangement of Eq. (4.27) yields an expression for the partial derivative $(\partial H/\partial P)_T$ that consists of experimentally accessible variables, such as volume, temperature and the thermal expansion coefficient, α, namely:

$$\left(\frac{\partial H}{\partial P}\right)_T = V - T\alpha V. \tag{4.28}$$

Eq. (4.22) can be used to derive the relationship between the molar heat at constant volume and the molar heat at constant pressure, given in Eq. (3.37). We start with the total differential of enthalpy that combines dH and dU, namely

$$dH = dU + PdV + VdP.$$

Dividing this equation by dT and keeping the pressure constant ($dP = 0$) yields:

$$\left(\frac{\partial H}{\partial T}\right)_P = \left(\frac{\partial U}{\partial T}\right)_P + P\left(\frac{\partial V}{\partial T}\right)_P. \tag{4.29}$$

In order to eliminate the partial derivative $(\partial U/\partial T)_P$ in Eq. (4.29), the total differential of internal energy

$$dU = \left(\frac{\partial U}{\partial T}\right)_V dT + \left(\frac{\partial U}{\partial V}\right)_T dV$$

is divided by dT at constant pressure. The result is:

$$\left(\frac{\partial U}{\partial T}\right)_P = \left(\frac{\partial U}{\partial T}\right)_V + \left(\frac{\partial U}{\partial V}\right)_T \left(\frac{\partial V}{\partial T}\right)_P. \tag{4.30}$$

Substituting Eq. (4.30) into Eq. (4.29) gives

$$\left(\frac{\partial H}{\partial T}\right)_P = \left(\frac{\partial U}{\partial T}\right)_V + \left[\left(\frac{\partial U}{\partial V}\right)_T + P\right]\left(\frac{\partial V}{\partial T}\right)_P$$

or

$$C_p = C_v + \frac{TV\alpha^2}{\beta}, \tag{4.31}$$

if one mole of substance is considered.

4.2.2 Adiabatic changes

If a system is completely thermally isolated, its internal energy can only be changed by the work done on or by the system. Thus,

$$(dU)_Q = -PdV. \tag{4.32}$$

Using Eq. (4.24) and the definition of the molar heat at constant volume as given in Eq. (3.31), the total differential of the internal energy reads:

$$dU = c_v dT + \left(T\frac{\alpha}{\beta} - P\right) dV. \tag{4.33}$$

In the case of an adiabatic process the change in the internal energy corresponds to the work increment, - PdV, (see Eq. (4.32)), therefore

$$c_v dT + T\frac{\alpha}{\beta} dV = 0. \tag{4.34}$$

Substituting dV by the expression given in Eq. (2.7) yields:

$$\left(c_v + \frac{\alpha^2 VT}{\beta}\right)dT = TV\alpha dP. \tag{4.35}$$

According to Eq. (4.31), the term in the parenthesis corresponds to the heat capacity of a system at constant pressure, c_p. Thus for one mole of a pure substance Eq. (4.35) is modified as follows:

$$\boldsymbol{C}_p dT = TV\alpha dP. \tag{4.36}$$

Dividing Eq. (4.36) by dP and $\boldsymbol{C}_p$ and considering the fact that this relationship holds for adiabatic processes yields:

$$\left(\frac{dT}{dP}\right)_Q = \frac{T\alpha V}{\boldsymbol{C}_p}. \tag{4.37}$$

Eq. (4.37) gives the change in temperature with pressure when the heat exchange between the system and its surroundings does not take place. This kind of processes play an important role in geosciences.

4.2.3 Temperature dependence of entropy

Entropy as a function of temperature is obtained by integrating the expressions in Eqs. (4.15) and (4.16) over the temperature range from 0 to T. One has

$$S = \int_0^T \frac{c_v}{T} dT \tag{4.38}$$

and

$$S = \int_0^T \frac{c_p}{T} dT. \tag{4.39}$$

Replacing the heat capacities c_v and c_p in Eqs. (4.38) and (4.39) by the corresponding molar quantities $\boldsymbol{C}_v$ and $\boldsymbol{C}_p$, respectively, yields the *molar entropy*, $\boldsymbol{S}$. It is

$$S = \int_0^T \frac{C_v}{T} dT \tag{4.40}$$

and

$$S = \int_0^T \frac{C_p}{T} dT. \tag{4.41}$$

The molar entropy has the same dimension as the molar heat capacity, namely $\mathrm{J mole^{-1} K^{-1}}$. The lower integration limit in Eqs. (4.40) and (4.41) corresponds with the entropy at absolute zero, S_0. It is in principle not known and is assumed to be *zero* for pure solids that are in *internal equilibrium*. Internal equilibrium means that all particles constituting the crystal are in their proper crystallographic positions. In other words, the crystal is perfect. This assumption is in accordance with Planck's version of the *third law of thermodynamics* that states that entropy of pure substances is de facto zero at 0 K. Thus Eqs. (4.40) and (4.41) give the 'absolute value' of the entropy. It is referred to as '*Third law*' or '*conventional*' entropy. The expression:

$$S_{298} = \int_0^{298} \frac{C_p}{T} dT \tag{4.42}$$

gives the entropy of a substance at room temperature. It is referred to as the *conventional standard entropy*. Note that the conventional standard entropy is not comparable to the standard enthalpy of formation, $\Delta_f H_{298}$, that refers to the reaction forming the compound from the elements under standard conditions. This quantity is zero for all elements. The conventional standard entropy of any element, however, is always a positive non-zero value.

The conventional standard entropy, together with the heat capacity polynomial, can be used to calculate the entropy at any temperatures of interest. Using Eq. (3.59) the calculation reads:

$$S_T = S_{298} + \int_{298}^{T} \frac{(a + bT + cT^{-2} + dT^{-1/2} + eT^2)}{T} dT \tag{4.43}$$

$$= S_{298} + a \times \ln\frac{T}{298} + b(T-298) - \frac{c}{2}\left(\frac{1}{T^2} - \frac{1}{298^2}\right)$$

$$-2d\left(\frac{1}{\sqrt{T}} - \frac{1}{\sqrt{298}}\right) + \frac{e}{2}(T^2 - 298^2).$$

Example: The conventional standard entropy of forsterite, Mg_2SiO_4, is 94.1 $Jmole^{-1}K^{-1}$ (Robie and Hemingway 1995). Using this value and the C_p polynomial given in Tab. 3.2 we can calculate the entropy for forsterite at 800 K as follows:

$$S_{800}^{fo} = 94.1 + 87.36\ln\frac{800}{298} + 87.17\times10^{-3}(800-298)$$

$$+\frac{3.699\times10^6}{2}\left(\frac{1}{800^2} - \frac{1}{298^2}\right) - 2\times 0.8436\times10^3$$

$$\times\left(\frac{1}{\sqrt{800}} - \frac{1}{\sqrt{298}}\right) - \frac{2.237\times10^{-5}}{2}(800^2 - 298^2)$$

$$= \underline{238.1\ \mathrm{Jmol^{-1}K^{-1}}}.$$

In the case where a phase transition occurs in the temperature range between 298.15 K and T, the entropy change associated with the transition has to be taken into account. Under these circumstances, the calculation of the entropy reads:

$$S_T = S_{298} + \int_{298}^{T_{tr}} \frac{C_{p,1}}{T} dT + \Delta_{tr} S + \int_{T_{tr}}^{T} \frac{C_{p,2}}{T} dT. \tag{4.44}$$

In Eq. (4.44) T_{tr}, $C_{p,1}$ and $C_{p,2}$ designate the transition temperature, the molar heat capacity of the low temperature phase and the molar heat capacity of the high temperature phase, respectively. $\Delta_{tr}S$ is the molar entropy of transition. It represents the difference between the entropy of phase 1 and phase 2 at the temperature of transition

$$\Delta_{tr}S = (S_2 - S_1)_{T_{tr}}. \tag{4.45}$$

There is a close relationship between the molar entropy of transition and molar enthalpy of transition, namely

$$\Delta_{tr}\boldsymbol{S} = \frac{\Delta_{tr}\boldsymbol{H}}{T_{tr}}. \tag{4.46}$$

4.2.4 Entropy changes associated with irreversible processes

The total entropy change associated with an irreversible process is according to Eq. (4.4) greater than zero. Thus, for an finite change of state it holds that

$$\Delta S_{tot} = \Delta S_{syst} + \Delta S_{surr} > 0. \tag{4.47}$$

The validity of this statement can be demonstrated with a simple example:

Consider the transformation of sillimanite to andalusite at 0.1 MPa and 900 K, which is, following Holdaway, within the stability field of andalusite. His phase diagram places the inversion curve andalusite/sillimanite at 1048 K and 0.1 MPa. The transformation at 900 K and 0.1 MPa is, thus, an irreversible process and infinitesimal changes in temperature will not change the direction of reaction. Hence, the entropy of this inversion can not be calculated using Eq. (4.46) that holds for reversible processes only. In order to solve the problem the process of transformation must be carried out over several reversible steps, such as:

a) temperature increase of sillimanite from 900 to 1048 K (the transformation temperature)
b) transformation of sillimanite to andalusite at 1048 K
c) temperature decrease of andalusite to 900 K.

The reversible steps and the associated changes in the entropy can be written as follows:

$$\begin{array}{lll}
a) & Al_2SiO_5^{sill}(900) \rightarrow Al_2SiO_5^{sill}(1048) & \Delta\boldsymbol{S}_a \\
b) & Al_2SiO_5^{sill}(1048) \rightarrow Al_2SiO_5^{and}(1048) & \Delta\boldsymbol{S}_b \\
c) & Al_2SiO_5^{and}(1048) \rightarrow Al_2SiO_5^{and}(900) & \Delta\boldsymbol{S}_c \\
\hline
d) & Al_2SiO_5^{sill}(1048) \rightarrow Al_2SiO_5^{and}(900) & \Delta\boldsymbol{S}_d
\end{array}$$

Reaction *d*) is the sum of reactions *a*) through *c*). It describes the transformation of sillimanite to andalusite at 900 K and 0.1 MPa. Consequently, $\Delta\boldsymbol{S}_d$ is the entropy change associated with this transition. It represents the sum of the entropy changes associated with the reversible steps *a*) through *c*) and is, therefore, identical to the change in the entropy of the system. On can write

$$\Delta S_d = \Delta S_{sys} = \Delta S_a + \Delta S_b + \Delta S_c.$$

The entropy change associated with each single step is calculated as follows:

$$\Delta S_a = \int_{900}^{1048} \frac{C_p^{sill}}{T} dT,$$

$$\Delta S_b = -\frac{\Delta_{tr} H_{eq}^{and/sill}}{1048},$$

$$\Delta S_c = -\int_{900}^{1048} \frac{C_p^{and}}{T} dT$$

and

$$\Delta_d S = \Delta S_{sys} = \int_{900}^{1048} \frac{(C_p^{sill} - C_p^{and})}{T} dT - \frac{\Delta_{tr} H_{eq}^{and/sill}}{1048},$$

where $\Delta_{tr} H_{eq}^{and/sill}$ is the heat of transformation at equilibrium temperature (1048 K). It is negative, because the tabulated values refer to the transformation from the low to the high-temperature form. In our case, however, the transformation occurs from the high-temperature to the low temperature form.

Assume that the system's surroundings is a heat reservoir that is large enough to maintain the same temperature when heat is transferred from it to the system or vice-versa. Assume further that the walls of the system are perfectly diathermic, such that the heat exchange occurs reversibly. Consider what happens in the surroundings in the course of the transformation process. Because the heat transfer between the system and the surroundings occurs reversibly, the entropy change of the surroundings is given by:

$$\Delta S_{surr} = \frac{\Delta_{tr} H_{900}^{and/sill}}{900},$$

where $\Delta_{tr}\boldsymbol{H}_{900}^{and/sill}$ is the heat of transformation for the inversion andalusite → sillimanite at 900 K. It has a positive sign, because during the course of the transition heat flows from the system to the surroundings. Normally, the heat of transformation is determined experimentally at the equilibrium temperature, which in our case would be 1048 K. Hence, the tabulated heat of transformation has to be transformed to 900 K. This transformation is carried out using Kirchhoff's law and the calculation reads:

$$\Delta_{tr}\boldsymbol{H}_{900}^{\mathrm{and/sill}} = \Delta_{tr}\boldsymbol{H}_{eq}^{and/sill} - \int_{900}^{1048} \Delta_{tr}\boldsymbol{C}_p^{and/sill}\, dT,$$

where $\Delta_{tr}\boldsymbol{C}_p^{and/sill}$ is the heat capacity change associated with the transition andalusite → sillimanite.

The total change in the entropy as a result of the irreversible transition of sillimanite to andalusite at 900 K and 0.1 MPa is given by the sum of the entropy change of the system and the surroundings.

$$\Delta\boldsymbol{S}_{tot} = \Delta\boldsymbol{S}_{sys} + \Delta\boldsymbol{S}_{surr}.$$

The data required to calculate the numerical values for this example are taken from Robie and Hemingway (1995), where

$$\boldsymbol{C}_p^{and} = 277.306 - 6.588\times10^{-3}T - 1.9141\times10^{6}T^{-2} - 2.2656\times10^{3}T^{-1/2}[\mathrm{Jmol^{-1}K^{-1}}],$$

$$\boldsymbol{C}_p^{sill} = 280.19 - 6.900\times10^{-3}T - 1.376\times10^{6}T^{-2} - 2.399\times10^{3}T^{-1/2}[\mathrm{Jmol^{-1}K^{-1}}]$$

and

$$\Delta_{tr}\boldsymbol{H}_{eq}^{and/sill} = 3.07\ \mathrm{kJmol^{-1}}.$$

Using these data, the change in the entropy of the system associated with the transition of sillimanite to andalusite at 900 K is calculated as follows:

$$\Delta S_{sys} = \int_{900}^{1048} \frac{(2.884 - 3.12\times10^{-4}T + 5.381\times10^{5}T^{-2} - 133.4T^{-1/2})}{T}dT - \frac{3070}{1048}$$

$$= -0.17\ \text{Jmol}^{-1}\text{K}^{-1} - 2.93\ \text{Jmol}^{-1}\text{K}^{-1} = \underline{-3.10\ \text{Jmol}^{-1}\text{K}^{-1}}.$$

The enthalpy of transformation for andalusite → sillimanite at 900 K, $\Delta_{tr}H_{900}^{and/sill}$, is:

$$\Delta_{tr}H_{900}^{and/sill} = 3070 - \int_{900}^{1048} (2.884 - 3.12\times10^{-4}T + 5.381\times10^{5}T^{-2}$$

$$- 133.4T^{-1/2})dT = \underline{2903\ \text{Jmol}^{-1}}.$$

Thus, 2903 Jmol^{-1} must be transferred from the system to the surroundings in order to keep the temperature of the system constant. The absorbance of this heat by the surroundings causes its entropy to change. It is:

$$\Delta S_{surr} = \frac{2903\ \text{Jmol}^{-1}}{900\ \text{K}} = \underline{3.23\ \text{Jmole}^{-1}\text{K}^{-1}}.$$

The total change in the entropy change is given by the sum of the entropy change in the system and in the surroundings, namely

$$\Delta S_{tot} = -3.10 + 3.23 = \underline{0.13\ \text{Jmol}^{-1}\text{K}^{-1}}.$$

It is positive, and is, thus, in accordance with the second law of thermodynamics.

In the case that the transformation for sillimanite → andalusite takes place at the equilibrium temperature of 1048 K, the entropy change of the system is

$$\Delta S_{sys} = -\frac{3070}{1048} = \underline{-2.93\ \text{Jmol}^{-1}\text{K}^{-1}}.$$

The heat that is transferred to the surroundings corresponds to the enthalpy of transformation for sillimanite → andalusite. It is positive, because from the standpoint of the surroundings, energy is absorbed. Its entropy change is

$$\Delta S_{surr} = \frac{3070}{1040} = \underline{2.93\ \mathrm{Jmol^{-1}K^{-1}}}.$$

The total change in the entropy is thus:

$$\Delta S_{tot} = -2.93\ \mathrm{Jmol^{-1}K^{-1}} + 2.93\ \mathrm{Jmol^{-1}K^{-1}} = 0.$$

At the equilibrium temperature the transformation is reversible. Infinitesimal changes in temperature cause the direction of the process to reverse, unless kinetic reasons hinder it.

4.3 Statistical interpretation of entropy

The atomistic basis for the second law lies in the fact that matter consists of particles moving around, mixing, colliding and exchanging kinetic energy with one another. Therefore, a *macrostate* of a system that is defined by the internal energy, U, the volume, V, and the number of the particles, N, is build up of a very large number of distinguishable *microstates*, which are characterized by the specific spatial and energetic arrangement of the particles. The total number of microstate corresponding to a particular macrostate is termed the *thermodynamic probability* and is designated W. The macrostate is the subject of classical thermodynamics. It is the most probable arrangement of particles that would be observed if it were possible to make an instantaneous observation of the state.

Because of the mobility of particles, it is assumed that each microstate is equally probable and the observable macrostate is the one with the greatest number of the microstates. In one mole of a substance with 6.022×10^{23} particles, the number of arrangements within the most probable distribution is much larger than the number of all other arrangements. It is, therefore, the only observable macrostate.

The number of microstates or *complexions* is a property of a system just as volume and energy. However, while energy and volume are additive properties, the number of microstates is multiplicative. For a system consisting of two subsystems, the number of microstates, Ω, is given by

$$\Omega_{12} = \Omega_1 \cdot \Omega_2,$$

because each microstate of system 1 can be combined with any microstate of system 2. On the other hand, entropy, being an extensive property, is additive. In order to establish a relationship between the randomness and entropy, the multiplicative property, Ω, has to be logarithmized, such that

$$S_{12} = S_1 + S_2 \propto \ln\Omega_{12} = \ln\Omega_1 + \ln\Omega_2.$$

Introducing the proportionality constant k, a quantitative relationship between the entropy of a system and its 'randomness' is obtained, namely

$$S = k\ln\Omega. \tag{4.48}$$

The quantity k is called *Boltzmann's constant.* It is obtained by dividing the universal gas constant, R, by Avogadro's number ($6.022\text{x}10^{23}$). Eq. (4.48) is known as *Boltzmann's equation.*

In a perfect crystal, where all constituents occupy proper crystallographic positions, the thermodynamic probability is one and thus the entropy zero, as required by the third law. Any possible randomness associated with electronic states or randomness within the nucleus are ignored in this connection.

4.3.1 Thermal entropy

Quantum mechanical theory is based on the principle of the quantization of energy. This means, if a particle is confined within a fixed volume, that it can only have certain discrete energy, ε. One says it occupies an allowed energy level. The particle with the lowest possible energy occupies the lowest energy level or ground state. It is designated ε_0. The particle occupying the next higher energy level has the energy ε_1, the particle in the succeeding level of increasing energy has the energy ε_2, etc. There are n_o particles in the ground state, n_1 particles have the energy ε_1, n_2 have the energy ε_2, etc. No particles have energies lying between the allowed levels. The relationship between the population of different energy levels by particles and the entropy can be best demonstrated on a simple example.

Consider a hypothetical perfect crystal in which all lattice sites are occupied by identical particles. The crystal contains N particles and has the energy U. The question to be addressed is: in how many ways can the N particles be distributed over the available energy levels such that the total energy of the crystal is U? The number of distinguishable arrangement is given by the thermodynamic probability Ω. In the case, where N particles are distributed over the energy levels ε_0 through ε_k the thermodynamic probability or the number of microstates is given as:

$$\Omega = \frac{N!}{n_o!n_1!n_2!\ldots n_i!\ldots n_k!}. \tag{4.49}$$

The most probable distribution of particles determines the macrostate of the system. It is characterized by the set of occupancies that give the maximum Ω. Because a function has a maximum when its first differential is zero, the condition:

$$d\Omega = 0 \tag{4.50}$$

must be fulfilled.

In an isolated system the total number of particles, N, and the internal energy, U, are fixed, that is

$$N = \sum_i n_i = \text{const} \tag{4.51}$$

and

$$U = \sum_i \varepsilon_i n_i = \text{const.} \tag{4.52}$$

From (4.51) and (4.52) follows further that

$$dN = \sum_i dn_i = 0 \tag{4.53}$$

and

$$dU = \sum_i \varepsilon_i dn_i = 0. \tag{4.54}$$

Entropy has a maximum value at equilibrium. Its differential is therefore zero and one can write:

$$dS = d(k\ln\Omega) = 0 \tag{4.55}$$

and

$$d\ln\Omega = 0. \tag{4.56}$$

Using Stirling's approximation ($\ln X! \approx X\ln X - X$) and considering the fact that N is constant, one obtains:

$$d\ln\Omega = d\left[\ln N! - \sum_i (n_i \ln n_i - n_i)\right] \tag{4.57}$$
$$= -\sum_i \frac{\partial}{\partial n_i}(n_i \ln n_i - n_i)dn_i$$
$$= -\sum_i \ln n_i dn_i.$$

Considering the constraints given in Eqs. (4.51) and (4.52) and applying Lagrange's method of undetermined multipliers, one obtains:

$$d\left(\ln\Omega + \alpha\sum_i n_i + \beta\sum_i \varepsilon_i n_i\right) = 0. \tag{4.58}$$

Because the multipliers α and β as well as ε_i are constants, differentiation of Eq. (4.58) yields:

$$-\sum_i \ln n_i dn_i + \alpha\sum_i dn_i + \beta\sum_i \varepsilon_i dn_i = 0 \tag{4.59}$$

or

$$\sum_i (-\ln n_i + \alpha + \beta)dn_i = 0. \tag{4.60}$$

The solution for Eq. (4.60) requires that each sum of the terms in the parenthesis is zero for all values of i, i.e.

$$-\ln n_i + \alpha + \beta\varepsilon_i = 0. \tag{4.61}$$

From Eq. (4.61) it follows that

$$n_i = e^{\alpha} \cdot e^{\beta\varepsilon_i}. \tag{4.62}$$

The total number of particles in a system is, therefore,

$$N = \sum_i n_i = e^{\alpha}\sum_i e^{\beta\varepsilon_i} \tag{4.63}$$

and consequently

$$\alpha = \ln N - \ln \sum_i e^{\beta \varepsilon_i}. \tag{4.64}$$

The thermodynamic meaning of the Lagrange multiplier, β, can be determined, as follows: Assume, that a small quantity of heat, δQ_{rev}, is transferred reversibly to the system. According to Eq. (4.1) a reversible heat exchange can be expressed in terms of the entropy change. Hence, the heat transfer can be represented as:

$$\delta Q_{rev} = TdS = kTd\ln\Omega. \tag{4.65}$$

Substituting Eq. (4.57) into Eq. (4.65) yields:

$$\delta Q_{rev} = -kT\sum_i \ln n_i dn_i. \tag{4.66}$$

Using the relationship given in Eq. (4.59), Eq. (4.66) can be rewritten:

$$\delta Q_{rev} = -kT\left(\alpha\sum_i dn_i + \beta\sum_i \varepsilon_i dn_i\right). \tag{4.67}$$

In the case of a closed system it holds that the total number of particles, N, remains constant, i.e.

$$\sum_i dn_i = 0. \tag{4.68}$$

Hence, Eq. (4.67) reduces to

$$\delta Q_{rev} = -kT\beta\sum_i \varepsilon_i dn_i. \tag{4.69}$$

According to Eq. (4.52)

$$\sum_i \varepsilon_i dn_i = dU \tag{4.70}$$

and in the case of an isochoric process $dU = \delta Q_{rev}$. Therefore,

$$\delta Q_{rev} = -kT\beta\delta Q_{rev}. \tag{4.71}$$

Hence, the second Lagrange multiplier, β, can be written in terms of reciprocal temperature

$$\beta = -\frac{1}{kT}. \tag{4.72}$$

Substituting α and β in Eq. (4.62) by the expressions given in Eqs. (4.64) and (4.72) yields

$$n_i = N\frac{e^{-\frac{\varepsilon_i}{kT}}}{\sum_i e^{-\frac{\varepsilon_i}{kT}}}. \tag{4.73}$$

Eq. (4.73) gives the distribution of the particles over the available energy levels. It is referred to as the *Maxwell-Boltzmann* distribution equations. If Eq. (4.73) is substituted for n_i in Eq. (4.49), the thermodynamic probability, Ω, is given in terms of the total number of particles and energy levels, ε_i.

The denominator in Eq. (4.73) is known as the *partition function* termed Z.

$$Z = \sum_i e^{-\frac{\varepsilon_i}{kT}}. \tag{4.74}$$

Using Eq. (4.74), the entropy of a system can be written in terms of Z, T and N as follows:

$$S = Nk\left(\ln Z + T\frac{\partial}{\partial T}\ln Z\right). \tag{4.75}$$

Substituting n_i in Eq. (4.52) by the expression given in Eq. (4.73), the internal energy, U, reads:

$$U = NkT^2\frac{\partial}{\partial T}\ln Z. \tag{4.76}$$

4.3.2 Configurational entropy (entropy of mixing)

In the preceding section we considered the entropy in terms of the numbers of ways in which particles can be distributed over available energy levels. Now, we consider the entropy in terms of the numbers of ways in which the particles distribute themselves over different sites in a crystal.

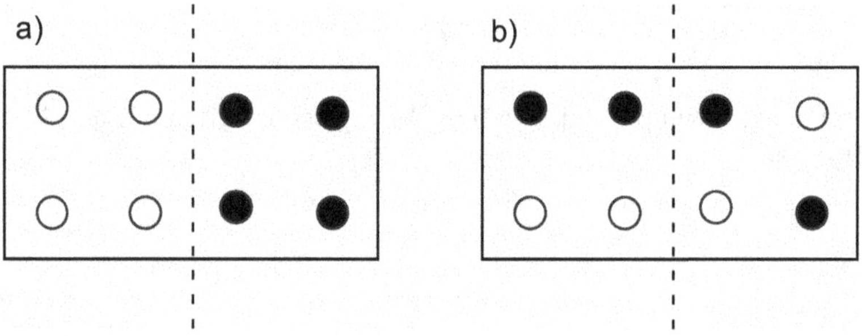

Fig. 4.1 Representation of atomic mixing. a) Two pure crystals consisting of *A* atoms (white) and *B* atoms (black) in close physical contact (initial state); b) mixing of *A* and *B* atoms.

For the sake of simplicity, suppose that the crystals contain four atoms each. Initially, all the *A* atoms are in the crystal located to the left of the dashed line and all the *B* atoms to the right (see Fig. 4.1a). The number of distinguishable atomic arrangement is one, because an interchange among the *A* atoms or among the *B* atoms does not produce a new configuration.

When one *A* atom on the left side of the dashed line interchanges with one *B* atom on the right side, the atom *B* can be placed in four different positions. Analogously, the atom *A* can take any of the four positions on the right side of the line. Hence there are $4^2 = 16$ possible arrangements.

When a second *A* atom is to be exchanged with *B* atom, there are 4 x 3 = 12 possibilities of arrangement. However, half of them are not distinguishable from one another, because they are the result of an interchange of B atoms among themselves. The number of complexions is therefore $(4 \times 3/2)^2 = 36$.

When a third *A* atom is to be exchanged with a *B* atom across the dashed line in Fig. 4.1, four possible sites are available. Two of them remain occupied by *B* atom and two of them are obtained by moving the first and second *A* atom successively onto these two positions and thus making their original positions available for the third *A* atom. Again, each atomic arrangement in crystal left to the dashed line can be combined with the same number of different occupations in the crystal to the right. Hence, there are 16 possible complexions.

For the fourth A atom there is only one possible configuration. All others are obtained by an interchange of *A* atoms among themselves and, therefore, do not represent new configuration. The same applies to the *B* atom diffusing into the crystal on the left side. The number of complexions is, therefore, $1^2 = 1$.

The configurations, discussed above, are shown in Tab. 4.1.

Table 4.1 Possible microstates and the probabilities for a macrostate.

Atoms in crystal *A*	Atoms in crystal *B*	Number of microstates	Probability of a macrostate
4A	4B	$1^2 = 1$	1/70
3A +1B	1A + 3B	$4^2 = 16$	16/70
2A + 2B	2A +2B	$(4 \times 3/2)^2 = 36$	36/70
1A + 3B	3A + 1B	$4^2 = 16$	16/70
4B	4A	$1^2 = 1$	1/70
		sum = 70	

The total number of microstates or complexions, Ω, is 70. This represents the number of ways in which two types of elements, four of each kind, can be distributed over 8 sites and is calculated:

$$\Omega = \frac{N!}{n_A! \cdot n_B!} . \tag{4.77}$$

In the case of the given example, the calculation reads:

$$\Omega = \frac{8!}{4! \cdot 4!} = \frac{8 \cdot 7 \cdot 6 \cdot 5 \cdot 4 \cdot 3 \cdot 2}{4 \cdot 3 \cdot 2 \cdot 4 \cdot 3 \cdot 2} = 70.$$

The last column in Tab. 4.1 gives the probability of encountering a given distribution. The distribution with the highest probability occurs most frequently. In our case, it is the distribution given in the third row. Its probability is 36/70.

In the case of a real crystal, the number of positions is multiple of Avogadro's number. The number of complexions in a mixed crystal consisting of the components *A* and *B* is calculated according to Eq. (4.77), i.e.

$$\Omega_{A,B} = \frac{(zN_A)!}{(x_A zN_A)!(x_B zN_A)!} ,$$

where z, N_A, x_A and x_B designate the number of equivalent crystallographic sites per formula unit, Avorgadro's number, the fraction of the component *A* and the fraction of the component *B* in the crystal, respectively.

In the case of an ideal solution, where the volume of mixing and the enthalpy of mixing are zero the change in the configurational entropy associated with the process of mixing is identical to the *entropy of mixing*. Hence, one can write:

$$\Delta_m \bar{S}^{id} = \bar{S}^{conf} = \bar{S}^{sol} - \bar{S}^{mm} = k[\ln\Omega_{A,B} - (\ln\Omega_A + \ln\Omega_B)]\,, \tag{4.78}$$

where k is Boltzmann's constant (R/N_A). The entropy of the mechanical mixture is represented by the sum of entropies of components A and B. Because A and B are pure phases, all sites are occupied by the particles constituting the corresponding component. The thermodynamic probabilities, Ω_A and Ω_B are, therefore, 1 and the corresponding configurational entropies zero. Hence, the configurational entropy of a mixed crystal (A,B) is given by

$$\bar{S}^{conf} = k\ln\Omega_{A,B} = k\ln\frac{(zN_A)!}{(x_A zN_A)!(x_B zN_A)!}\,. \tag{4.79}$$

Applying the Stirling's approximation yields:

$$\begin{aligned}\bar{S}^{conf} &= k[(zN_A)\ln(zN_A) - (zN_A) - (x_A zN_A)\ln(x_A zN_A) \\ &+ (x_A zN_A) - (x_B zN_A)\ln(x_B zN_A) + (x_B zN_A)] \\ &= zkN_A[\ln(zN_A) - 1 - x_A\ln(x_A zN_A) + x_A - x_B\ln(x_B zN_A) + x_B] \\ &= zkN_A[\ln(zN_A) - 1 - x_A\ln x_A - x_A\ln(zN_A) \\ &+ x_A - x_B\ln x_B - x_B\ln(zN_A) + x_B].\end{aligned}$$

Considering that

$$x_A + x_B = 1$$

and therefore,

$$\ln(zN_A) - x_A\ln(zN_A) - x_B\ln(zN_A) = 0$$

and

$$x_A + x_B - 1 = 0$$

one obtains

$$\bar{S}^{conf} = -zkN_A[x_A \ln x_A + x_B \ln x_B]. \quad (4.80)$$

Substituting the universal gas constant R for kN_A in Eq. (4.80) gives

$$\bar{S}^{conf} = -zR[x_A \ln x_A + x_B \ln x_B]. \quad (4.81)$$

In Eq. (4.81) x_A and x_B represent the mole fractions of atom A and B, respectively, in the solid solution (A,B). z gives the number of the thermodynamically equivalent sites in the sublattice where mixing takes place.

Example: In olivine, $(Mg,Fe)_2SiO_4$, there are two different sixfold coordinated sites termed *M1* and *M2* that are occupied by magnesium and iron. If neither of the two sites is preferred by either cation, the occupation of both sites is equal and their number per mole olivine is two times N_A. The configurational part of the entropy of mixing for olivine reads:

$$\bar{S}^{conf} = -2R[x_{Mg}^{ol} \ln x_{Mg}^{ol} + x_{Fe}^{ol} \ln x_{Fe}^{ol}],$$

where

$$x_{Mg}^{ol} = \frac{[Mg]}{[Mg]+[Fe]} \quad \text{and} \quad x_{Fe}^{ol} = \frac{[Fe]}{[Mg]+[Fe]}$$

with [Mg] and [Fe] giving the atomic fractions of magnesium and iron ions, respectively. The values of these fractions are equal to the mole fraction of the forsterite and fayalite components constituting the olivine solid solution.

If, however, cation ordering takes place such that one of the sites is preferred by one of the two cations, *M1* and *M2* belong to two different sublattices and the mixing entropy for olivine is given by:

$$\bar{S}^{conf} = -R[(1-x_{Fe,M1}^{ol})\ln(1-x_{Fe,M1}^{ol}) + x_{Fe,M1}^{ol} \ln x_{Fe,M1}^{ol} + (1-x_{Fe,M2}^{ol})\ln(1-x_{Fe,M2}^{ol}) + x_{Fe,M2}^{ol} \ln x_{Fe,M2}^{ol}],$$

where $x_{Fe,M1}^{ol}$ and $x_{Fe,M2}^{ol}$ give the mole fractions for Fe at *M1* and *M2* site, respectively.

In the general case where i ions occupy j sublattices with z equivalent sites, the enthalpy of mixing reads:

$$\bar{S}^{conf} = -R\sum_i z\sum_j x_{ij}\ln x_{ij}. \tag{4.82}$$

In the preceding sections it was shown that volume and enthalpy showed no effects on mixing in the case of an ideal mixing. This does not hold for entropy. Even an ideal solutions shows changes in the entropy function because mixing is a natural process.

The entropy of mixing represents the difference between the entropy of the solution and the entropy of the mechanical mixture In other words, it gives the difference between the entropy of a system before and after the mixing at atomic level.

In a two component system, *A-B*, the entropy for a mechanical mixture is given by

$$\bar{S}^{mm} = x_A\mathbf{S}_A + x_B\mathbf{S}_B, \tag{4.83}$$

where $\mathbf{S}_A$ and $\mathbf{S}_B$ are the *molar entropies* of the pure components *A* and *B*, respectively.

The entropy for a solution *A-B*, however, reads:

$$\bar{S}^{sol} = x_A S_A + x_B S_B, \tag{4.84}$$

where S_A and S_B are the *partial molar entropies* of the components.

The entropy change due to the solution of *A* in *B* or vice versa at constant temperature and pressure is obtained by subtracting Eq. (4.83) from Eq. (4.84), that is:

$$(\bar{S}^{sol} - \bar{S}^{mm}) = \Delta_m\bar{S} = x_A(S_A - \mathbf{S}_A) + x_B(S_B - \mathbf{S}_B). \tag{4.85}$$

Comparing the coefficients from Eqs. (4.81) and (4.85) shows that

$$(S_i - \mathbf{S}_i) = -zR\ln x_i \tag{4.86}$$

or

$$S_i = \mathbf{S}_i - zR\ln x_i. \tag{4.87}$$

In a solution, the mole fraction of the solute, x_i is smaller than 1. Its logarithm is, therefore, negative. Hence, the partial molar entropy, S_i, is larger than the molar entropy of the pure component, $\mathbf{S}_i$. This is in accordance with the second law of thermodynamics that states that the entropy of a system increases during the course of a natural process.

Example 1: In orthopyroxene, $(Mg,Fe)_2Si_2O_6$, magnesium and iron occupy two crystallographically different sixfold coordinated sites termed *M1* and *M2*. The larger and more strongly distorted *M2* site is preferred by the iron ion. If we want to calculate the configurational entropy of a Mg-Fe orthopyroxene, the mole fractions for magnesium and iron on both sites must be known. The calculation of the entropy then reads:

$$\bar{S}^{conf} = -R[x_{Mg,\,M1}^{opx} \ln x_{Mg,\,M1}^{opx} + x_{Fe,\,M1}^{opx} \ln x_{Fe,\,M1}^{opx} + x_{Mg,\,M2}^{opx} \ln x_{Mg,\,M2}^{opx} + x_{Fe,\,M2}^{opx} \ln x_{Fe,\,M2}^{opx}].$$

Because the sum of the mole fractions for the atoms on each lattice site is unity, only the concentration for one of the two ions needs to be determined experimentally. It is the concentration of the iron that is normally measured. In this case, the equation for the calculation of the entropy of mixing must be modified to:

$$\bar{S}^{conf} = -R[(1 - x_{Fe,\,M1}^{opx})\ln(1 - x_{Fe,\,M1}^{opx}) + x_{Fe,\,M1}^{opx} \ln x_{Fe,\,M1}^{opx} + (1 - x_{Fe,\,M2}^{opx})\ln(1 - x_{Fe,\,M2}^{opx}) + x_{Fe,\,M2}^{opx} \ln x_{Fe,\,M2}^{opx}].$$

Let us now calculate the entropy of mixing, or more precisely, the configurational entropy for an orthopyroxene solid solution containing 57.31 mol% enstatite, $Mg_2Si_2O_6$, and 42.69 mol% ferrosilite, $Fe_2Si_2O_6$. The mole fractions of the iron on *M2* and *M1* sites, at 973 K are 0.6488 and 0.2051, respectively (modified after Saxena et al., 1987). The configurational entropy of the orthopyroxene is given by

$$\begin{aligned}\bar{S}^{conf} &= -8.3144[(1-0.2051)\ln(1-0.2051) + 0.2051\ln 0.2051 \\ &\quad + (1-0.6488)\ln(1-0.6488) + 0.6488\ln 0.6488] \\ &= \underline{9.61\ \mathrm{Jmol^{-1}K^{-1}}}.\end{aligned}$$

If the iron ions were distributed identically over *M1* and *M2* sites, $x_{Fe,\,M1}^{opx}$ and $x_{Fe,\,M2}^{opx}$ would be equal and their values would correspond to the mole fraction of ferrosilite. The calculation of the entropy of mixing would then read:

$$\bar{S}^{conf} = -2 \times 8.3144[0.5731 \ln 0.5731 + 0.4269 \ln 0.4269]$$
$$= \underline{11.35 \text{ Jmol}^{-1}\text{K}^{-1}}.$$

In the latter case, where the iron is distributed statistically over the two crystallographic sites the "randomness" is higher than in the case, where iron cations show some preference for the *M2* site. Consequently, the configurational entropy value is more positive.

Example 2: In garnet with the general chemical formula $X_3Y_2Si_3O_{12}$, the cation mixing takes place primarily on the *X* and *Y* sublattices. *X* represents the dodecahedrally coordinated sites that are occupied by divalent ions such as magnesium, ferrous iron, calcium and manganese. *Y* is an octahedrally coordinated site that contains trivalent ions such aluminium, ferric iron, chromium, etc. There are three dodecahedral and two octachedral cations per formula unit. In the case that mixing occurs on both *X* and *Y* sites the configurational entropy is calculated as follows:

$$\bar{S}^{conf} = -3R\sum_i x_i^{[8]} \ln x_i^{[8]} - 2R\sum_i x_j^{[6]} \ln x_j^{[6]},$$

where $x_i^{[8]}$ and $x_j^{[6]}$ designate the mole fractions of the cations occupying the *X* and *Y* site, respectively.

For example, a garnet of composition $(Mg,Fe^{2+},Ca,Mn)_3(Al,Fe^{3+},Cr^{3+})_2Si_3O_{12}$ contains: 67.1 mol% pyrope, $Mg_3Al_2Si_3O_{12}$, 18.0 mol% almandine, $Fe_3Al_2Si_3O_{12}$, 13.0 mol% grossular, $Ca_3Al_2Si_3O_{12}$, 1.1 mol% andradite, $Ca_3Fe_2Si_3O_{12}$, 0.7 mol% spessartine, $Mn_3Al_2Si_3O_{12}$, and 0.1 mol% uvarovite, $Ca_3Cr_2Si_3O_{12}$. What is the configurational entropy of this garnet?

First, we calculate the mole fractions of the cations:

$$x_{Mg}^{[8]} = \frac{[Mg]}{[Mg]+[Fe^{2+}]+[Ca]+[Mn]} = \frac{67.1}{67.1+18.0+(13.0+1.1+0.1)+0.7}$$
$$= 0.671,$$

$$x_{Fe^{2+}}^{[8]} = \frac{[Fe^{2+}]}{[Mg]+[Fe^{2+}]+[Ca]+[Mn]} = \frac{18.0}{67.1+18.0+(13.0+1.1+0.1)+0.7}$$
$$= 0.180,$$

$$x_{Ca}^{[8]} = \frac{[Ca]}{[Mg]+[Fe^{2+}]+[Ca]+[Mn]} = \frac{13.0+1.1+0.1}{67.1+18.0+(13.0+1.1+0.1)+0.7}$$
$$= 0.142,$$

$$x_{Mn}^{[8]} = \frac{[Mn]}{[Mg]+[Fe^{2+}]+[Ca]+[Mn]} = \frac{0.7}{67.1+18.0+(13.0+1.1+0.1)+0.7}$$
$$= 0.007,$$

$$x_{Al}^{[6]} = \frac{[Al]}{[Al]+[Fe^{3+}]+[Cr^{3+}]} = \frac{67.1+18.0+13.0+0.7}{(67.1+18.0+13.0+0.7)+1.1+0.1}$$
$$= 0.988,$$

$$x_{Fe^{3+}}^{[6]} = \frac{[Fe^{3+}]}{[Al]+[Fe^{3+}]+[Cr^{3+}]} = \frac{1.1}{(67.1+18.0+13.0+0.7)+1.1+0.1}$$
$$= 0.011$$

and

$$x_{Cr^{3+}}^{[6]} = \frac{[Cr^{3+}]}{[Al]+[Fe^{3+}]+[Cr^{3+}]} = \frac{0.1}{(67.1+18.0+13.0+0.7)+1.1+0.1}$$
$$= 0.001.$$

Using these site fractions the configurational entropy is calculated as follows:

$$\begin{aligned}\bar{S}^{conf} &= -3\times 8.3144[0.671\ln 0.671\\ &+0.18\ln 0.18+0.142\ln 0.142+0.007\ln 0.007]\\ &-2\times 8.3144[0.988\ln 0.988+0.011\ln 0.011+0.001\ln 0.001]\\ &= \underline{23.29\ \mathrm{Jmol^{-1}K^{-1}}}.\end{aligned}$$

Mixing behavior of the type presented in the above examples is termed *mixing on sites*.

Example 3: Consider a diopside ($CaMgSi_2O_6$) - jadeite ($NaAlSi_2O_6$) solid solution. In this pyroxene, calcium and sodium occupy the larger *M2* sites and magnesium and aluminium the smaller *M1* sites. In order to preserve electroneutrality each calcium atom has to be in close vicinity of a magnesium atom and the same for a sodium and aluminium atom. The solid solution can therefore be modeled as a mixture of *CaMg* and *NiAl* pairs. Their proportions correspond to the mole fractions of the components, $x^{cpx}_{CaMgSi_2O_6}$ and $x^{cpx}_{NaAlSi_2O_6}$, respectively. Hence, the calculation of the configurational entropy of a diopside-jadeite solid solution reads:

$$\bar{S}^{conf} = -R[(1-x^{cpx}_{NaAlSi_2O_6})\ln(1-x^{cpx}_{NaAlSi_2O_6}) + x^{cpx}_{NaAlSi_2O_6}\ln x^{cpx}_{NaAlSi_2O_6}].$$

For a clinopyroxene consisting of 80 mol% diopside and 20% jadeite a configurational entropy of

$$\bar{S}^{conf} = -8.4144[0.8\ln 0.8 + 0.2\ln 0.2] = \underline{4.16\ \mathrm{Jmol^{-1}K^{-1}}}$$

is obtained.

This type of mixing behavior is referred to as *molecular mixing*. It requires complete ordering of cations on a local scale.

Eq. (4.82) was derived assuming that the pure components are perfect ordered crystals whose configurational entropies are zero. The entropy of mixing is therefore identical to the configurational entropy of the mixed crystal. In other words, the configurational entropy of an ideal solid solution is determined only by the 'randomness' of the atoms or ions over the different crystallographic sites. Therefore, Eq. (4.82) can, as well, be used to calculate the configurational entropy of pure phases exhibiting cation disorder, for example alkali feldspar, $KAlSi_3O_8$. In the high temperature modification, sanidine, *Al* and *Si* are randomly distributed over two crystallographically distinct tetrahedral sites termed t_1 and t_2. Sanidine's configurational entropy is therefore given by

$$S^{conf} = -4 \times R[x_{Al}\ln x_{Al} + x_{Si}\ln x_{Si}].$$

In sanidine the total number of tetrahedrally coordinated cations per formula unit is four ($1Al + 3Si$). Thus, one quarter of the available sites is occupied by aluminium and three quarters of them by silicon. Because the atomic fractions for aluminium and silicon are equal for all sites, no distinction is made between t_1 and t_2 and we have:

$$S^{conf} = -4 \times 8.3144[0.25\ln 0.25 + 0.75\ln 0.75] = 18.70\ \underline{\mathrm{Jmol^{-1}K^{-1}}}.$$

In partially ordered monoclinic orthoclase the distribution of aluminium over the t_1 site differs from that over t_2 site. In order to calculated the configurational entropy, each site has to be accounted for by individual terms. The configurational entropy of an orthoclase is therefore calculated as follows:

$$S^{conf} = -2R[x_{Al}^{t_1}\ln x_{Al}^{t_1} + x_{Si}^{t_1}\ln x_{Si}^{t_1} + x_{Al}^{t_2}\ln x_{Al}^{t_2} + x_{Si}^{t_2}\ln x_{Si}^{t_2}].$$

Example: Consider an orthoclase with the following *Al* distribution: $t_1 = 0.433$ and $t_2 = 0.067$ (Hovis 1986). Its configurational entropy is

$$\begin{aligned} S^{conf} &= 2 \times 8.3144[0.433\ln 0.433 + 0.067\ln 0.067 \\ &+ (1-0.433)\ln(1-0.433) + (1-0.067)\ln(1-0.067)] \\ &= \underline{15.46\ \text{Jmol}^{-1}\text{K}^{-1}}. \end{aligned}$$

In triclinic microcline *four* distinguishable sites are available for *Al* and *Si*, because the two t_1 and t_2 sites are split into t_{1o} and t_{1m} and t_{2o} and t_{2m} sites, respectively. The mathematical expression for the calculation of microcline's configurational entropy reads:

$$\begin{aligned} S^{conf} &= -R[x_{Al}^{t_{1o}}\ln x_{Al}^{t_{1o}} + x_{Al}^{t_{1m}}\ln x_{Al}^{t_{1m}} + x_{Al}^{t_{2o}}\ln x_{Al}^{t_{2o}} + x_{Al}^{t_{2m}}\ln x_{Al}^{t_{2m}} \\ &+ x_{Si}^{t_{1o}}\ln x_{Si}^{t_{1o}} + x_{Si}^{t_{1m}}\ln x_{Si}^{t_{1m}} + x_{Si}^{t_{2o}}\ln x_{Si}^{t_{2o}} + x_{Si}^{t_{2m}}\ln x_{Si}^{t_{2m}}]. \end{aligned}$$

Example: Consider a microcline with the following distribution of Al over the four sites: $x_{Al}^{t_{1o}} = 0.425$, $x_{Al}^{t_{1m}} = 0.350$, $x_{Al}^{t_{2o}} = 0.110$ and $x_{Al}^{t_{2m}} = 0.110$ (Kroll and Ribbe 1983). The calculation of its configurational entropy yields:

$$\begin{aligned} S^{conf} &= -8.3144[0.425\ln 0.425 + 0.350\ln 0.350 \\ &+ 0.110\ln 0.110 + 0.110\ln 0.110 \\ &+ (1-0.425)\ln(1-0.425) + (1-0.350)\ln(1-0.350) \\ &+ (1-0.110)\ln(1-0.110) + (1-0.110)\ln(1-0.110)] \\ &= \underline{16.81\ \text{Jmol}^{-1}\text{K}^{-1}}. \end{aligned}$$

In a hypothetical completely ordered microcline Al would occupy only the t_{1o} sites, while the remaining three sites would be occupied exclusively by Si. The configurational entropy of such a microcline crystal would be zero. For kinetical reasons, complete cation order is never attained in natural feldspars.

In addition to changes in the configurational entropy, changes in the thermal entropy arise from the mixing process. This part of the total entropy is termed the *par-*

tial molar excess entropy, S_i^{ex}, and it occurs in non-ideal solutions. Following Eq. (4.87) the partial molar excess entropy of a component i is given by

$$S_i^{ex} = S_i - (\boldsymbol{S}_i + zR\ln x_i), \tag{4.88}$$

where S_i and $\boldsymbol{S}_i$ designate the partial molar entropy and the molar entropy of the component i, respectively. The term in the parenthesis gives the partial molar entropy of the component i for the case of ideal mixing.

4.4 Entropy of reaction

Consider a system consisting of four pure phases A, B, C and D once again, and let a chemical reaction of the type

$$\nu_a A + \nu_b B = \nu_c C + \nu_d D$$

occur. In order to keep the temperature constant, heat must be exchanged between the system and its surroundings. As discussed in the preceding chapter, this heat is referred to as the heat of reaction, $\Delta_r\boldsymbol{H}$. In the case where the heat exchange occurs reversibly it holds:

$$\frac{\Delta_r\boldsymbol{H}}{T} = \Delta_r\boldsymbol{S}, \tag{4.89}$$

where T is the temperature at which the reaction takes place. $\Delta_r\boldsymbol{S}$ is called the *entropy of reaction*. The bold type characters designate a reaction between pure phases. The entropy of reaction can be calculated using the molar entropies of the phases participating in the reaction at the temperature in question. For a generalized reaction, the calculation of the reaction entropy reads:

$$\Delta_r\boldsymbol{S} = -\nu_a\boldsymbol{S}_A - \nu_b\boldsymbol{S}_B + \nu_c\boldsymbol{S}_C + \nu_d\boldsymbol{S}_D. \tag{4.90}$$

In Eq. (4.90), $\boldsymbol{S}_A$, $\boldsymbol{S}_B$, $\boldsymbol{S}_C$ and $\boldsymbol{S}_D$ designate the molar entropies of phases A, B, C and D, respectively.

4.4.1 Temperature dependence of the entropy of reaction

The temperature dependence of the entropy of reaction is given by:

$$\left(\frac{\partial \Delta_r \boldsymbol{S}}{\partial T}\right)_P = \frac{\Delta_r \boldsymbol{C}_p}{T}, \tag{4.91}$$

where $\Delta_r \boldsymbol{C}_p$ is the change in heat capacity as defined in Eq. (3.90).

Hence, the entropy of reaction at temperature T is obtained by integrating Eq. (4.91) between the temperature limits T_o and T:

$$\Delta_r \boldsymbol{S}_T = \Delta \boldsymbol{S}_{T_o} + \int_{T_o}^{T} \frac{\Delta_r \boldsymbol{C}_p}{T} dT. \tag{4.92}$$

Generally, in the field of geosciences the lower integration limit is taken to be room temperature (298.15 K). In this case, the integration constant, $\Delta_r \boldsymbol{S}_{298}$, is the stoichiometric sum of the conventional standard entropies, $\boldsymbol{S}_{298,i}$, of the reactants. If any of the reactants undergoes a phase transformation in the temperature interval in question, the entropy associated with the transformation must be taken into account.

Example: Consider the reaction:

$$NaAlSi_3O_8 = NaAlSi_2O_6 + SiO_2.$$

If one wants to calculate the entropy of this reaction at 298 K, one needs only the third law entropies of the reactants. These values are tabulated in various thermodynamic tables. In our example, we take the data given by Robie and Hemingway (1995):

$$\boldsymbol{S}^{ab}_{298,\,NaAlSi_3O_8} = 207.4 \text{ Jmol}^{-1}\text{K}^{-1},$$

$$\boldsymbol{S}^{jd}_{298,\,NaAlSi_2O_6} = 133.5 \text{ Jmol}^{-1}\text{K}^{-1} \text{ and}$$

$$\boldsymbol{S}^{\alpha-qtz}_{298,\,SiO_2} = 41.5 \text{ Jmol}^{-1}\text{K}^{-1}.$$

Using these values the calculation yields:

$$\Delta_r \boldsymbol{S}_{298} = -207.4 + 133.5 + 41.5 = \underline{-32.4 \text{ JK}^{-1}}.$$

In order to calculate the entropy of reaction at 800 K the molar heat capacities of the reactants as a function of temperature are required. Robie and Hemingway (1995) give the following $\boldsymbol{C}_p$-polynomials for the phases in question:

$$C_{p,\,NaAlSi_3O_8}^{ab} = 583.9 - 9.285\times10^{-2}T + 1.678\times10^{6}T^{-2} - 6424T^{-1/2} + 2.272\times10^{-5}T^{2}[\ \mathrm{Jmol^{-1}K^{-1}}],$$

$$C_{p,\,NaAlSi_2O_6}^{jd} = 3.011\times10^{2} + 1.014\times10^{-2}T - 2.239\times10^{6}T^{-2} - 2.055\times10^{3}T^{-1/2}[\ \mathrm{Jmol^{-1}K^{-1}}]$$

and

$$C_{p,\,SiO_2}^{\alpha-qtz} = 81.145 + 1.828\times10^{-2}T - 1.81\times10^{5}T^{-2} - 698.5\mathrm{T}^{-1/2} + 5.406\times10^{-6}T^{2}[\ \mathrm{Jmol^{-1}K^{-1}}].$$

A summation of the polynomials according to the stoichiometric proportions of the reactants yields:

$$\Delta_r C_p = -201.655 + 0.1213T - 4.098\times10^{6}T^{-2} + 3.67\times10^{3}T^{-1/2} - 1.731\times10^{-5}T^{2}[\ \mathrm{Jmol^{-1}K^{-1}}].$$

Using this polynomial, the entropy of reaction at 800 K is calculated as follows:

$$\Delta_r S_{800} = -32.4 + \int_{298}^{800}\left(\frac{-201.655 + 0.1213T - 4.098\times10^{6}T^{-2}}{T} + \frac{3.67\times10^{3}T^{-1/2} - 1.731\times10^{-5}T^{2}}{T}\right)dT$$

$$= -32.4 - 201.655\times\ln\frac{800}{298} + 0.1213\,(800 - 298)$$

$$+\frac{4.098\times10^{6}}{2}\left(\frac{1}{800^{2}} - \frac{1}{298^{2}}\right) - 2\times3.67\times10^{3}\left(\frac{1}{\sqrt{800}} - \frac{1}{\sqrt{298}}\right)$$

$$-\frac{1.731\times10^{-5}}{2}(800^{2} - 298^{2})$$

$$= \underline{-29.6\ \mathrm{Jmol^{-1}K^{-1}}}.$$

4.5 Problems

1. The third law entropy of hercynite, $\boldsymbol{S}_{298}$, equals 106.274 $Jmol^{-1}K^{-1}$ (Knacke et al. 1991).

- Calculate the entropy of hercynite for 1000°C.

$$C_{p,\,FeAl_2O_4} = 155.310 + 26.150\times10^{-3}T - 3.523\times10^{6}T^{-2}[\mathrm{Jmol}^{-1}\mathrm{K}^{-1}].$$

2. The melting temperature of diopside, $CaMgSi_2O_6$, at 0.1 MPa is 1665 K. Its enthalpy of melting equals 128.448 $kJmol^{-1}$.

- Show that the crystallization of diopside at 0.1 MPa and 1500 K is an irreversible process.

$$C^{di}_{p,\,CaMgSi_2O_6} = 186.021 + 123.763\times10^{-3}T - 5.590\times10^{6}T^{-2} - 43.932T^{2}[\mathrm{Jmol}^{-1}\mathrm{K}^{-1}],$$

$$C^{melt}_{p,\,CaMgSi_2O_6} = 355.640\ [\mathrm{Jmol}^{-1}\mathrm{K}^{-1}]$$

3. In high temperature cordierite (i.e., indialite), $Mg_2Al_4Si_5O_{18}$, Al and Si, are tetrahedrally coordinated and located in two crystallographically different T_1 and T_2 sites.

- Calculate the configurational entropy for the case that 1 Si and 2 Al are disordered over 3 T_1 sites and 4 Si and 2 Al are disordered over 6 T_2 sites.
- Calculate the configurational entropy of indialite for the case that Si and Al are randomly distributed over 9 tetrahedral sites.

$R = 8.3144$ $Jmol^{-1}K^{-1}$

4. Calculate the configurational entropy of a clinopyroxene containing 10 mol% enstatite, $Mg_2Si_2O_6$, 70 mol% diopside, $CaMgSi_2O_6$, and 20 mol% Ca-Tschermak, $CaAl_2SiO_8$.

5. Calculate the change in entropy associated with the reaction:

$$9\,\text{talc} + 4\,\text{forsterite} \rightarrow 5\,\text{anthophyllite} + 4H_2O$$

taking place at 527°C and 0.1 MPa.

$$C^{tlc}_{p,\,Mg_3Si_4O_{10}(OH)_2} = 0.5343 + 3.7416\times10^{-5}T - 8805.2T^{-2} - 2.1532T^{-1/2},$$

$$C^{fo}_{p,\,Mg_2SiO_4} = 0.2349 + 0.1069\times10^{-5}T - 542.9T^{-2} - 1.9064T^{-1/2};$$

$$C^{anth}_{p,\,Mg_7Si_8O_{22}(OH)_2} = 1.2773 + 2.5825\times10^{-5}T - 9704.6T^{-2} - 9.0747T^{-1/2},$$

$$C^{steam}_{p,\,H_2O} = 0.0401 + 0.8656\times10^{-5}T + 487.5T^{-2} - 0.2512T^{-1/2}$$

C_p[kJmol^{-1}K^{-1}]

$$S^{tlc}_{298,\,Mg_3Si_4O_{10}(OH)_2} = 260.80\ [\text{Jmol}^{-1}\text{K}^{-1}],$$

$$S^{fo}_{298,\,Mg_2SiO_4} = 94.10\ [\text{Jmol}^{-1}\text{K}^{-1}],$$

$$S^{anth}_{298,\,Mg_5Si_8O_{22}(OH)_2} = 537.00\ [\text{Jmol}^{-1}\text{K}^{-1}]\ \text{and}$$

$$S^{steam}_{298,\,H_2O} = 188.80\ [\text{Jmol}^{-1}\text{K}^{-1}].$$

Data: Holland and Powell (1990)

Chapter 5 Gibbs free energy and Helmholtz free energy

In chapter 4 it was shown that the total entropy change for an irreversible process is positive. For a reversible processes it is zero. If the entropy function is used to address the question whether or not a given process proceeds spontaneously in a certain direction, the *system and its surroundings* must be taken into consideration. This, however, is inconvenient situation. It would be better to have a function just for the system that would indicate whether or not a process is potentially spontaneous without a need for considering changes in the surroundings.

In our example in section 4.2.4 it was shown that sillimanite transformed to andalusite irreversibly. The heat exchange between the system and its surroundings, however, took place reversibly. In the course of the transformation of sillimanite to andalusite at 900 K an amount of heat equal to 2903 $Jmol^{-1}$ was transferred from the system to the surroundings in order to keep the temperature of the system constant. Due to the heat absorption, the entropy of the surroundings increased by 3.23 $Jmol^{-1}K^{-1}$ (see page 143). From the standpoint of the system, this entropy change is, however, *negative*. It is smaller than the change in the entropy of the system (-3.10 $Jmol^{-1}K^{-1}$) that was calculated using the reversible path. Eq. (4.4) can, thus, be rewritten as follows:

$$dS_{syst} - \frac{\delta Q_{rev}}{T} > 0 \tag{5.1}$$

for an irreversible process and

$$dS_{syst} - \frac{\delta Q_{rev}}{T} = 0 \tag{5.2}$$

for a reversible process.

Under isochoric conditions one has $\delta Q_{\mathrm{rev}} = dU$. Eq. (5.1) can be rewritten, dropping the subscript '*syst*', as

$$dU - TdS < 0 \tag{5.3}$$

and Eq. (5.2) as

$$dU - TdS = 0. \tag{5.4}$$

In the case of an isobaric processes, δQ_{rev} can be replaced by dH and one obtains:

$$dH - TdS \leq 0, \tag{5.5}$$

where the sign 'smaller than' designates an irreversible process and the sign 'equal to' a reversible one.

Eqs. (5.3) through (5.5) can be used to determine whether or not a process is potentially spontaneous without any consideration of any changes in the surroundings. If (dU - TdS) at constant volume or (dH - TdS) at constant pressure is smaller than zero, a process is potentially spontaneous. If (dU - TdS) or (dH - TdS) equal zero, there is thermodynamic equilibrium, and if (dU - TdS) or (dH - TdS) are greater than zero, a process does not proceed in the given direction. For the sake of convenience, we introduce dF for (dU - TdS) and dG for (dH - TdS), and we can then write:

$$dF = dU - TdS \tag{5.6}$$

and

$$dG = dH - TdS. \tag{5.7}$$

Integrating Eqs. Eq. (5.6) and Eq. (5.7) yields:

$$F = U - TS \tag{5.8}$$

and

$$G = H - TS. \tag{5.9}$$

The new functions F and G are termed the *Helmholtz free energy* and the *Gibbs free energy*, respectively. Their values are normally expressed in Joules or calories. Both are functions of state and both can be used to characterize the nature of a processes. Their total differentials read:

$$dF = dU - TdS - SdT \tag{5.10}$$

and

$$dG = dH - TdS - SdT. \tag{5.11}$$

Substituting Eq. (4.5) into equation Eq. (5.10) and Eq. (4.6) into Eq. (5.11) yields:

$$dF = -SdT - PdV \tag{5.12}$$

and

$$dG = -SdT + VdP. \tag{5.13}$$

In Eqs. (5.12) and (5.13) the Helmholtz free energy and the Gibbs free energy are given as functions of temperature and volume and temperature and pressure, respectively. These two equations are more convenient than Eqs. (4.5) and (4.6) that can, in principle, also be used to describe changes in a system. Eqs. (4.5) and (4.6) contain entropy as a variable which is difficult to control experimentally.

Eqs. (4.5), (4.6), (5.12) and (5.13) can be used to derive several important relationships, namely

$$\left(\frac{\partial U}{\partial V}\right)_S = \left(\frac{\partial F}{\partial V}\right)_T = -P, \tag{5.14}$$

$$\left(\frac{\partial H}{\partial P}\right)_S = \left(\frac{\partial G}{\partial P}\right)_T = V \tag{5.15}$$

and

$$\left(\frac{\partial F}{\partial T}\right)_V = \left(\frac{\partial G}{\partial T}\right)_P = -S. \tag{5.16}$$

An additional relationship is:

$$\frac{\partial}{\partial T}\left(\frac{G}{T}\right)_P = -\frac{H}{T^2}. \tag{5.17}$$

Eq. (5.17) is obtained using the quotient rule of differentiation. According to this rule, a function of the type $y = u/v$ is differentiated as follows:

$$y' = \frac{u'v - uv'}{v^2}. \tag{5.18}$$

Applying it to the function (G/T), one obtains:

$$\left[\frac{\partial(G/T)}{\partial T}\right]_P = \frac{T\left(\frac{\partial G}{\partial T}\right)_P - G}{T^2} = \frac{-TS-G}{T^2} = -\frac{H}{T^2}. \tag{5.19}$$

5.1 Chemical potential of pure phases

If the Gibbs free energy, G, of a pure phase is divided by its number of moles, an intensive function of state termed the *chemical potential*, μ, is obtained:

$$\frac{G}{n} = \mu. \tag{5.20}$$

The chemical potential is thus the *molar Gibbs free energy*. Accordingly, its magnitude is given in Joule mol^{-1} or cal mol^{-1}.

The total differential of the chemical potential reads:

$$d\mu = \left(\frac{\partial\mu}{\partial T}\right)_P dT + \left(\frac{\partial\mu}{\partial P}\right)_T dP. \tag{5.21}$$

Considering Eqs. (5.15) and (5.16), Eq. (5.21) can be rewritten as:

$$d\mu = -\boldsymbol{S}dT + \boldsymbol{V}dP, \tag{5.22}$$

where $\boldsymbol{S}$ and $\boldsymbol{V}$ designate the molar entropy and molar volume of the pure phases, respectively.

5.1.1 Chemical potential of ideal gases

A mathematical expression for the chemical potential of an ideal gas can be derived using the relationship:

$$\left(\frac{\partial\mu}{\partial P}\right)_T = \boldsymbol{V}, \tag{5.23}$$

which describes the pressure dependence of the chemical potential at constant temperature. Separating the variables in Eq. (5.23) yields:

$$d\mu = \boldsymbol{V}dP. \tag{5.24}$$

In the case of an ideal gas, the molar volume, V, in Eq. (5.24) can be replaced by RT/P and one obtains:

$$d\mu = \frac{RT}{P}dP = RTd\ln\boldsymbol{P}. \tag{5.25}$$

Integrating Eq. (5.25) between the initial pressure P_0 and the final pressure P gives

$$\int_{\mu(P_o)}^{\mu(P)} d\mu = RT\int_{P_o}^{P} d\ln P = \mu(P) - \mu(P_o) = RT\ln\frac{\boldsymbol{P}}{\boldsymbol{P}_o}. \tag{5.26}$$

Rearranging Eq. (5.26) and remembering that the temperature remains constant yields:

$$\mu(P, T) = \mu(P_o, T) + RT\ln\frac{\boldsymbol{P}}{\boldsymbol{P}_o}, \tag{5.27}$$

where $\mu(P_0,T)$ designates the *standard chemical potential* that refers to the temperature of the experiment, T, and to some arbitrarily chosen *standard pressure*, P_0. The latter corresponds generally to the ambient pressure of 0.1 MPa. Because of this widely accepted convention, pressure and temperature specifications for the standard potential are omitted and the standard potential is designated as μ^{o}. Introducing the superscript '*id*' to designate the ideal behavior of a gas gives:

$$\mu^{id} = \mu^{\text{o}} + RT\ln\frac{\boldsymbol{P}[Pa]}{0.1\times10^{6}[Pa]}. \tag{5.28}$$

In the case that pressure is given in *bar*, Eq. (5.28) simplifies to

$$\mu^{id} = \mu^{\text{o}} + RT\ln\boldsymbol{P}[bar], \tag{5.29}$$

because the reference pressure is then 1 bar and ln 1 = 0. One should, however, keep in mind that P[bar] is a dimensionless quantity (bar/bar) that is numerically equal to the pressure given in bars.

5.1.2 Chemical potential of non-ideal gases

In order to derive an expression analogous to Eq. (5.28) for non-ideal gases, the pressure in Eq. (5.25) has to be replaced by the *fugacity*, f, so that

$$d\mu = RTd\ln f \tag{5.30}$$

is obtained.

The relationship between the pressure and fugacity is given as:

$$f = \varphi P. \tag{5.31}$$

The proportionality factor, φ, is termed the *fugacity coefficient*. It fulfills the following boundary condition:

$$\lim_{P \to 0} \varphi = 1. \tag{5.32}$$

Thus, the fugacity of a gas equals pressure as pressure decreases and approaches 0. This means that at some sufficiently low pressures all gases behave ideally.

Integrating Eq. (5.30) between the limits f_o and f yields:

$$\mu^{real} = \mu^{o} + RT\ln\frac{f}{f_o}. \tag{5.33}$$

The standard chemical potential, μ^o, refers to the standard fugacity f_o whose value is 0.1 MPa for all temperatures. Using the relationship between the pressure and fugacity, given in Eq. (5.31), Eq. (5.33) can be rewritten to obtain:

$$\mu^{real} = \mu^{o} + RT\ln\frac{P}{P_o} + RT\ln\frac{\varphi}{\varphi_o}. \tag{5.34}$$

Because P_o and f_o, equal 0.1 MPa, φ_o must have a value of 1, and it follows that

$$\mu^{real} = \mu^{id} + RT\ln\varphi. \tag{5.35}$$

If pressure is given in bars, f_o equals 1 bar and Eq. (5.33) simplifies to:

$$\mu^{real} = \mu^{o} + RT\ln f. \tag{5.36}$$

Accordingly, Eq. (5.34) simplifies to:

$$\mu^{real} = \mu^{o} + RT\ln\boldsymbol{P} + RT\ln\varphi. \tag{5.37}$$

The relationship between the fugacity of a real gas, its molar volume and its fugacity coefficient can be derived using Eqs. (5.29) and (5.36). Differentiating Eq. (5.36) with respect to pressure at constant temperature, yields:

$$V^{real}dP = RTd\ln\boldsymbol{f}. \tag{5.38}$$

An analogous relationship is obtained for an ideal gas by differentiating Eq. (5.29), namely:

$$V^{id}dP = RTd\ln\boldsymbol{P}. \tag{5.39}$$

Subtracting Eq. (5.39) from Eq. (5.38) gives:

$$(V^{real} - V^{id})dP = RTd\ln\frac{\boldsymbol{f}}{\boldsymbol{P}}. \tag{5.40}$$

Substituting the term $RT/\boldsymbol{P}$ for V^{id} in Eq. (5.40) and then integrating it between the limits $P = 0$ and P yields:

$$\int_0^P \left(V^{real} - \frac{RT}{\boldsymbol{P}}\right)dP = RT\left[\ln\left(\frac{\boldsymbol{f}}{\boldsymbol{P}}\right)_P - \ln\left(\frac{\boldsymbol{f}}{\boldsymbol{P}}\right)_{P=0}\right] \tag{5.41}$$

or

$$RT\ln\varphi = \int_0^P \left(V^{real} - \frac{RT}{\boldsymbol{P}}\right)dP. \tag{5.42}$$

The term $\ln(\boldsymbol{f}/\boldsymbol{P})_{P=0}$ equals zero, since all gases behave ideally at pressures close to zero.

V^{real} as well as $RT/\boldsymbol{P}$, approach infinity as $\boldsymbol{P}$ approaches 0. Therefore, the two terms cannot be integrated separately. The difference between them, however, is definite and a nonzero quantity.

The integration procedure can be best solved graphically. Using experimentally determined values for the volume of the gas, the difference V^{real} - $RT/\boldsymbol{P}$ is calculated and plotted versus pressure. The area under the curve gives, then, the fugacity coefficient.

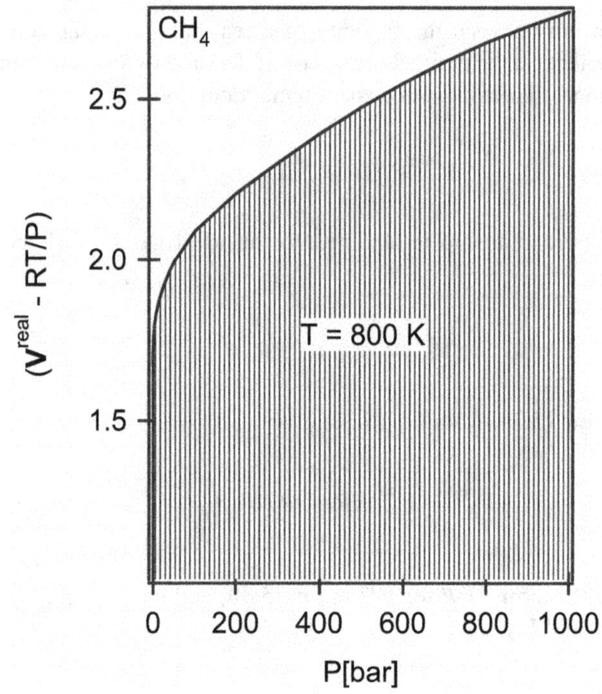

Fig. 5.1 Graphical representation of the difference V^{real} - RT/P as a function of pressure for methane, CH_4, at 800 K, using volume data of Grevel and Chatterjee 1992.

Fig. 5.1 shows a plot of the term V^{real} - RT/P versus P for methane, CH_4, as a function of pressure at 800 K. The calculated areas and the corresponding fugacity coefficient determined from the graph are summarized in Tab. 5.1.

Table 5.1 Fugacity coefficient and fugacity of methane, CH_4, determined by graphical integration of the V^{real} - RT/P vs. P curve, depicted in Fig. 5.1. (Data: Grevel and Chatterjee 1992)

P[bar]	area	$\ln\varphi$	φ	f[bar]
0.5	0.59	$8.86\text{x}10^{-5}$	1.000	0.5
1.0	1.44	$2.16\text{x}10^{-4}$	1.000	1.0
10.0	17.77	$2.67\text{x}10^{-3}$	1.003	10.0

Table 5.1 Fugacity coefficient and fugacity of methane, CH_4, determined by graphical integration of the V^{real} - RT/P vs. P curve, depicted in Fig. 5.1. (Data: Grevel and Chatterjee 1992)

P[bar]	area	lnφ	φ	f[bar]
50.0	94.93	0.01427	1.014	50.7
100.0	196.94	0.02961	1.030	103.0
500.0	1114.76	0.16759	1.182	591.2
1000.0	2433.76	0.36589	1.442	1441.8

5.2 The chemical potential of components in solutions

Considering Eq. (5.13), the total differential of the Gibbs free energy, dG, of a multi component phase can be written as follows:

$$dG = -SdT + VdP + \sum_i \left(\frac{\partial G}{\partial n_i}\right)_{P,\,T,\,n_{j \neq i}} dn_i. \tag{5.43}$$

The differential in Eq. (5.43) gives the change in the total Gibbs free energy of the system when an infinitesimal amount of component i is added to it, while pressure, temperature and the concentrations of the remaining components are kept constant. It thus represents the partial molar Gibbs free energy of component i in the solution. It is referred to as the *chemical potential of the component i*, μ_i, i.e.

$$\left(\frac{\partial G}{\partial n_i}\right)_{P,\,T,\,n_{j \neq i}} = \mu_i. \tag{5.44}$$

Using this definition of the chemical potential, Eq. (5.43) can be rewritten as:

$$dG = -SdT + VdP + \sum_i \mu_i dn_i. \tag{5.45}$$

At constant temperature and pressure Eq. (5.45) reduces to:

$$dG = \sum_i \mu_i dn_i. \tag{5.46}$$

Assume, that a solution is made by continuously adding small increments, all of the same composition, so that the bulk composition of the solution remains constant. In this case, the values of the chemical potentials of the various components remain constant and Eq. (5.46) can be integrated such that

$$G = \sum_i n_i \mu_i \tag{5.47}$$

is obtained. Thus, the total Gibbs free energy of a mixed phase at some pressure and temperature is given by the sum of the chemical potentials of the components multiplied by the respective number of moles.

Example: The total Gibbs free energy of an orthopyroxene crystal with $n^{opx}_{MgSiO_3}$ mole enstatite and $n^{opx}_{FeSiO_3}$ mole ferrosilite is calculated as follows:

$$G^{opx} = n^{opx}_{MgSiO_3}\mu^{opx}_{MgSiO_3} + n^{opx}_{FeSiO_3}\mu^{opx}_{FeSiO_3}.$$

Dividing Eq. (5.46) by the total number of moles yields the *molar Gibbs free energy*, $\overline{G}$. It is

$$\frac{G}{\sum_i n_i} = \overline{G} = \sum_i \frac{n_i}{\sum_i n_i}\mu_i = \sum_i x_i \mu_i. \tag{5.48}$$

Example: In the case of orthopyroxene, the molar Gibbs free energy is calculated according to

$$\begin{aligned}\overline{G}^{opx} &= \frac{n^{opx}_{MgSiO_3}}{n^{opx}_{MgSiO_3} + n^{opx}_{FeSiO_3}}\mu^{opx}_{MgSiO_3} + \frac{n^{opx}_{FeSiO_3}}{n^{opx}_{MgSiO_3} + n^{opx}_{FeSiO_3}}\mu^{opx}_{FeSiO_3} \\ &= x^{opx}_{MgSiO_3}\mu^{opx}_{MgSiO_3} + x^{opx}_{FeSiO_3}\mu^{opx}_{FeSiO_3},\end{aligned}$$

and because the mole fractions of the components composing a mixed phase sum to 1, one can also write:

$$\overline{G}^{opx} = (1 - x^{opx}_{FeSiO_3})\mu^{opx}_{MgSiO_3} + x^{opx}_{FeSiO_3}\mu^{opx}_{FeSiO_3}.$$

As shown previously, the chemical potential of a component in a solution is equal to its partial molar Gibbs free energy at the given pressure, temperature and composition of the solution. The value of the chemical potential depends not only on the concentration of the component in the solution but also on the solution itself. That is, the concentrations and the chemical composition of the remaining components composing the solution also come into play. A component of the same concentration yields different chemical potentials in different solutions.

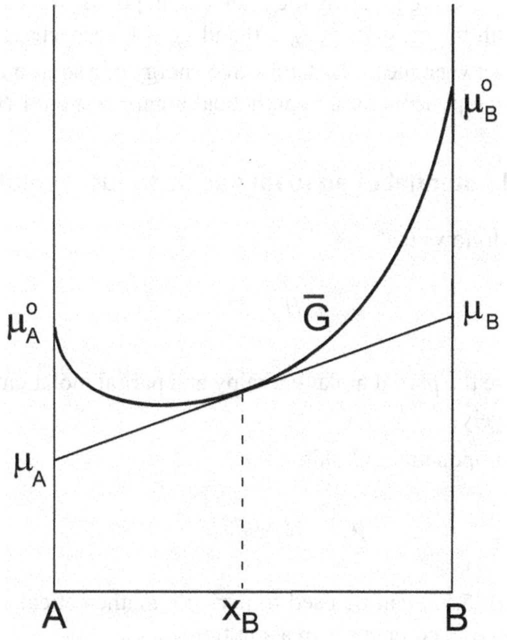

Fig. 5.2 Graphical representation of the relationship between the molar Gibbs free energy, $\overline{G}$, and the chemical potentials of the components A and B in the solution of the composition x_B. $\overset{o}{\mu}_A$ and $\overset{o}{\mu}_B$ are the chemical potentials of the pure components A and B, respectively.

The chemical potentials of the components can be calculated using the integral Gibbs free energy of the solution. For a binary system, A-B, the calculation gives:

For the component A:

$$\mu_A = \overline{G} - x_B\left(\frac{\partial \overline{G}}{\partial x_B}\right)_{P,T} \tag{5.49}$$

and

$$\mu_B = \overline{G} + (1 - x_B)\left(\frac{\partial \overline{G}}{\partial x_B}\right)_{P,\,T} \tag{5.50}$$

for the component B.

According to Eqs. (5.49) and (5.50), the chemical potentials of the components A and B in the solution are given by the intersection points of the tangent line to the $\overline{G}$ vs. x_B curve with the ordinate at $x_B = 0$ and $x_B = 1$, respectively. Fig. 5.2 shows the relationships between the molar Gibbs free energy of a solution and the chemical potentials of the components for a hypothetical binary system A-B.

5.2.1 Chemical potential of an ideal gas in an ideal solution

From Eq. (5.9) it follows that

$$\mu_i = H_i - TS_i, \tag{5.51}$$

where H_i and S_i are the partial molar enthalpy and partial molar entropy of the component i, respectively.

For the pure component i, it holds:

$$\boldsymbol{\mu}_i = \boldsymbol{H}_i - T\boldsymbol{S}_i. \tag{5.52}$$

Eqs. (5.51) and (5.52) can be used to derive a mathematical expression for the chemical potential of a component in a solution.

Consider an ideal single component gas. At temperature T and pressure P, its chemical potential equals:

$$\boldsymbol{\mu}_i = \boldsymbol{H}_i - T\boldsymbol{S}_i.$$

If this gas is mixed with another ideal gas at the same temperature and pressure, its chemical potential is then given by:

$$\mu_i = H_i - TS_i.$$

The difference between the chemical potential of the gas i before and after the mixing is then:

$$\mu_i - \boldsymbol{\mu}_i = H_i - \boldsymbol{H}_i - T(S_i - \boldsymbol{S}_i). \tag{5.53}$$

In an ideal mixture, the molar enthalpy and the partial molar enthalpy of a component are equal, and therefore $H_i - \boldsymbol{H}_i = 0$. The difference $S_i - \boldsymbol{S}_i$, however, is equal to $- R \ln x_i$. Thus, instead of Eq. (5.53) one can write:

$$\mu_i - \boldsymbol{\mu}_i = RT\ln x_i \tag{5.54}$$

or

$$\mu_i = \boldsymbol{\mu}_i + RT\ln x_i. \tag{5.55}$$

Substituting the expression given in Eq. (5.28) for $\boldsymbol{\mu}_i$ and adding the superscript 'id' to μ_i in order to stress the ideal behavior of the mixture, yields:

$$\mu_i^{id} = \boldsymbol{\mu}_i^{o} + RT\ln\frac{\boldsymbol{P}[Pa]}{0.1\times10^6[Pa]} + RT\ln x_i. \tag{5.56}$$

According to Dalton's law, the product $x_i P$ equals the partial pressure, P_i, of gas i in an ideal gas mixture. Hence, Eq. (5.56) can be rewritten as follows:

$$\mu_i^{id} = \mu_i^{o} + RT\ln\frac{P_i[Pa]}{0.1\times10^6[Pa]}. \tag{5.57}$$

If the pressure is given in bars, Eq. (5.57) simplifies to:

$$\mu_i^{id} = \mu_i^{o} + RT\ln P_i. \tag{5.58}$$

In order to avoid any confusion, it should be noted that $\boldsymbol{P}$ designates the total pressure, that in the case of a pure gas is identical to the gas pressure. No subscript is therefore needed. In the case of an ideal gas mixture, P, however, represents the sum of the partial pressures of the gases participating in the mixture, i.e. $P = \Sigma x_i P_i$. The partial pressure of each different gas must then be defined by an appropriate subscript.

5.2.2 Chemical potential of a real gas in ideal and non-ideal gas mixtures

Suppose that two non-ideal gases mix ideally at constant pressure and temperature.

The chemical potential of each gas in such a mixture reads:

$$\mu_i^{real} = \boldsymbol{\mu}_i^{real} + RT\ln x_i, \tag{5.59}$$

where $\boldsymbol{\mu}_i^{real}$ is the chemical potential of the pure gas i at the given temperature and pressure, and x_i is its mole fraction in the mixture. Substituting the expression given in Eq. (5.35) for $\boldsymbol{\mu}_i^{real}$ yields:

$$\mu_i^{real} = \boldsymbol{\mu}_i^{id} + RT\ln\boldsymbol{\varphi}_i + RT\ln x_i \tag{5.60}$$

or

$$\mu_i^{real} = \boldsymbol{\mu}_i^{\mathrm{o}} + RT\ln\frac{P}{P_o} + RT\ln\boldsymbol{\varphi}_i + RT\ln x_i \tag{5.61}$$

if $\boldsymbol{\mu}^{id}$ is replaced by the expression giving the chemical potential of an ideal gas at pressure, P, and temperature, T. The product x_iP corresponds to the partial pressure, P_i, of the gas i in the mixture. Using this relationship, Eq. (5.61) obtains the following form:

$$\mu_i^{real} = \boldsymbol{\mu}_i^{\mathrm{o}} + RT\ln\frac{P_i}{P_o} + RT\ln\boldsymbol{\varphi}_i. \tag{5.62}$$

In this equation $\boldsymbol{\varphi}_i$ represents the fugacity coefficient for a pure gas i. The product

$$P_i\boldsymbol{\varphi}_i = f_i^{id} \tag{5.63}$$

is frequently termed the *partial fugacity* of the gas i in an ideal gas mixture.

In the case of ideal mixing, the partial molar volume of a non-ideal gas, V_i^{real}, and its molar volume, $\boldsymbol{V}_i^{real}$, are the same. In real mixtures, however, the two differ one from another. Therefore, it holds:

$$RT\ln\varphi_i = \int_0^P\left(V_i^{real} - \frac{RT}{P}\right)dP \neq RT\ln\boldsymbol{\varphi}_i = \int_0^P\left(\boldsymbol{V}_i^{real} - \frac{RT}{P}\right)dP, \tag{5.64}$$

where φ_i designates the fugacity coefficient of a non-ideal gas in a non-ideal gas mixture at the given temperature, pressure and concentration of the gas i in the mixture. It is referred to as the *partial fugacity coefficient.*

Following Eqs. (5.34) and (5.35), the chemical potential of a pure non-ideal gas i, $\boldsymbol{\mu}_i^{real}$ equals:

$$\boldsymbol{\mu}_i^{real} = \boldsymbol{\mu}_i^{\mathrm{o}} + RT\ln\frac{P}{P_o} + RT\ln\boldsymbol{\varphi}_i. \tag{5.65}$$

The chemical potential of the same gas in a non-ideal gas mixture, however, reads:

$$\mu_i^{real} = \boldsymbol{\mu}_i^{\mathrm{o}} + RT\ln\frac{P}{P_o} + RT\ln\varphi_i + RT\ln x_i, \tag{5.66}$$

where φ_i represents the partial fugacity coefficient as defined in Eq. (5.64). It depends on pressure, temperature and the composition of the mixture.

If Eq. (5.65) is subtracted from Eq. (5.66), the change in the chemical potential due to the non ideal mixing at constant pressure and temperature is obtained. One has:

$$\mu_i^{real} - \boldsymbol{\mu}_i^{real} = RT\ln\frac{\varphi_i}{\boldsymbol{\varphi}_i} + RT\ln x_i \tag{5.67}$$

or

$$\mu_i^{real} = \boldsymbol{\mu}_i^{real} + RT\ln\frac{\varphi_i}{\boldsymbol{\varphi}_i} + RT\ln x_i. \tag{5.68}$$

The second term on the right side of Eq. (5.68) gives the relationship between the partial fugacity coefficient of a non-ideal gas in a non-ideal gas mixture, φ_i, and its fugacity coefficient, $\boldsymbol{\varphi}_i$, as a pure gas. It is termed the *activity coefficient*, γ. Eq. (5.68) can thus be rewritten to obtain:

$$\mu_i^{real} = \boldsymbol{\mu}_i^{real} + RT\ln\gamma_i + RT\ln x_i. \tag{5.69}$$

Combining the two last terms in Eq. (5.69) yields:

$$\mu_i^{real} = \mu_i^{real} + RT\ln a_i, \tag{5.70}$$

where

$$a_i = \gamma_i \cdot x_i. \tag{5.71}$$

The term a_i is the *activity* of the non-ideal gas i in an non-ideal mixture at the given temperature, pressure and concentration of the gas, x_i. It is, thus, a 'corrected mole fraction' of a gas that accounts for the non-ideal behavior of the mixture. The 'correction' factor is the activity coefficient γ_i.

Replacing μ_i^{real} in Eq. (5.69) by the expression given in Eq. (5.65) gives:

$$\mu_i^{real} = \mu_i^{o} + RT\ln\frac{P_i}{P_o} + RT\ln\varphi_i + RT\ln\gamma_i + RT\ln x_i. \tag{5.72}$$

Considering the definition of the activity coefficient, the term $RT\ln\varphi_i$ in Eq. (5.72) cancels and one obtains:

$$\mu_i^{real} = \mu_i^{o} + RT\ln\frac{P}{P_o} + RT\ln\varphi_i + RT\ln x_i. \tag{5.73}$$

Because the product $P_i{\cdot}x_i$ corresponds to the partial pressure of the gas, P_i, Eq. (5.73) can be rewritten as

$$\mu_i^{real} = \mu_i^{o} + RT\ln\frac{P_i}{P_o} + RT\ln\varphi_i \tag{5.74}$$

or

$$\mu_i^{real} = \mu_i^{o} + RT\ln\frac{f_i}{P_o}, \tag{5.75}$$

where

$$f_i = P_i\varphi_i \tag{5.76}$$

represents the fugacity of the non-ideal gas i in a non-ideal mixture at some given concentration of the gas. f_i is referred to as the *partial fugacity*.

The following relationships exist between the partial fugacity of a non-ideal gas in a non-ideal mixture, f_i, the fugacity of a pure non-ideal gas, $\boldsymbol{f}_i$, and the fugacity of a non-ideal gas in a ideal gas mixture, f_i^{id}:

$$f_i = a_i \cdot \boldsymbol{f}_i = \gamma_i \cdot f_i^{id}. \qquad (5.77)$$

The level of scientific knowledge has now reached the point whereby the mixing properties of many geologically relevant and important binary gas mixtures are known (see Kerrick and Jacobs 1981; Grevel and Chatterjee 1992; Aranovich and Newton 1999). Hence, the partial fugacities of the gases participating in such mixtures have been determined too. In those cases where the thermodynamic properties of mixtures are not known, the fugacity of a real gas in a non-ideal gas mixture is determined by using the *fugacity rule of Lewis and Randall.* According to this rule, the fugacity of a real gas in a non-ideal gas mixture can be estimated as follows:

$$f_i \approx f_i^{id} = x_i \cdot \boldsymbol{f}_i. \qquad (5.78)$$

5.2.3 Chemical potential of components in ideal solid solutions

In analogy to Eq. (5.59), the chemical potential of a component in an ideal solid solution is given by:

$$\mu_i = \mu_i^{\mathrm{o}} + RT\ln x_i, \qquad (5.79)$$

where μ_i^{o} designates the standard potential of the component, *i*. Its numerical value depends on pressure, temperature and the value of the standard state chosen. In most cases the standard potential refers to the pure phase at the pressure and temperature conditions of interest. It thus differs from the standard potential of a gas, which normally refers to the actual temperature and *ambient* pressure (0.1 MPa).

In deriving the expression for the chemical potential of an ideal gas in an ideal gas mixture, it was shown that the term $RT \ln x_i$ accounts for the change in the entropy of the gas due to the mixing process. The same holds for a solid component in a solid solution. However, the entropy change of crystalline solid is given by Eq. (4.86):

$$S_i - \boldsymbol{S}_i = -z_i R \ln x_i,$$

where z_i refers to the number of crystallographic sites per formula unit containing the same fraction of atoms of the component i. If, however, the mixing atoms are distributed non-equally over different crystallographic sites, each site must be given separately. The more general form of Eq. (5.79), therefore, reads:

$$\mu_i = \mu_i^{o} + RT\sum_i z_i \ln x_i. \tag{5.80}$$

Example 1: Let us consider an olivine solid solution once again. In this solution magnesium and ferrous iron occupy the sixfold coordinated *M1* and *M2* sites. Although the two sites are crystallographically non-equivalent, neither cation shows a clear preference for either site. Therefore, the chemical potential of fayalite, Fe_2SiO_4, in an olivine solid solution, $(Mg,Fe)_2SiO_4$, can be written as follows:

$$\mu_{Fe_2SiO_4}^{ol} = \mu_{Fe_2SiO_4}^{o,ol} + 2RT\ln x_{Fe}^{ol},$$

where x_{Fe}^{ol} gives the atomic fraction of iron in olivine. Because $x_{Fe}^{ol} = x_{Fe_2SiO_4}^{ol}$, the expression for the chemical potential of fayalite can be rewritten replacing the atomic fraction, x_{Fe}^{ol}, by the mole fraction, $x_{Fe_2SiO_4}^{ol}$:

$$\mu_{Fe_2SiO_4}^{ol} = \mu_{Fe_2SiO_4}^{o,ol} + 2RT\ln x_{Fe_2SiO_4}^{ol}.$$

If one wants to eliminate the multiplier in front of the term that accounts for the concentration of the component in the solution, one has to divide the above given equation by 2 and one obtains:

$$\frac{1}{2}\mu_{Fe_2SiO_4}^{ol} = \frac{1}{2}\mu_{Fe_2SiO_4}^{o,ol} + RT\ln x_{Fe_2SiO_4}^{ol}$$

or

$$\mu_{FeSi_{0.5}O_2}^{ol} = \mu_{FeSi_{0.5}O_2}^{o,ol} + RT\ln x_{FeSi_{0.5}O_2}^{ol}.$$

In this case, one has to keep in mind that the chemical potential refers to one half of the formula unit.

Example 2: As discussed previously, in orthopyroxene, $(Mg,Fe)_2Si_2O_6$, ferrous iron occupies preferably the larger and more distorted *M2* sites. Hence, its site fractions for *M1* and *M2* are different and must, therefore, appear as separate terms. The chemical potential of ferrosilite, $Fe_2Si_2O_6$, in an orthopyroxene solid solution reads:

$$\mu_{Fe_2Si_2O_6}^{opx} = \mu_{Fe_2Si_2O_6}^{o,opx} + RT\ln x_{Fe,\,M1}^{opx} + RT\ln x_{Fe,\,M2}^{opx},$$

where $x_{Fe,\,M1}^{opx}$ and $x_{Fe,\,M2}^{opx}$ are the atomic fractions of iron on the *M1* and *M2* sites, respectively. Note, that the mineral name ferrosilite is used to designate a component. In a strict sense this is not correct because a mineral name designates a phase of definite composition and definite crystal structure. Nonetheless, in the mineralogical literature mineral names are often assigned to components. This practice, however, becomes problematic when a component undergoes a phase transitions in the course of a thermodynamic process. One should always bear in mind that a component is a chemical entity that should be expressed in terms of a chemical formula. A mineral name can be used only when confusion with the phase is excluded.

Example 3: In section 4.3.2 it was shown that the configurational entropy of the diopside-jadeite solid solution can be expressed using the mole fractions of the components. Therefore, the chemical potential of diopside in the diopside-jadeite solid solution can be written as follows:

$$\mu_{CaMgSi_2O_6}^{cpx} = \mu_{CaMgSi_2O_6}^{o,cpx} + RT\ln x_{CaMgSi_2O_6}^{cpx}.$$

The term containing mole fraction accounts for the change in the entropy of a component as a result of the process of mixing. If there are different crystallographic sites in a structure that are available for some atoms or ions, their distribution over these sites must be known in order to formulate the chemical potential.

5.2.4 Chemical potential of components in non-ideal solid solutions

Similarly as in the case of non-ideal gas mixtures, the mole fraction must be replaced by the activity in order to express the chemical potential of a component in a non-ideal solid mixture. At constant pressure and temperature the chemical potential of the component i then reads:

$$\mu_i = \mu_i^{o} + z_i RT\ln a_i. \tag{5.81}$$

In Eq. (5.81) μ_i^{o} and z_i designate the standard potential of the component i and z_i gives the number of crystallographic sites per formula unit with the same atomic fraction of the component i. Generally μ_i^{o} is independent of composition, but it depends on temperature and pressure.

The activity of a component is proportional to its mole fraction. The proportionality factor is referred to as the *activity coefficient*. Thus, it holds:

$$a_i = \gamma_i \cdot x_i \tag{5.82}$$

and the chemical potential reads:

$$\mu_i = \mu_i^{\mathrm{o}} + z_i RT\ln x_i + z_i RT\ln\gamma_i. \tag{5.83}$$

The activity coefficient is a function of pressure, temperature and composition of the solution. It is an empirical quantity and is often evaluated from the phase equilibrium studies. In some cases it can be determined by different electrochemical methods, e.g. emf-measurements or calorimetric studies.

Activity and activity coefficient as a function of composition

The relationship between the concentration of a component and its activity in a solution is generally complex. In every solution, however, there are concentration regions where the relationship is relatively simple.

At high concentration of a component when its mole fraction in the solution approaches 1, the activity coefficient also approaches 1 and the activity becomes equal to the mole fraction, i.e.

$$a_i = x_i. \tag{5.84}$$

The chemical potential is, according to Eq. (5.81), then:

$$\mu_i = \mu_i^{\mathrm{o}} + z_i RT\ln x_i. \tag{5.85}$$

The compositional region, where a component behaves this way is referred to as the region of *Raoult's law*. Within this region the chemical potential of a component is directly proportional to the logarithm of its mole fraction. Its extension to lower concentrations depends on the interaction between the components in the solution. The smaller the interaction, the larger is the region in which Raoult's law is obeyed. In Fig. 5.3 Raoult's region, R, extends from x_B^R to $x_B = 1$.

At low concentrations of a component, when its mole fraction approaches 0, the chemical potential varies linearly with the logarithm of the mole fraction. The proportionality factor, however, differs from 1. It can be greater, as well as smaller, than one. The activity is directly proportional to the mole fraction, i.e:

$$a_i = h_i \cdot x_i. \tag{5.86}$$

The proportionality factor, h_i is called *Henry constant.* Its value depends not only on the nature of the solute but also on the solution. The compositional region in which a component displays this kind of behavior is referred to as the region of *Henry's law* and the chemical potential is given as:

$$\mu_i = \mu_i^{\mathrm{o}} + z_i RT \ln x_i + z_i RT \ln h_i. \tag{5.87}$$

In Fig. 5.3 the region of Henry's law is designated by the letter *H.* It extends from $x_B = 0$ to x_B^H. The extension of Henry's law region toward higher concentrations depends on the nature of mixing in the solution. The smaller the interaction between the components in the solution, the larger is this region. In ideal mixtures, where no or only negligible interactions between the components exist, Henry's line and Raoult's' line coincide and the activity equals the mole fraction from $x_B = 0$ to $x_B = 1$.

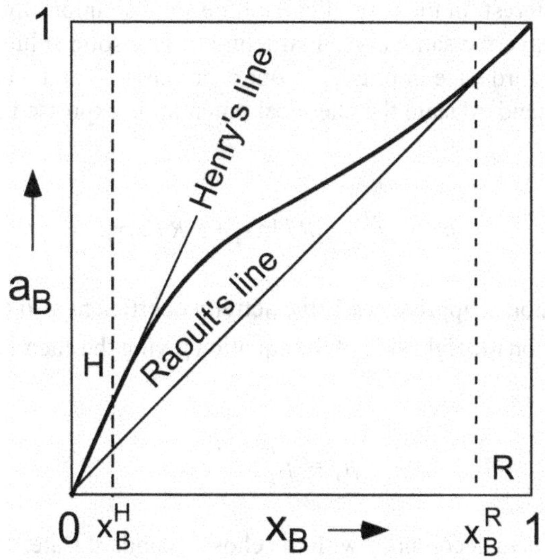

Fig. 5.3 Schematic representation of the activity of the component B, a_B, as a function of the mole fraction x_B in the solution A-B at constant temperature and pressure. R = region of Raoult's law and H = region of Henry's law. Raoult's and Henry's line are tangents to the $a_B(x_B)$ curves at low and high concentration of B, respectively.

The region of Henry's law and the one of Raoult's law are separated by the so-called *intermediate region*. Here, the activity coefficient is a function of solution composition and the expression for the chemical potential has the general form given in Eq. (5.83), namely:

$$\mu_i = \mu_i^{\text{o}} + z_i RT\ln x_i + z_i RT\ln \gamma_i.$$

In the mineralogical literature there are different approaches to solving the '$\gamma(x)$ problem'. Some of them will be discussed later.

5.2.5 The standard state

The classical thermodynamics does not give any information about the absolute value of the chemical potential. Therefore an arbitrary standard state has to be chosen. One type of a standard state was already presented in the discussion of the chemical potential of ideal and non-ideal gases. In both cases, a state of an ideal gas at 0.1 MPa and the temperature of interest was chosen as a standard. In the case of condensed phases other standard states are normally used. Here we want to present three of them.

a) The most widely used standard state is that of a *pure phase* at pressure and temperature of interest. In the case of crystalline solid solutions the pure end-member phase must have the same crystal structure as any solid solution composition. We used this standard state implicitly in our examples 1, 2 and 3 on pages 182 and 183. Using this standard state the chemical potential is expressed according to Eq. (5.83) as:

$$\mu_i = \mu_i^{\text{o}} + z_i RT\ln x_i + z_i RT\ln \gamma_i.$$

As the mole fraction x_i approaches 1, the activity coefficient also approaches 1 and the two last terms on the right side of the equation giving the chemical potential vanish, so that

$$\mu_i = \mu_i^{\text{o}} \tag{5.88}$$

is obtained. This is in accordance with the chosen standard state.

b) The standard state presented in the previous section could be difficult to apply, if a component does not occur as a pure end-member phase (e.g. pyroxene with Rare Earth cation only). In this case the state of *infinite dilution* is chosen as the standard. The corresponding standard potential consists of the chemical potential that consists of the hypothetical pure end-member phase and Henry's constant, h_i, such that:

$$\mu_i^{\infty} = \mu_i^{o} + RT\ln h_i. \tag{5.89}$$

Hence, this standard potential refers to a 'pure end-member phase' having the thermodynamic properties extrapolated from the state of infinite dilution.

Using the standard potential defined in Eq. (5.89), the chemical potential of a component i can be written as follows:

$$\mu_i = \mu_i^{\infty} + RT\ln x_i. \tag{5.90}$$

Eq. (5.90) holds only in the region of Henry's law. At higher concentration of the component, the mole fraction must be corrected introducing a new activity coefficient, γ_i^{∞}. The chemical potential, therefore, reads:

$$\mu_i = \mu_i^{\infty} + RT\ln x_i + RT\ln\gamma_i^{\infty}. \tag{5.91}$$

The activity coefficient γ_i^{∞} approaches 1 as the mole fraction approaches 1 such at $x_i = 1$ it holds that:

$$\mu_i = \mu_i^{\infty}. \tag{5.92}$$

In order to determine the relationship between the activity coefficient related to the pure phase and the one related to infinite dilution, we take the expression given in Eq. (5.83) and first add and then subtract the term $RT\ln h_i$:

$$\mu_i = \mu_i^{o} + z_iRT\ln h_i - z_iRT\ln h_i + z_iRT\ln x_i + z_iRT\ln\gamma_i. \tag{5.93}$$

The first two terms on the right side of the 'equal' sign of Eq. (5.93) give the standard potential at infinite dilution, μ_i^{∞}. Hence, one can write:

$$\mu_i = \mu_i^{\infty} + z_iRT\ln x_i + z_iRT\ln\frac{\gamma_i}{h_i}. \tag{5.94}$$

A comparison of the terms in Eq. (5.94) to those in Eq. (5.91) shows that

$$\gamma_i^{\infty} = \frac{\gamma_i}{h_i}. \tag{5.95}$$

The two different activity coefficients are proportional to one another. As can be seen in Fig. 5.4, the Henry's constant corresponds with the activity coefficient γ_i at $x_i = 0$ (double arrow). At this concentration, γ_i^{∞} has the value of 1. Generally it holds: If γ_i has a value greater than one, γ_i^{∞} is smaller than one and vice versa.

In the case that the pure component and the solid solution have different crystal structures, the mixing process is associated with a phase transition. Hence, the Gibbs' free energy of transition must be included in the expression for the chemical potential. Assume that a component i crystallizes in an α-structure, while the solutions has the β-structure. The chemical potential will then read:

$$\mu_i = \mu_i^{0,\alpha} + z_i RT\ln x_i + z_i RT\ln\gamma_i^{\beta} + (\mu_i^{0,\beta} - \mu_i^{0,\alpha}), \tag{5.96}$$

where the difference $\mu_i^{0,\beta} - \mu_i^{0,\alpha}$ gives the Gibbs free energy of the phase transition from α to β.

The last two terms in Eq. (5.96) can be combined to give a new activity coefficient, γ_i^{α}, namely:

$$\gamma_i^{\alpha} = \gamma_i \exp\left(\frac{\mu_i^{0,\beta} - \mu_i^{0.\alpha}}{RT}\right). \tag{5.97}$$

Using this activity coefficient, Eq. (5.96) simplifies to

$$\mu_i = \mu_i^{0,\alpha} + z_i RT\ln x_i + z_i RT\ln\gamma_i^{\alpha}. \tag{5.98}$$

Fig. 5.4 shows graphically the three different standard states of the chemical potential. In order to simplify the image, a binary system consisting of the hypothetical components A and B was chosen, where the component B forms a complete solid solution with A. The heavy line gives the chemical potential of the component B as a function of its mole fraction. The double arrow at $x_B = 0$ corresponds to the term $RT\ln h_B$. Note that the arrow giving the activity coefficient related to the state of infinite dilution and the one giving the activity coefficient related to the state of a pure phase point in opposite directions.

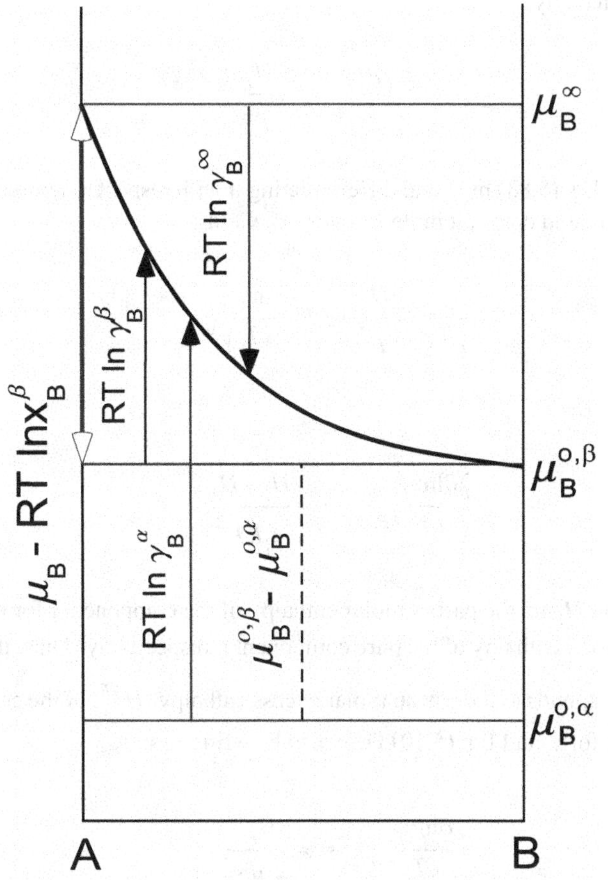

Fig. 5.4 Relationships between different standard states of the chemical potential in the binary system *A-B*. μ_B^{∞}, $\mu_B^{o,\beta}$ and $\mu_B^{o,\alpha}$ are the standard potentials of the component *B* referenced to the state of infinite dilution of *B* in *A*, to the pure component *B* occurring as the phase β, and to the pure component *B* occurring as the phase α, respectively. γ_B^{∞}, γ_B^{β} and γ_B^{α} are the corresponding activity coefficients. $\mu_B^{o,\beta} - \mu_B^{o,\alpha}$ gives the Gibbs free energy of the phase transformation from α to β.

The activity coefficient as a function of temperature and pressure

The activity and the activity coefficient of a component depend not only on composition but also on temperature and pressure.

In order to derive a mathematical expression for the *temperature dependence* of the activity coefficient, we make use of a relationship analogues to the one given in Eq. (5.17), namely:

$$\frac{\partial}{\partial T}\left(\frac{\mu_i}{T}\right)_P = -\frac{H_i}{T^2}. \tag{5.99}$$

Dividing Eq. (5.83) by T and differentiating it with respect to temperature at constant pressure and constant mole fraction, x_i, yields:

$$-\frac{H_i}{T^2} = -\frac{\boldsymbol{H}_i}{T^2} + z_i R\left(\frac{\partial \ln\gamma_i}{\partial T}\right)_{P, x_i} \tag{5.100}$$

or

$$\left(\frac{\partial \ln\gamma_i}{\partial T}\right)_{P, x_i} = -\frac{H_i - \boldsymbol{H}_i}{z_i R T^2}, \tag{5.101}$$

where H_i and $\boldsymbol{H}_i$ are the partial molar enthalpy of the component i for mole fraction x_i and the molar enthalpy of the pure component i, respectively. Thus, the difference H_i - $\boldsymbol{H}_i$ corresponds to the partial molar excess enthalpy, H_i^{ex}, of the component i in the solution for x_i and Eq. (5.101) can also be written as:

$$\left(\frac{\partial \ln\gamma_i}{\partial T}\right)_{P, x_i} = -\frac{H_i^{ex}}{z_i R T^2}. \tag{5.102}$$

Hence, the partial molar excess enthalpy of a component determines the temperature dependence of the activity coefficient.

The *pressure dependence* of the activity coefficient can be derived analogously. In this case, the relationship corresponding to the one given in Eq. (5.14) is used, namely:

$$\left(\frac{\partial \mu_i}{\partial P}\right)_T = V_i. \tag{5.103}$$

Differentiating Eq. (5.83) with respect to pressure at constant temperature and composition of the solution gives:

$$V_i = V_i + z_i RT\left(\frac{\partial \ln \gamma_i}{\partial P}\right)_{T, x_i} \tag{5.104}$$

or upon rearranging terms

$$\left(\frac{\partial \ln \gamma_i}{\partial P}\right)_{T, x_i} = \frac{V_i - V_i}{z_i RT}. \tag{5.105}$$

In Eq. (5.105), V_i and V_i designate the partial molar volume of the component i for the mole fraction x_i and the molar volume of the pure component i, respectively. The difference V_i - V_i, therefore, corresponds to the partial molar excess volume of the component, V_i^{ex}, for given concentration of i in the solution. One can also write:

$$\left(\frac{\partial \ln \gamma_i}{\partial P}\right)_{T, x_i} = \frac{V_i^{ex}}{z_i RT}. \tag{5.106}$$

Hence, the pressure dependence of the activity coefficient of a component in a solution is determined by its partial molar excess volume.

5.3 Gibbs free energy of mixing

At constant temperature and pressure, the *Gibbs free energy of mixing*, $\Delta_m \overline{G}$, is given by the difference between the molar Gibbs free energy of the system before and after mixing, i.e.

$$\Delta_m \overline{G} = \overline{G}^{after} - \overline{G}^{before}. \tag{5.107}$$

Only in the case of a mechanical mixture does it hold that $\Delta_m \overline{G} = 0$. In solutions, the Gibbs free energy of mixing always differs from zero, because mixing is a natural process that generates an increase in entropy.

Using the definition of the molar Gibbs free energy given in Eq. (5.48), Eq. (5.107) reads:

$$\Delta_m \overline{G} = \sum_i x_i \mu_i - \sum_i x_i \mu_i, \tag{5.108}$$

where μ_i and μ_i are the chemical potentials of the pure component i and in the solution, respectively.

Expanding μ_i in Eq. (5.108), according to Eq. (5.83), yields:

$$\Delta_m\overline{G} = \sum_i x_i(\mu_i^o + z_iRT\ln x_i + z_iRT\ln\gamma_i) - \sum_i x_i\mu_i^o. \tag{5.109}$$

Because the standard potential, μ_i^o, in Eq. (5.109) refers to the pure component i, the first and the last terms cancel and one obtains:

$$\Delta_m\overline{G} = RT\sum_i x_i\ln x_i^{z_i} + RT\sum_i x_i\ln\gamma_i^{z_i} \tag{5.110}$$

or

$$\Delta_m\overline{G} = RT\sum_i x_i\ln a_i, \tag{5.111}$$

where

$$a = (x_i\gamma_i)^{z_i}. \tag{5.112}$$

In Eq. (5.110), the term $RT\sum_i x_i\ln x_i^{z_i}$ gives the change in the Gibbs free energy due to ideal mixing and the term $RT\sum_i x_i\ln\gamma_i^{z_i}$ accounts for the non-ideal behavior of the solution. The latter is referred to as the *excess Gibbs free energy of mixing*. Using these definitions one can write:

$$\Delta_m\overline{G} = \Delta_m\overline{G}^{id} + \Delta_m\overline{G}^{ex}. \tag{5.113}$$

If one defines the *excess chemical potential* of a component as:

$$\mu_i^{ex} = z_iRT\ln\gamma_i \tag{5.114}$$

the excess Gibbs free energy of mixing reads:

$$\Delta_m \overline{G}^{ex} = \sum_i x_i \mu_i^{ex}. \tag{5.115}$$

Analogously to Eqs. (5.49) and (5.50), the excess chemical potentials of the various components in a binary solution can be calculated from the excess Gibbs free energy as follows:

$$\mu_A^{ex} = RT\ln\gamma_A = \Delta_m \overline{G}^{ex} - x_B\left(\frac{\partial \Delta_m \overline{G}^{ex}}{\partial x_B}\right)_{P,T} \tag{5.116}$$

and

$$\mu_B^{ex} = RT\ln\gamma_B = \Delta_m \overline{G}^{ex} + (1-x_B)\left(\frac{\partial \Delta_m \overline{G}^{ex}}{\partial x_B}\right)_{P,T}. \tag{5.117}$$

According to Eq. (5.16), the differentiation of the excess Gibbs free energy of mixing with respect to temperature at constant pressure yields:

$$\left(\frac{\partial \Delta_m \overline{G}^{ex}}{\partial T}\right)_P = -\Delta_m \overline{S}^{ex}, \tag{5.118}$$

where $\Delta_m \overline{S}^{ex}$ is the *excess entropy of mixing*. Replacing $\Delta_m \overline{G}^{ex}$ by $RT\sum_i x_i \ln\gamma_i^{z_i}$ gives:

$$\begin{aligned} \Delta_m \overline{S}^{ex} &= -\frac{\partial}{\partial T}\left(RT\sum_i x_i \ln\gamma_i^{z_i}\right)_{P,x_i} \\ &= -R\sum_i x_i \ln\gamma_i^{z_i} - RT\sum_i x_i z_i\left(\frac{\partial \ln\gamma_i}{\partial T}\right)_{P,x_i}. \end{aligned} \tag{5.119}$$

Substituting Eq. (5.102) into Eq. (5.119) yields:

$$\Delta_m \overline{S}^{ex} = -R\sum_i x_i \ln\gamma_i^{z_i} - \frac{1}{T}\sum_i x_i H_i^{ex}. \tag{5.120}$$

Multiplying both sides of Eq. (5.120) by T and considering that $RT\sum_i x_i \ln\gamma_i^{z_i}$

and $\sum_i x_i H_i^{ex}$ correspond to $\Delta_m\overline{G}^{ex}$ and $\Delta_m\overline{H}^{ex}$, respectively, the following relationship between the excess Gibbs free energy of mixing, the excess enthalpy of mixing and the excess entropy of mixing is obtained:

$$\Delta_m\overline{G}^{ex} = \Delta_m\overline{H}^{ex} - T\Delta_m\overline{S}^{ex}. \tag{5.121}$$

So, the Gibbs-Helmholtz relation holds also for the excess functions of mixing.

5.3.1 Mixing and activity models

In a binary system *A-B* the excess Gibbs free energy of mixing is zero at both ends of a composition line and it is a function of mole fraction for the intermediate compositions. In order to describe $\Delta_m\overline{G}^{ex}(x)$ mathematically, Guggenheim (1937) suggested a power series having the argument $(x_A - x_B)$, so that:

$$\Delta_m\overline{G}^{ex} = x_A x_B[K_o + K_1(x_A - x_B)^1 + K_2(x_A - x_B)^2 + \ldots], \tag{5.122}$$

where x_A and x_B are the mole fractions of the components *A* and *B*, respectively and K_o, K_1, and K_2 are empirical parameters that are generally functions of pressure and temperature.

Using the relationship $(x_A - x_B) = [(1 - x_B) - x_B] = (1 - 2x_B)$, Eq. (5.122) can be rewritten as follows:

$$\Delta_m\overline{G}^{ex} = (1 - x_B)x_B[K_o + K_1(1 - 2x_B) + K_2(1 - 2x_B)^2 + \ldots]. \tag{5.123}$$

Depending on the number of parameters that are necessary to describe the excess Gibbs free energy, different types of mixtures are distinguishable.

Solutions whose excess Gibbs free energy can be represented using only parameters with an even number as a subscript are called *symmetrical*. If all constants but the first are zero, the solution is called *simple* (Guggenheim 1967).

From relations (5.116) and (5.117) and (5.123) it follows that:

$$\mu_A^{ex} = RT\ln\gamma_A = x_B^2[K_o + K_1(3 - 4x_B) + K_2(1 - 2x_B)(5 - 6x_B)\ldots] \tag{5.124}$$

and

$$\mu_B^{ex} = RT\ln\gamma_B = (1-x_B)^2[K_o + K_1(1-4x_B) + K_2((2x_B-1)(6x_B-1)\ldots)]. \quad (5.125)$$

These expressions are also termed the Redlich-Kister equations (Redlich and Kister 1948).

In the case of a simple solution, Eqs. (5.124) and (5.125) simplify to:

$$\mu_A^{ex} = RT\ln\gamma_A = x_B^2 K_o \quad (5.126)$$

and

$$\mu_B^{ex} = RT\ln\gamma_B = (1-x_B)^2 K_o. \quad (5.127)$$

In order to determine the physical meaning behind K_o, the values of the excess chemical potential for components at infinite dilution must be considered. If component A is infinitely diluted in B, x_B approaches 1 and Eq. (5.126) reads:

$$\mu_A^{ex,\infty} = RT\ln\gamma_A^{\infty} = K_o. \quad (5.128)$$

In the case that component B is infinitely diluted in A, x_B approaches 0 and Eq. (5.127) obtains the following form:

$$\mu_B^{ex,\infty} = RT\ln\gamma_B^{\infty} = K_o. \quad (5.129)$$

In Eqs. (5.128) and (5.129) $\mu_A^{ex,\infty}$ and $\mu_B^{ex,\infty}$ designate the excess chemical potential of component A and B at infinite dilution, respectively. γ_A^{∞} and γ_B^{∞} are the corresponding activity coefficients.

Eqs. (5.128) and (5.129) can be combined to give:

$$\mu_A^{ex,\infty} = \mu_B^{ex,\infty} = RT\ln\gamma_A^{\infty} = RT\ln\gamma_B^{\infty} = K_o. \quad (5.130)$$

Hence, in the case of simple solutions the excess chemical potentials of the components at infinite dilution are equal.

The constant K_o is often replaced by W^G (e.g. Thompson 1967) or G^{∞} (Froese and Gunter 1976). Both W^G and G^{∞} refer to the physical meaning behind this quantity, which is, as demonstrated above, the partial molar excess Gibbs free energy of the component at infinite dilution.

The partial molar excess Gibbs free energy, G^{∞}, consists, according to Eqs. (3.21) and (5.9), of the partial molar excess internal energy, U^{∞}, the partial molar excess entropy, S^{∞}, and the partial molar excess volume, V^{∞}, as follows:

$$G^{\infty} = U^{\infty} - TS^{\infty} + PV^{\infty}. \tag{5.131}$$

Because U^{∞} and H^{υ} are nearly equal at ambient pressure, Eq. (5.131) is normally written as

$$G^{\infty} = H^{\infty} - TS^{\infty} + PV^{\infty} \tag{5.132}$$

or

$$W^{G} = W^{H} - TW^{S} + PW^{V} \tag{5.133}$$

for the case where W^{G} is used instead of G^{∞}. W^{G} is frequently referred to as an *interaction parameter*. W^{H}, W^{S} and W^{V} depend on temperature and pressure. Their dependence, however, is small and can be ignored in most cases. Solutions where all terms except W^{H} are zero are termed *regular* (see Hildebrand 1929).

Substituting W^{G} for K_{o} in Eq. (5.22) and truncating the $\Delta_m\overline{G}^{ex}$-polynomial after the first term yields:

$$\Delta_m\overline{G}^{ex} = (1 - x_B)x_B W^{G}_{A,\,B}. \tag{5.134}$$

Accordingly it holds that:

$$\Delta_m\overline{V}^{ex} = (1 - x_B)x_B W^{V}_{A,\,B}, \tag{5.135}$$

$$\Delta_m\overline{S}^{ex} = (1 - x_B)x_B W^{S}_{A,\,B} \tag{5.136}$$

and

$$\Delta_m\overline{H}^{ex} = (1 - x_B)x_B W^{H}_{A,\,B}. \tag{5.137}$$

If two parameters are required to describe the excess Gibbs free energy of mixing, Eq. (5.123) becomes:

$$\Delta_m \overline{G}^{ex} = (1-x_B)x_B[K_o + K_1(1-2x_B)] \tag{5.138}$$

and the excess chemical potentials and the activity coefficients of the component A and B are given by:

$$\mu_A^{ex} = RT\ln\gamma_A = x_B^2[K_o + K_1(3-4x_B)] \tag{5.139}$$

and

$$\mu_B^{ex} = RT\ln\gamma_B = (1-x_B)^2[K_o + K_1(1-4x_B)]. \tag{5.140}$$

The relationship between the parameters K_0 and K_1 and the excess chemical potentials of the components can be determined in the same way as in the case of a simple solution. One has to set the mole fraction x_B successively to one and to zero. For $x_B = 1$, Eq. (5.139) reads:

$$\mu_A^{ex,\infty} = RT\ln\gamma_A^{\infty} = K_o - K_1 \tag{5.141}$$

and similarly for $x_B = 0$, Eq. (5.140) becomes:

$$\mu_B^{ex,\infty} = RT\ln\gamma_B^{\infty} = K_o + K_1. \tag{5.142}$$

From Eqs. (5.141) and (5.142) it follows that

$$K_o = \frac{\mu_A^{ex,\infty} + \mu_B^{ex,\infty}}{2} \tag{5.143}$$

and

$$K_1 = \frac{\mu_B^{ex,\infty} - \mu_A^{ex,\infty}}{2} \tag{5.144}$$

or

$$K_o = \frac{W^G_{A-B} + W^G_{B-A}}{2} \tag{5.145}$$

and

$$K_1 = \frac{W^G_{B-A} - W^G_{A-B}}{2}, \tag{5.146}$$

if W^G is used instead of the chemical excess potentials of the components at infinite dilution.

When the K parameters with both odd and even subscripts are required to describe the $\Delta_m\overline{G}^{ex}$ as a function of composition, the solution is called *asymmetric*.

Substituting the expressions given in Eqs. (5.145) and (5.146) for K_o and K_1 into Eq. (5.138) yields:

$$\Delta_m\overline{G}^{ex} = x_B(1-x_B)[W^G_{B-A} + (W^G_{A-B} - W^G_{B-A})x_B]. \tag{5.147}$$

Rearranging terms in Eq. (5.147) gives the so-called *sub-regular* or *Margules formulation* of the excess Gibbs free energy of mixing (see Margules 1895). It holds that:

$$\Delta_m\overline{G}^{ex} = x_B(1-x_B)[x_B W^G_{A-B} + (1-x_B)W^G_{B-A}] \tag{5.148}$$

or

$$\Delta_m\overline{G}^{ex} = x_B^2(1-x_B)W^G_{A-B} + x_B(1-x_B)^2 W^G_{B-A}. \tag{5.149}$$

In this context the W^G's are termed *Margules Parameters*.

Using the relationship between the excess Gibbs free energy of mixing and the activity coefficients of the components, as given in Eqs.(5.116) and (5.117), yields:

$$RT\ln\gamma_A = x_B^2[W^G_{A-B} + 2(W^G_{B-A} - W^G_{A-B})(1-x_B)] \tag{5.150}$$

and

$$RT\ln\gamma_B = (1-x_B)^2[W^G_{B-A} + 2(W^G_{A-B} - W^G_{B-A})x_B]. \tag{5.151}$$

Eqs. (5.149) through (5.151) are frequently used expressions to describe the subregular mixing behavior in mineral solutions. Their usefulness has been discussed a number of times e.g. Carlson and Colburn (1942), Hardy (1953) and Thompson (1967).

Two additional expressions for the activity coefficients in binary solutions have been developed by Carlson and Colburn (1942), namely:

$$RT\ln\gamma_A = (2W^G_{B-A} - W^G_{A-B})x_B^2 + 2(W^G_{A-B} - W^G_{B-A})x_B^3 \tag{5.152}$$

and

$$RT\ln\gamma_B = (2W^G_{A-B} - W^G_{B-A})(1-x_B)^2 + 2(W^G_{B-A} - W^G_{A-B})(1-x_B)^3. \tag{5.153}$$

They are derived from Eqs. (5.150) and (5.151) by rearranging of terms and factoring out the mole fractions.

Following Eqs. (5.147) through (5.149), the $\Delta_m\overline{G}^{ex}$ polynomial for regular solutions reduces to the one for simple mixtures if $W^G_{B-A} = W^G_{B-A} = W^G_{A,B}$.

Example 1: According to Hovis (1995) the heat of mixing for the sanidine-analbite solid solution series is a symmetric function of composition. The enthalpic interaction parameter, W^H, at 977 K, has a value of 17.0 kJmol^{-1}. The solution can be described as being regular which means that both W^V and W^S are zero and, consequently, $W^H = W^G$. The excess Gibbs free energy of mixing is given, according to (5.134), by:

$$\Delta_m\overline{G}^{ex} = (1 - x_{KAlSi_3O_8})x_{KAlSi_3O_8} \times 17.0\ \text{kJmol}^{-1}.$$

The resulting excess Gibbs free energy of mixing is shown graphically as a function of mole fraction of sanidine in Fig. 5.5. The two dotted lines are tangents to the $\Delta_m\overline{G}^{ex}$ curve at $x_{KAlSi_3O_8} = 0$ and 1, respectively. Their intersections with the ordinate at $x_{KAlSi_3O_8} = 0$ and 1 give the excess chemical potentials of the components at infinite dilution and correspond to the interaction parameter, W^G. Because the mixture is symmetric, the interaction parameters of the two components are equal.

In order to calculate the activity coefficients of the two components Eqs. (5.126) and (5.127) are used. Rearranging terms and inserting the numerical value for W^G gives for analbite:

$$\gamma_{NaAlSi_3O_8} = \exp\left[(x_{KAlSi_3O_8})^2 \cdot \frac{17000\,\text{Jmol}^{-1}}{RT}\right]$$

and for sanidine:

$$\gamma_{KAlSi_3O_8} = \exp\left[(1 - x_{KAlSi_3O_8})^2 \cdot \frac{17000\,\text{Jmol}^{-1}}{RT}\right].$$

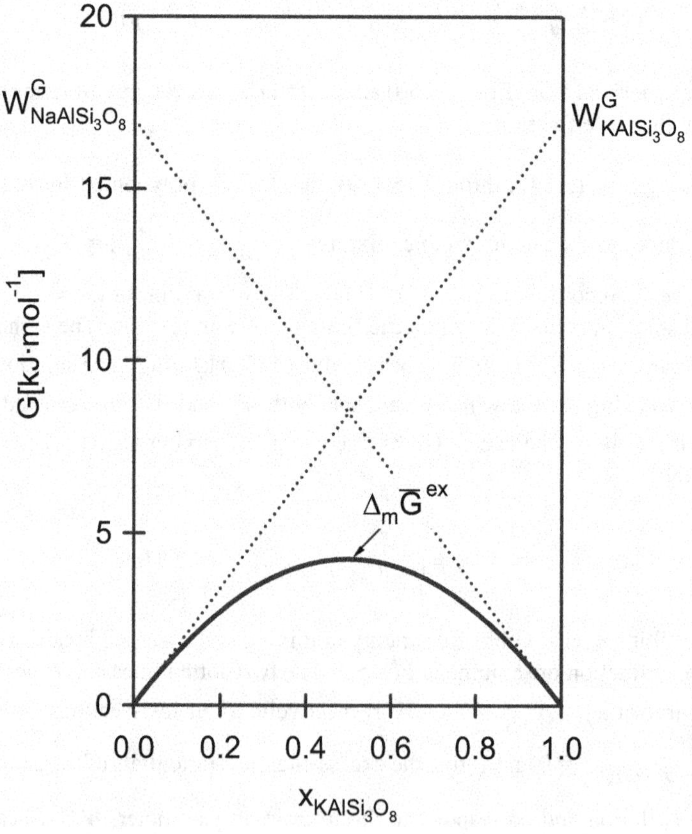

Fig. 5.5 Excess Gibbs free energy $\Delta_m\overline{G}^{ex}$ as a function of the composition in the system sanidine-analbite. $W^G_{NaAlSi_3O_8}$ and $W^G_{KAlSi_3O_8}$ are the interaction parameters of the corresponding components. They are equal because of the symmetric nature of the solution. (Data: Hovis 1995)

Multiplying the activity coefficients by the corresponding mole fractions yields the activities of the components, namely:

$$a_{NaAlSi_3O_8} = (x \cdot \gamma)_{NaAlSi_3O_8}$$

and

$$a_{KAlSi_3O_8} = (x \cdot \gamma)_{KAlSi_3O_8}.$$

The results of the calculations for a temperature of 1173 K are given in Tab. 5.2.

Table 5.2 Activity coefficients and activities of the components in the system analbite-sanidine at $T = 1173$ K. (Data: Hovis 1995)

$x_{KAlSi_3O_8}$	$\gamma_{NaAlSi_3O_8}$	$a_{NaAlSi_3O_8}$	$\gamma_{KAlSi_3O_8}$	$a_{KAlSi_3O_8}$
0.000	1.000	1.000	5.715	0.000
0.100	1.018	0.916	4.104	0.410
0.200	1.072	0.858	3.051	0.610
0.300	1.170	0.819	2.349	0.704
0.400	1.322	0.793	1.873	0.749
0.500	1.546	0.773	1.546	0.773
0.600	1.873	0.749	1.322	0.793
0.700	2.349	0.704	1.170	0.819
0.800	3.051	0.610	1.072	0.858
0.900	4.104	0.410	1.018	0.916
1.000	5.715	0.000	1.000	1.000

Fig. 5.6 shows the activities of analbite and sanidine as a function of the mole fraction of sanidine.

Example 2: Using the experimental data on the (Na,K)Cl solvus, Chatterjee (1991) derived polythermal-polybaric expressions for the Margules parameters.

They are:

$$W_{Na-K}^{G} = 23730 - 20.527\,T + 0.52\times10^{-6}P\ [\mathrm{Jmol}^{-1}]$$

and

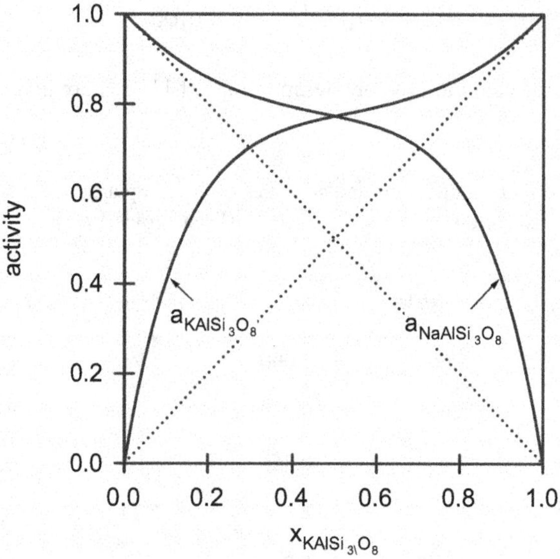

Fig. 5.6 Activities of analbite and sanidine as a function of the mole fraction of sanidine. Dotted lines give the activities for the case of ideal mixing. Note that the activities of the components are equal for $x_{KAlSi_3O_8} = 0.5$.

$$W_{K-Na}^{G} = 37959 - 30.954\,T + 0.52\times10^{-6}P\ [\mathrm{Jmol}^{-1}],$$

with T expressed in K and P in Pascal. In these equations the numbers refer, according to Eq. (5.131), to the partial molar excess internal energy, U^{∞}, the partial molar excess entropy, S^{∞}, and the partial molar excess volume, V^{∞} of the respective component. Because V^{∞} is small, U^{∞} can be replaced by H^{∞} at ambient pressure (P = 0.1 MPa).

In order to calculate the excess Gibbs free energy of mixing as a function of composition we use Eq. (5.149). At 873 K and 0.1 MPa the calculation reads:

$$\Delta_m \overline{G}^{ex} = (x_{KCl})^2 (1 - x_{KCl})(23730 - 20.527 \times 873) \\ + (x_{KCl})(1 - x_{KCl})^2 (37959 - 30.954 \times 873)$$

and the results are shown in Fig. 5.7.

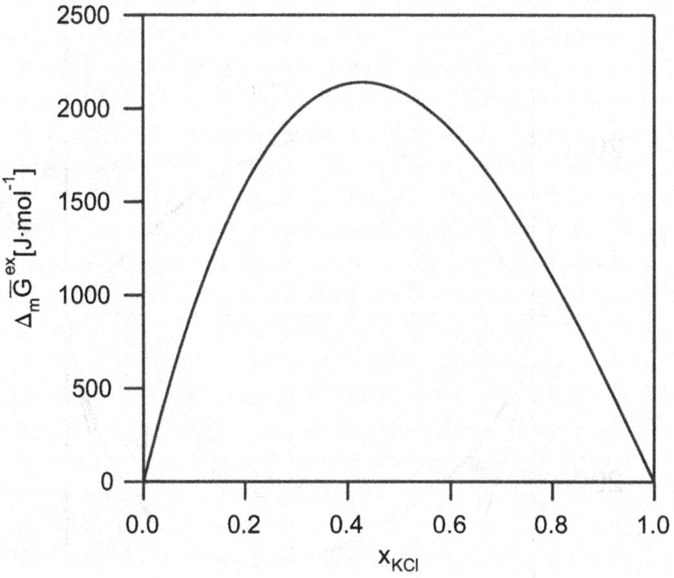

Fig. 5.7 Excess Gibbs free energy of mixing for the halite-sylvite solid solution as a function of the mole fraction x_{KCl} at 873 K and 0.1 MPa (Data: Chatterjee 1991).

Fig. 5.8 shows the excess Gibbs free energy of mixing, $\Delta_m \overline{G}^{ex}$, the ideal Gibbs free energy of the ideal mixing, $\Delta_m \overline{G}^{id}$, and the total Gibbs free energy of mixing, $\Delta_m \overline{G}^{real}$ as a function of the concentration of sylvite in the halite-sylvite solid solution.

Using Eqs. (5.150) and (5.151) the activity coefficients of halite and sylvite in the NaCl-KCl solid solution are calculated as follows:

$$RT\ln\gamma_{NaCl} = (x_{KCl})^2[W^G_{Na-K} + 2(W^G_{K-Na} - W^G_{Na-K})(1 - x_{KCl})]$$

and

$$RT\ln\gamma_{KCl} = (1 - x_{KCl})^2[W^G_{K-Na} + 2(W^G_{Na-K} - W^G_{K-Na})x_{KCl}].$$

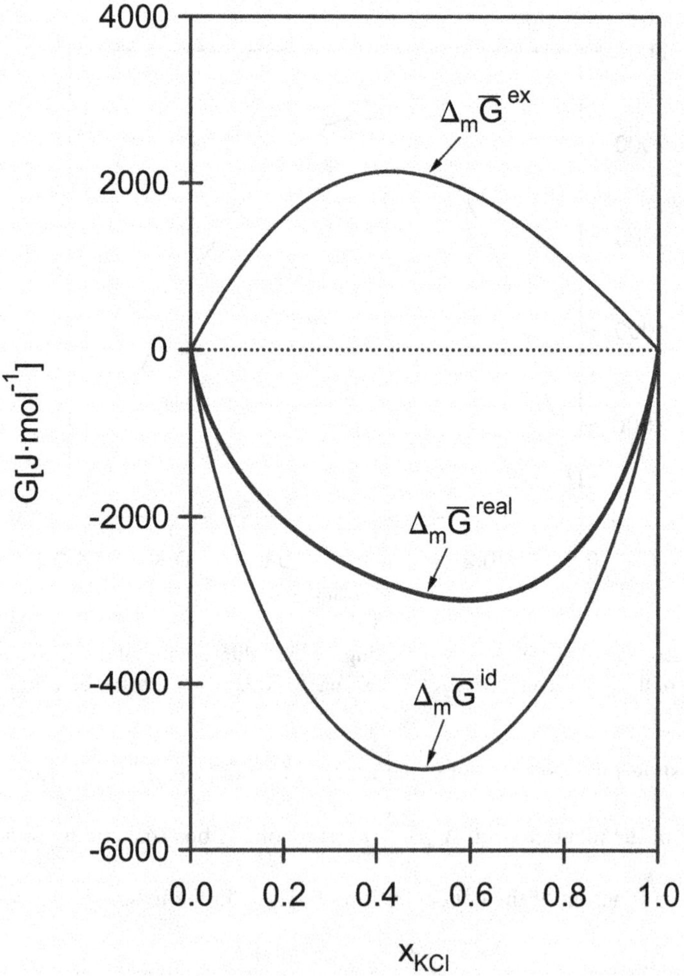

Fig. 5.8 The Gibbs free energy of mixing in the system halite-sylvite as a function of composition of the solution. (Calculated using data of Chatterjee 1991).

Following Chatterjee (1991), the Margules parameters for $T = 873$ K and 0.1 MPa are: $W^G_{Na-K} = 5810$ Jmol^{-1} and $W^G_{K-Na} = 10936$ Jmol^{-1}.

Inserting these values into the equations given above, yields the following expressions for the activity coefficients of NaCl and KCl:

$$RT\ln\gamma_{NaCl} = (x_{KCl})^2[5810 + 2(10936 - 5810)(1 - x_{KCl})]$$

and

$$RT\ln\gamma_{KCl} = (1 - x_{KCl})^2[10936 + 2(5810 - 10936)x_{KCl}].$$

The results of the calculation are presented graphically in Fig. 5.9.

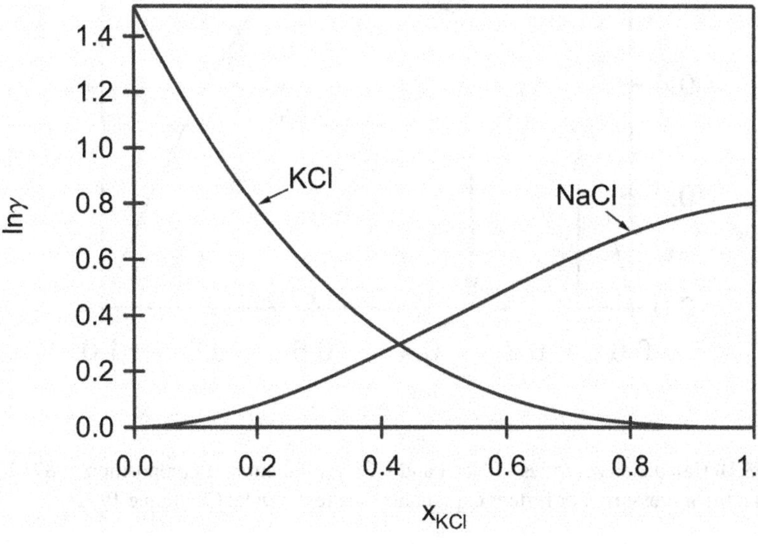

Fig. 5.9 The logarithms of activity coefficients for NaCl and KCl in the halite-sylvite solid solution as a function of composition at 873 K and 0.1 MPa. (Data: Chatterjee 1991)

Fig. 5.10 shows the activities of the components NaCl and KCl as a function of the mole fraction of sylvite, x_{KCl}, which were calculated using Eq. (5.82). They are:

$$a_{\mathrm{NaCl}} = (\gamma \cdot x)_{NaCl}$$

and

$$a_{\mathrm{KCl}} = (\gamma \cdot x)_{KCl} \cdot$$

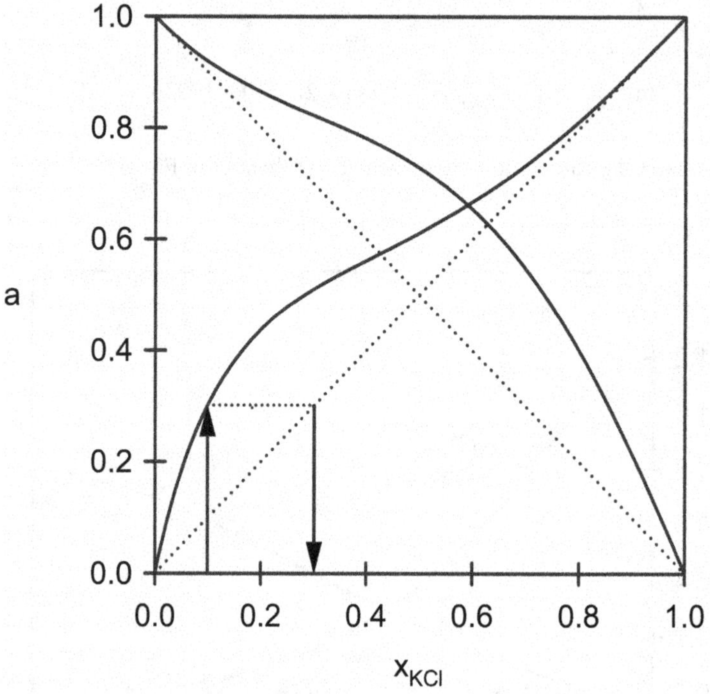

Fig. 5.10 Calculated activities of NaCl and KCl as a function of composition at 873 K and ambient pressure. For further explanation see text. (Data: Chatterjee 1991)

As shown in Fig. 5.10, at $x_{NaCl} = 0.1$ the value of the NaCl activity is 0.3022. This means that NaCl with a mole fraction of 0.1 'behaves' thermodynamically as if it had a mole fraction of 0.3022.

The expression in Eq. (5.122) can be extended to multicomponent systems (Redlich and Kister 1948). By truncating after the third degree-term, the excess Gibbs free energy of mixing of a ternary solution is given as (King 1969):

$$\Delta_m \bar{G}^{ex} = x_A x_B[K_o + K_1(x_A - x_B)] + x_A x_C[L_o + L_1(x_A - x_C)] \qquad (5.154)$$
$$+ x_B x_C[M_o + M_1(x_B - x_C)] + x_A x_B x_C N,$$

where the x_is are the mole fractions of the components A, B and C. K, L and M are empirical terms and are constant for constant P and T. The constant N describes the ternary interaction between the components. As demonstrated by Cheng and Ganguly (1994), Eq. (5.154) is equivalent to that of Wohl (1953), which in the case of a ternary solution reads:

$$\Delta_m \bar{G}^{ex} = x_A x_B[x_A W^G_{B-A} + x_B W^G_{A-B}] + x_A x_C[x_A W^G_{C-A} + x_C W^G_{A-C}] \qquad (5.155)$$
$$+ x_B x_C[x_B W^G_{C-B} + x_C W^G_{B-C}] + x_A x_B x_C\Big[\frac{1}{2}(W^G_{A-B} + W^G_{B-A}$$
$$+ W^G_{A-C} + W^G_{C-A} + W^G_{B-C} + W^G_{C-B}) + N_{A-B-C}\Big].$$

Eq. (5.155) is obtained by replacing the constants K_i, L_i and M_i by the corresponding interaction parameters, W^G_{i-j}, which, in the case of a binary solution, are termed Margules parameters, namely:

$$K_o = \frac{W^G_{A-B} + W^G_{B-A}}{2}, \qquad (5.156)$$

$$K_1 = \frac{W^G_{B-A} - W^G_{A-B}}{2}, \qquad (5.157)$$

$$L_o = \frac{W^G_{A-C} + W^G_{C-A}}{2}, \qquad (5.158)$$

$$L_1 = \frac{W^G_{C-A} - W^G_{A-C}}{2}, \qquad (5.159)$$

$$M_o = \frac{W^G_{B-C} + W^G_{C-B}}{2} \qquad (5.160)$$

and

$$M_1 = \frac{W^G_{C-B} - W^G_{B-C}}{2}. \tag{5.161}$$

In the case that the bounding binaries behave as simple solutions Eq. (5.155) reduces to:

$$\Delta_m \bar{G}^{ex} = x_A x_B W^G_{A,B} + x_A x_C W^G_{A,C} + x_B x_C W^G_{B,C} + x_A x_B x_C D \tag{5.162}$$

with $D = (W_{A,B} + W_{A,C} + W_{B,C} + N_{A,B,C})$. $W^G_{A,B}$, $W^G_{A,C}$ and $W^G_{B,C}$ are the interaction parameters of the bounding binaries A-B, A-C and B-C, respectively.

The activity coefficients of the components in ternary systems are calculated similar to the case for binary solutions. It holds that:

$$RT\ln\gamma_A = \Delta_m \bar{G}^{ex} - x_B\left(\frac{\partial \Delta_m \bar{G}^{ex}}{\partial x_B}\right)_C - x_C\left(\frac{\partial \Delta_m \bar{G}^{ex}}{\partial x_C}\right)_B, \tag{5.163}$$

$$RT\ln\gamma_B = \Delta_m \bar{G}^{ex} - x_A\left(\frac{\partial \Delta_m \bar{G}^{ex}}{\partial x_A}\right)_C - x_C\left(\frac{\partial \Delta_m \bar{G}^{ex}}{\partial x_C}\right)_A \tag{5.164}$$

and

$$RT\ln\gamma_C = \Delta_m G^{ex} - x_A\left(\frac{\partial \Delta_m \bar{G}^{ex}}{\partial x_A}\right)_B - x_B\left(\frac{\partial \Delta_m \bar{G}^{ex}}{\partial x_B}\right)_A. \tag{5.165}$$

If the excess Gibbs free energy of mixing can be described by the expression given in Eq. (5.162), the following activity coefficients are obtained for the components A, B and C:

$$RT\ln\gamma_A = x_B^2 W^G_{A,B} + x_C^2 W^G_{A,C} + x_B x_C (W^G_{A,B} + W^G_{A,C} - W^G_{B,C}) + x_B x_C (1 - 2x_A) D, \tag{5.166}$$

$$RT\ln\gamma_B = x_A^2 W^G_{A,B} + x_C^2 W^G_{B,C} + x_A x_C (W^G_{A,B} + W^G_{B,C} - W^G_{A,C}) + x_A x_C (1 - 2x_B) D \tag{5.167}$$

and

$$RT\ln\gamma_C = x_A^2 W^G_{A,C} + x_B^2 W^G_{B,C} + x_A x_B (W^G_{A,C} + W^G_{B,C} - W^G_{A,B}) + x_A x_C (1 - 2x_B) D. \quad (5.168)$$

In the case that all three bounding binaries of the system *A*-*B*-*C* behave as sub-regular solutions the activity coefficient of, for example component *A*, reads:

$$\begin{aligned} RT\ln\gamma_A &= x_B^2[W^G_{A-B} + 2x_A(W^G_{B-A} - W^G_{A-B})] \\ &+ x_C^2[W^G_{A-C} + 2x_A(W^G_{C-A} - W^G_{A-C})] \\ &+ 2x_A x_B x_C(W^G_{B-A} + W^G_{C-A}) - 2x_B x_C(x_B W^G_{C-B} + x_C W^G_{B-C}) \\ &+ x_B x_C\Big[\frac{1}{2}(W^G_{A-B} + W^G_{B-A} + W^G_{A-C} \\ &+ W^G_{C-A} + W^G_{B-C} + W^G_{C-B}) + N_{A,B,C}\Big](1 - 2x_A). \end{aligned} \quad (5.169)$$

The expressions for the activity coefficient of the two remaining components are analogous and are obtained by the cyclic permutation of the indices.

The use of Eqs. (5.166) through (5.169) is only possible if the ternary term $N_{A,B,C}$ is known. This, however, is rarely the case. To overcome the problem, the assumption is frequently made that the ternary interaction is negligibly small and it can be ignored.

Example: If Eq. (5.169) is applied to the ternary system albite (ab)-orthoclase (or)-anorthite (an), the following expressions for the activity coefficients of the components albite, orthoclase and anorthite are obtained:

$$\begin{aligned} RT\ln\gamma_{ab} &= x_{or}^2[W^G_{ab-or} + 2x_{ab}(W^G_{or-ab} - W^G_{ab-or})] \\ &+ x_{an}^2[W^G_{ab-an} + 2x_{ab}(W^G_{an-ab} - W^G_{ab-an})] \\ &+ 2x_{ab}x_{or}x_{an}(W^G_{or-ab} + W^G_{an-ab}) \\ &- 2x_{or}x_{an}(x_{or}W^G_{an-or} + x_{an}W^G_{or-an}) \\ &+ x_{or}x_{an}\Big[\frac{1}{2}(W^G_{ab-or} + W^G_{or-ab} + W^G_{ab-an} \\ &+ W^G_{an-ab} + W^G_{or-an} + W^G_{an-or}) + W^G_{ab,or,an}\Big](1 - 2x_{ab}) \end{aligned} \quad (5.170)$$

for the albite component,

$$
\begin{aligned}
RT\ln\gamma_{or} = {} & x_{ab}^2[W^G_{or-ab} + 2x_{or}(W^G_{ab-or} - W^G_{or-ab})] \\
& + x_{an}^2[W^G_{or-an} + 2x_{or}(W^G_{an-or} - W^G_{or-an})] \\
& + 2x_{ab}x_{or}x_{an}(W^G_{ab-or} + W^G_{an-or}) \\
& - 2x_{ab}x_{an}(x_{ab}W^G_{or-ab} + x_{an}W^G_{ab-an}) \\
& + x_{ab}x_{an}\left[\frac{1}{2}(W^G_{ab-or} + W^G_{or-ab} + W^G_{ab-an}\right. \\
& \left. + W^G_{an-ab} + W^G_{or-an} + W^G_{an-or}) + W^G_{ab,\,or,\,an}\right](1 - 2x_{or})
\end{aligned}
\tag{5.171}
$$

for the orthoclase component and

$$
\begin{aligned}
RT\ln\gamma_{an} = {} & x_{ab}^2[W^G_{an-ab} + 2x_{an}(W^G_{ab-an} - W^G_{an-ab})] \\
& + x_{or}^2[W^G_{an-or} + 2x_{an}(W^G_{or-an} - W^G_{an-or})] \\
& + 2x_{ab}x_{or}x_{an}(W^G_{ab-an} + W^G_{or-an}) \\
& - 2x_{ab}x_{or}(x_{ab}W^G_{or-ab} + x_{or}W^G_{ab-or}) \\
& + x_{ab}x_{or}\left[\frac{1}{2}(W^G_{ab-or} + W^G_{or-ab} + W^G_{ab-an}\right. \\
& \left. + W^G_{an-ab} + W^G_{or-an} + W^G_{an-or}) + W^G_{ab,\,or,\,an}\right](1 - 2x_{an})
\end{aligned}
\tag{5.172}
$$

for anorthite.

Consider now a ternary feldspar solid solution that contains 63.2 mol% albite, 6.6 mol% orthoclase and 30.2 mol% anorthite. In order to calculate the activity coefficients of the three components the binary and ternary Margules parameters are required.

Tab. 5.3 gives the Margules parameters that were derived from phase equilibria data for coexisting alkali feldspar and plagioclase (Fuhrman and Lindsley 1988). The values presented in the last column give the Gibbs free energy interaction parameters. They are calculated for 1073 K and 0.1 MPa using the relationship: $W^G = W^H - TW^S + PW^V$.

Table 5.3 Margules parameters for ternary phases in the system albite-orthoclase-anorthite (Fuhrman and Lindsley 1988).

	W^H[Jmol^{-1}]	W^S[Jmol^{-1}K^{-1}]	W^V[cm^{3}mol^{-1}]	W^G[Jmol^{-1}]
$W_{\text{ab-or}}$	18810	10.3	3.94	7758
$W_{\text{or-ab}}$	27320	10.3	3.94	16268
$W_{\text{ab-an}}$	28226	-	-	28226
$W_{\text{an-ab}}$	8471	-	-	8471
$W_{\text{or-an}}$	47396	-	-	47396
$W_{\text{an-or}}$	52468	-	-1.20	52468
$W_{\text{ab-or-an}}$	8700	-	-10.94	8700

Inserting the various W^G values into Eqs. (5.170) through (5.172) yields:

$$\gamma_{ab} = \exp\left\{\left\{0.066^2[7758 + 2 \times 0.632(16268\right.\right.$$

$$- 7758)] + 0.302^2[28226 + 2 \times 0.632(8471 - 28226)]$$

$$+ 2 \times 0.632 \times 0.066 \times 0.302 \times (16268 + 8471)$$

$$- 2 \times 0.066 \times 0.302 \times (0.066 \times 52468 + 0.302 \times 47396)$$

$$+ 0.066 \times 0.302 \times \left[\frac{1}{2}(7758 + 16268 + 28226 + 8471\right.$$

$$\left.\left.\left. + 47396 + 52468) + 8700\right](1 - 2 \times 0.632)\right\}/(8.3144 \times 1073)\right\} = \underline{0.981}$$

for the activity coefficient of albite,

$$\gamma_{or} = \exp\Big\{\Big\{0.632^2[16268 + 2 \times 0.066(7758 - 16268)]$$

$$+ 0.302^2[47396 + 2 \times 0.066 \times (52468 - 47396)]$$

$$+ 2 \times 0.632 \times 0.066 \times 0.302 \times (7758 + 52468)$$

$$- 2 \times 0.632 \times 0.302 \times (0.632 \times 16268 + 0.302 \times 28226)$$

$$+ 0.632 \times 0.302\Big[\frac{1}{2}(7758 + 16268 + 28226 + 8471 + 47396$$

$$+ 52468) + 8700\Big](1 - 2 \times 0.066)\Big\}/(8.3144 \times 1073)\Big\} = \underline{8.913}$$

for that of orthoclase and

$$\gamma_{an} = \exp\Big\{\Big\{0.632^2[8471 + 2 \times 0.302(28226 - 8471)]$$

$$+ 0.066^2[52468 + 2 \times 0.302(47396 - 52468)]$$

$$+ 2 \times 0.632 \times 0.066 \times 0.302(28226 - 47396)$$

$$- 2 \times 0.632 \times 0.302(0.632 \times 16268 + 0.066 \times 7758) + 0.632$$

$$\times 0.066\Big[\frac{1}{2}(7758 + 16268 + 28226 + 8471 + 47396 + 52468)$$

$$+ 8700\Big](1 - 2 \times 0.302)\Big\}/(1073 \times 8.3144)\Big\} = \underline{2.349}$$

for the activity coefficient of anorthite.

In order to account for the configurational entropy resulting from the aluminium-silicon order, which is governed by the Al-avoidance rule, the so-called ideal activities must be calculated. They are, according to Ghiorso (1984), given by:

$$a_{ab}^{id} = x_{ab}(1 - x_{an}^2), \tag{5.173}$$

$$a_{or}^{id} = x_{or}(1 - x_{an}^2) \tag{5.174}$$

and

$$a_{an}^{id} = \frac{1}{4}x_{an}(1 + x_{an})^2. \tag{5.175}$$

Inserting the numerical values for the mole fractions of albite, orthoclase and anorthite into Eqs. (5.173) through (5.175) yields:

$$a_{ab}^{id} = 0.632 \times (1 - 0.302^2) = \underline{0.574},$$

$$a_{or}^{id} = 0.066 \times (1 - 0.302^2) = \underline{0.060}$$

and

$$a_{an}^{id} = \frac{1}{4} \times 0.302 \times (1 + 0.302)^2 = \underline{0.128}.$$

In order to obtain the activities of the components in a non-ideal feldspar solid solution, the ideal activities are to be multiplied by the respective activity coefficients, i.e.:

$$a_{ab} = a_{ab}^{id} \times \gamma_{ab} = 0.574 \times 0.981 = \underline{0.563},$$

$$a_{or} = a_{or}^{id} \times \gamma_{or} = 0.060 \times 8.913 = \underline{0.535}$$

and

$$a_{an} = a_{an}^{id} \times \gamma_{an} = 0.128 \times 2.349 = \underline{0.301}.$$

The activity models, which are presented above, were derived from an analysis of the compositional behavior of the excess Gibbs free energy of mixing. Models that were developed considering interactions between the atoms constituting the solution are rarely used and will not be discussed here.

5.4 Gibbs free energy of reaction

Consider a closed thermodynamic system consisting of components A, B, C and D. If a chemical reaction of the type

$$\nu_a A + \nu_b B = \nu_c C + \nu_d D$$

takes place at constant pressure and temperature, the change in the Gibbs free energy, dG, is given by:

$$dG = \sum_A^D \left(\frac{\partial G}{\partial n_i}\right)_{P,\,T,\,n_{j \neq i}} dn_i, \tag{5.176}$$

where n_i designates the number of moles of the component i. Considering the definition of the chemical potential (Eq. (5.43)), Eq. (5.176) can be rewritten as follows:

$$dG = \sum_A^D \mu_i dn_i. \tag{5.177}$$

In the case of a chemical reaction the molar increments, dn_i's are related to one another through the stoichiometric coefficients and the extent of reaction, ξ, as given in Eq. (2.113). So, the change in the Gibbs free energy of a system is given by:

$$dG = \sum_A^D \nu_i \mu_i d\xi. \tag{5.178}$$

Dividing Eq. (5.178) by $d\xi$ and assuming that the reaction occurs at constant pressure and temperature, yields:

$$\left(\frac{\partial G}{\partial \xi}\right)_{P,\,T} = \sum_A^D \nu_i \mu_i = \Delta_r G. \tag{5.179}$$

$\Delta_r G$ is called the *Gibbs free energy of reaction*. It corresponds to the change in the Gibbs free energy of a system with respect to the extent of reaction at constant pressure and temperature.

The relationship between the Gibbs free energy of reaction, the enthalpy of reaction and the entropy of reaction is given by the *Gibbs-Helmholtz equation*, namely:

$$\Delta_r G = \Delta_r H - T\Delta_r S. \tag{5.180}$$

5.4.1 Standard Gibbs free energy of reaction

If the chemical potential, μ_i, in Eq. (5.177) is split into standard and rest potential, the Gibbs free energy of reaction reads:

$$\Delta_r G = \sum_A^D \nu_i(\mu_i^o + RT\ln a_i). \tag{5.181}$$

In the case that reactants, as well as products, occur in their standard states, all activities in Eq. (5.181) are 1 and the second term within the parentheses vanishes. The Gibbs free energy of reaction is then equal to the stoichiometric sum of the standard potentials, and Eq. (5.181) obtains the form:

$$\Delta_r \boldsymbol{G}^o = \sum_A^D \nu_i \mu_i^o. \tag{5.182}$$

In this case $\Delta_r \boldsymbol{G}^o$ is referred to as the *standard Gibbs free energy of reaction* and the corresponding reaction is called the *standard reaction*. The choice of standard conditions depends on the definition of the standard potential of the phase under consideration. For gases, the state of an ideal gas at 0.1 MPa and the temperature of interest are frequently chosen as the standard state. The standard potential of solid phases, however, generally refers to the pure phase at the temperature and pressure of interest. In the case of aqueous solutions, the state of infinite dilution is normally chosen.

The second term within the parentheses in Eq. (5.181) accounts for the concentration of the components participating in the reaction. Combining Eqs. (5.181) and (5.182) yields:

$$\Delta_r G = \Delta_r \boldsymbol{G}^o + RT\sum_A^D \nu_i \ln a_i. \tag{5.183}$$

A special type of reactions is that one in which a compound is formed from the elements at standard conditions. The Gibbs free energy associated with this reaction is referred to as the *standard Gibbs free energy of formation*, $\Delta_f \boldsymbol{G}_{298}$.

Using the Gibbs-Helmholtz equation, the standard Gibbs free energy of formation can be calculated from the standard enthalpy of formation, $\Delta_f \boldsymbol{H}_{298}$, and the conventional standard entropies of the components, as follows:

$$\Delta_f \boldsymbol{G}_{298} = \Delta_f \boldsymbol{H}_{298} - 298.15\sum_i \nu_i \boldsymbol{S}_{298} \tag{5.184}$$

or

$$\Delta_f G_{298} = \Delta_f H_{298} - 298.15\Delta_f S_{298}, \qquad (5.185)$$

where $\Delta_f S_{298}$ designates the change in the entropy associated with the formation of the compound from the elements at standard conditions. In analogy to the standard enthalpy of formation, $\Delta_f H_{298}$, $\Delta_f S_{298}$ can be called the *standard entropy of formation.*

5.5 Problems

1. Consider a pyrope-almandine garnet solid solution, $(Mg,Fe)_3Al_2Si_3O_{12}$, containing 70 mol% pyrope, $Mg_3Al_2Si_3O_{12}$.

- Calculate the Gibbs free energy of mixing at 1000 K assuming ideal mixing behavior for garnet.

2. Calculate the change in the chemical potential of an ideal gas in the case that at a constant temperature of 800 K, the pressure of the gas increases from 0.1 MPa to 0.5 GPa.

3. The binary solid solution muscovite, $KAl_2[AlSi_3O_{10}](OH)_2$ - paragonite, $NaAl_2[AlSi_3O_{10}](OH)_2$, shows asymmetric behavior. The corresponding Gibbs free interaction parameters, according to Chatterjee and Froese (1975), read:

$$W^G_{pg-mu} = 12230 + 0.71\,T[\mathrm{K}] + 6.65\times10^{-6}P[\mathrm{Pa}] \text{ and}$$

$$W^G_{mu-pg} = 19456 + 1.654\,T[\mathrm{K}] + 4.56\times10^{-6}P[\mathrm{Pa}]$$

- Express the excess Gibbs free energy of mixing as a function of composition.
- Calculate the molar excess enthalpy of a muscovite-paragonite solid solution containing 15 mol% paragonite for 0.2 GPa.
- Calculate the molar excess volume and the molar excess entropy of the muscovite-paragonite solid solution containing 10 mol% muscovite.
- Calculate the activity coefficients and activities of the components in a solid solution containing 15 mol% paragonite at 0.2 GPa and 600°C.
- Calculate the molar Gibbs free energy of mixing for a muscovite-paragonite solid solution containing 15 mol% paragonite at 0.2 GPa and 600°C using the activities calculated above.

4. The heat of formation of grossular, $Ca_3Al_2Si_3O_{12}$, equals -6638.30 kJmol^{-1}. Its third law entropy and molar volume are equal to 256.00 Jmol^{-1}K^{-1} and 125.35 cm^{3}mol^{-1}, respectively. The thermal expansion $\alpha = 2.39 \times 10^{-5}$ K^{-1}, the compressibility coefficient $\beta = 6.3 \times 10^{-12}$ Pa^{-1} and C_p[Jmol^{-1}K^{-1}] = 728.6 - $40.986 \times 10^{-3} T - 3.128 \times 10^6 T^{-2} - 6077.4\, T^{-0.5}$(Holland and Powell 1990).

- Calculate the chemical potential (apparent Gibbs free energy) of grossular at 2.0 GPa and 800°C. Assume that the thermal expansion as well as the compressibility coefficient are pressure and temperature independent.

5. The dehydration of phlogopite, $KMg_3[AlSi_3O_{10}](OH)_2$, in the presence of quartz can be described by the following chemical reaction:

$$\text{Phlogopite} + 3\,\text{Quartz} \rightarrow \text{Sanidine} + 3\ \text{Enstatite} + H_2O\ (\text{steam}).$$

- Calculate the standard Gibbs free energy of the reaction at 0.3 GPa and 600 K using the data in Tab. 5.4. Assume that steam behaves ideally, that the enthalpy, the entropy and the volume of reaction are temperature and pressure independent, and that total pressure equals the pressure of the steam.

Table 5.4 Thermodynamic data of the components involved in the reaction of phlogopite with quartz (Holland and Powell 1996)

Phase	$\Delta_f H_{298}$[kJmol^{-1}]	S_{298}[Jmol^{-1}K^{-1}]	V_{298}[cm^{3}mol^{-1}]
Phlogopite	-6219.44	328.00	149.64
Quartz	-910.88	41.50	22.69
Sanidine	-3964.90	230.00	109.00
Enstatite ($MgSiO_3$)	-1545.13	61.25	31.31
H_2O (steam)	-241.81	188.80	-

- How do you interpret the result?

Chapter 6 Thermal equilibrium

An isolated system is in the state of thermal equilibrium if the total entropy has a maximum value, i.e.

$$\sum_i dS_i = 0. \tag{6.1}$$

From Eq. (6.1) it follows that the temperature, as well as the pressure, are the same throughout the system. If this is not the case, irreversible equalization processes take place. This can be verified by a simple thought experiment. Consider an isolated system that is in equilibrium. If an infinitesimal amount of heat is transferred from one part of the system to another, then, in order to maintain thermal equilibrium, the condition

$$dS = dS_1 + dS_2 = \delta Q\left(\frac{1}{T_1} < \frac{1}{T_2}\right) = 0 \tag{6.2}$$

must be fulfilled. This is only the case if the temperature in all parts of the system is the same, that is

$$T_1 = T_2. \tag{6.3}$$

An analogous relationship holds for pressure.

In order to derive the equilibrium condition for a closed system, we start with the statement that says that a system is in equilibrium when all infinitesimal changes of state occur reversibly. Excluding chemical reactions from the consideration, these changes are given by Eqs. (4.5) and (4.6). Hence, the equilibrium conditions for *adiabatic* ($dS = 0$) and *isochoric* ($dV = 0$) changes of state read:

$$(dU)_{V,S} = 0 \tag{6.4}$$

and for *adiabatic* ($dS = 0$) and *isobaric* ($dP = 0$) changes:

$$(dH)_{P,S} = 0. \tag{6.5}$$

Expressions, analogous to Eq. (6.5) are obtained from Eqs. (5.12) and (5.13). For *isothermal* ($dT = 0$) and *isochoric* changes it holds:

$$(dF)_{T,\,V} = 0 \tag{6.6}$$

and for *isothermal* ($dT = 0$) and *isobaric* ($dP = 0$) changes:

$$(dG)_{T,\,P} = 0. \tag{6.7}$$

According to Eqs. (6.4) through (6.7), the functions of state U, H, F, and G and the total entropy, S, have extreme values at equilibrium. Because all states of a system in the neighborhood of equilibrium can be attained only by work, the values of the functions U, H, F and G must be at a minimum.

In the case that a phase undergoes a phase transition from α to β at constant P and T, the equilibrium is, according to Eq. (5.176), given by:

$$(dG)_{P,\,T} = \mu_\alpha dn_\alpha + \mu_\beta dn_\beta = 0. \tag{6.8}$$

The molar increments in Eq. (6.8) are interdependent, because the amounts of the disappearing and the appearing phase must be the same. Therefore, one can write:

$$dn_\beta = -dn_\alpha = dn \tag{6.9}$$

and

$$(\mu_\beta - \mu_\alpha)dn = 0. \tag{6.10}$$

Eq. (6.10) holds only if

$$\mu_\beta = \mu_\alpha. \tag{6.11}$$

Following Eq. (6.11), equilibrium between phases exists when the chemical potentials of the component in the coexisting phases are equal. This condition is not restricted to the case of two phases only. It holds for any number of phases that coexist at a given temperature and pressure.

For reacting systems containing many components, the condition of chemical equilibrium is given by:

$$(dU)_{S,\,V} = (dH)_{S,\,P} = (dF)_{V,\,T} = (dG)_{P,\,T} = \sum \nu_i \mu_i d\xi = 0. \tag{6.12}$$

For practical use only those functions whose variables are P, T and V are useful, because the use of entropy as an independent variable is inconvenient.

6.1 Stability conditions for phases in one-component systems

Because heat capacity can not have a negative value, the entropy of a stable phase can only increase with increasing temperature. This fact can be expressed mathematically as follows:

$$\left(\frac{\partial S}{\partial T}\right)_P = \frac{C_P}{T} > 0. \tag{6.13}$$

On the other hand, the derivative of the Gibbs free energy with respect to temperature at constant pressure gives negative entropy, Eq. (5.16):

$$\left(\frac{\partial G}{\partial T}\right)_P = -S.$$

Combining the two equations yields:

$$\left(\frac{\partial^2 G}{\partial T^2}\right)_P < 0. \tag{6.14}$$

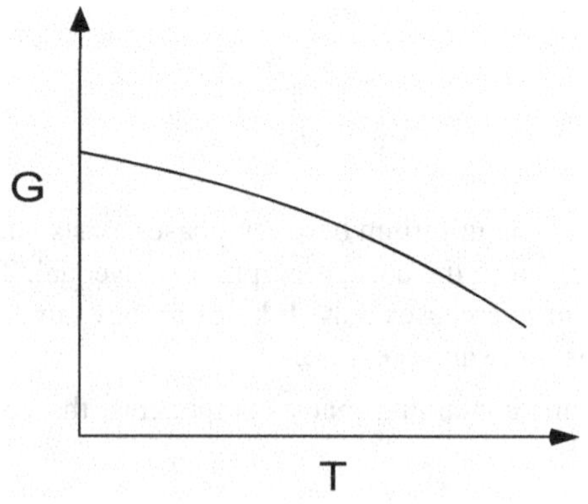

Fig. 6.1 Schematic representation of the Gibbs free energy of a stable phase as a function of temperature at constant pressure.

Thus, according to Eq. (6.14), the Gibbs free energy curve of a stable phase as a function of temperature at constant pressure is *concave* with respect to the T axis. Moreover, because of the relationship: $G = H - TS$, the $G(T)$-curve has a negative slope. The temperature dependence of the Gibbs free energy at constant pressure is shown schematically in Fig. 6.1.

According to Eq. (5.14) the volume corresponds to the derivative of the Gibbs free energy with respect to pressure, namely:

$$\left(\frac{\partial G}{\partial P}\right)_T = V.$$

Because the volume of a stable phase decreases with increasing pressure, its derivative with respect to temperature at constant temperature must be negative i.e.:

$$\left(\frac{\partial V}{\partial P}\right)_T < 0. \qquad (6.15)$$

Considering this two relationships, one can write:

$$\left(\frac{\partial^2 G}{\partial T^2}\right)_T < 0. \qquad (6.16)$$

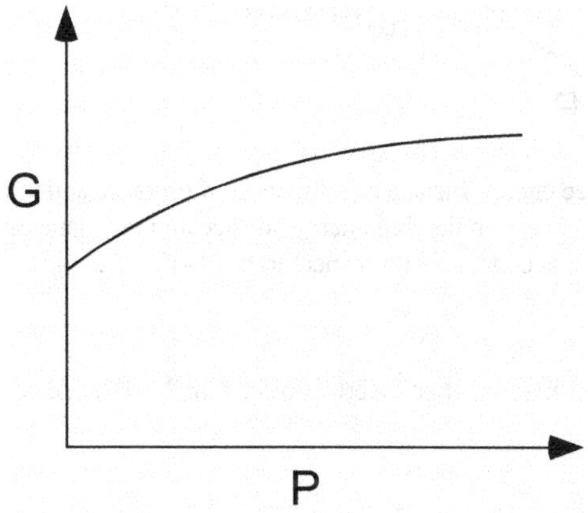

Fig. 6.2 Schematic representation of Gibbs free energy of a stable phase as a function of pressure.

Hence, the curve of Gibbs free energy of a stable phase as a function of pressure is concave with respect to the P axis. The mathematical expression for the $G(P)$ curve is $G = U - TS + PV$. Because volume can assume positive values only, the slope of the curve is normally positive. The run of the $G(P)$ curve is presented graphically in Fig. 6.2.

In an orthogonal coordinate system a $G(P,T)$ function of a stable phase is represented by a curved surface. The intersection between the surface and any plane, that stands perpendicular on the P-T-plane results in a curve which is *concave* with respect to the P-T-plane (see Fig. 6.3).

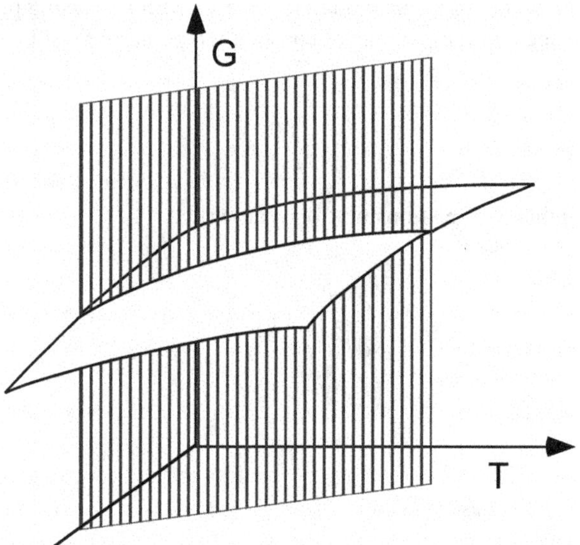

Fig. 6.3 Gibbs free energy surface as a function of pressure and temperature. The trace of the intersection between the free energy surface and the plane that is parallel to the G axis is a curve that is concave with respect to the P-T plane.

The stability conditions, discussed above, can be described mathematically as follows:

$$\left(\frac{\partial^2 G}{\partial T^2}\right)_P \left(\frac{\partial^2 G}{\partial P^2}\right)_T - \left(\frac{\partial^2 G}{\partial T \partial P}\right)^2 > 0. \tag{6.17}$$

In the case of polymorphism, every phase has its own *G(T)* and *G(P)* curve. Fig. 6.4 shows an example for a trimorphic substance with three phases α, β and γ. At a given temperature, the most stable phase has the lowest Gibbs free energy. The temperatures at which two *G(T)* curve cross is termed the *transformation temperature* and here, the chemical potential of the transforming phases are equal, as required by Eq. (6.11). The temperature of the intersection of the Gibbs free energy curves for the phases α and γ lies above the *G(T)*-curve of the phase β. This means that phase β is stable at this temperature, while the coexistence between α and γ is metastable.

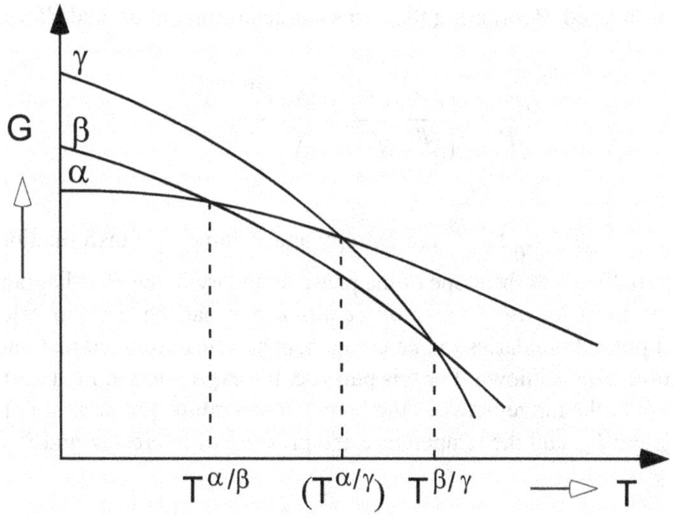

Fig. 6.4 *G(T)*-curves of a trimorphic substance with the phases α, β and γ. $T^{\alpha/\beta}$, $T^{\alpha/\gamma}$ and $T^{\beta/\gamma}$ are the transformation temperatures. The parenthesis designate the metastability of the phase transformation.

6.1.1 Phase equilibria in one-component systems.

At the transition point the coexisting phases are, according to Eq. (6.11), in equilibrium with one another, if the chemical potentials of the component in both phases are equal. Should equilibrium persist after infinitesimal changes in pressure and temperature, it must hold that the incremental changes in the chemical potentials are also equal i.e.

$$d\mu_i^{\alpha} = d\mu_i^{\beta}, \qquad (6.18)$$

where i designates the component and α and β the two coexisting phases.

In the case of pure components, the pressure and temperature dependence of the chemical potential is given by Eq. (5.22). Thus, Eq. (6.18) can be written as follows:

$$-S_i^{\alpha} dT + V_i^{\alpha} dP = -S_i^{\beta} dT + V_i^{\beta} dP. \tag{6.19}$$

In Eq. (6.19) S_i^{α} and S_i^{β} are the molar entropies and V_i^{α} and V_i^{β} the molar volumes of the component i in the phases α and β at P and T of interest, respectively. Because the system contains only one component, the subscript i is not necessary and can be dropped. Reordering the terms and factoring out dP and dT yields:

$$\frac{dT}{dP} = \frac{(V^{\beta} - V^{\alpha})}{(S^{\beta} - S^{\alpha})} = \frac{\Delta_{tr} V^{\alpha/\beta}}{\Delta_{tr} S^{\alpha/\beta}}, \tag{6.20}$$

where $\Delta_{tr} S^{\alpha/\beta}$ and $\Delta_{tr} V^{\alpha/\beta}$ are entropy and volume of transformation, respectively. Eq. (6.20) gives the slope of the phase boundary in the *P-T* diagram. It is referred to as the *Clausius-Clapeyron equation* and can be used to calculate the position of phase boundaries in one component systems provided that one point of the transformation is known. For this purpose, the expression in Eq. (6.20) must be integrated over the range between the known temperature and pressure of transformation, T_0 and P_0, and the temperature and pressure of interest, T and P, i.e.

$$\int_{T_0}^{T} dT = \int_{P_0}^{P} \frac{\Delta_{tr} V^{\alpha/\beta}}{\Delta_{tr} S^{\alpha/\beta}} dP. \tag{6.21}$$

In the case of solid-solid transitions the pressure and temperature dependence of the entropy and volume of transformation is small and can be neglected without introducing considerable error, provided the pressure and temperature intervals are not to large. Integrating Eq. (6.21) yields:

$$T = \frac{\Delta_{tr} V^{\alpha/\beta}}{\Delta_{tr} S^{\alpha/\beta}} (P - P_o) + T_o. \tag{6.22}$$

Example: Consider the one component system Al_2SiO_5. It consists of three phases: kyanite, andalusite and sillimanite. In order to calculate the boundaries between the three stability fields using the Clausius-Clapeyron equation the molar volumes, the standard entropies of the phases and the *P-T*-coordinates of two different transitions are required. We take 454 K and 950 K (derived from thermodynamic data given by Robie and Hemingway 1995) as the transition temperatures for the transformation of kyanite to andalusite and for andalusite to sillimanite at ambient pressure, respectively. Using these data and the data from Tab. 6.1 all three phase transition boundaries can be calculated.

Table 6.1 Thermodynamic data of the phases in the system Al_2SiO_5. (Robie and Hemingway 1995)

Phase	V[cm^{3}mol^{-1}]	S[Jmol^{-1}K^{-1}]
Andalusite	51.52	91.4
Kyanite	44.15	82.8
Sillimanite	49.86	95.4

We begin with the phase transition boundary between kyanite and andalusite. Because of the linear relationship between the temperature and pressure of transformation, the phase transition boundary can be defined by two transition points. The first one, which holds for the ambient pressure, is at 454 K. For the second one, a pressure of 0.3 GPa is chosen and the calculation of the corresponding temperature reads:

$$T_{0.3GPa}^{ky/and} = \frac{51.52\times10^{-6}\,\mathrm{m^3mol^{-1}} - 44.15\times10^{-6}\,\mathrm{m^3mol^{-1}}}{91.4\,\mathrm{Jmol^{-1}K^{-1}} - 82.8\,\mathrm{Jmol^{-1}K^{-1}}}(3.0\times10^{8}\,\mathrm{Pa}$$

$$-1.0\times10^{5}\,\mathrm{Pa}) + 454\,\mathrm{K} = \underline{711\ \mathrm{K}}.$$

The transition temperature for the transformation of andalusite to sillimanite at ambient pressure is 950 K. The transition temperature at 0.3 GPa is then:

$$T_{0.3GPa}^{and/sil} = \frac{49.86\times10^{-6}\,\mathrm{m^3mol^{-1}} - 51.52\times10^{-6}\,\mathrm{m^3mol^{-1}}}{95.4\ \mathrm{Jmol^{-1}K^{-1}} - 91.4\ \mathrm{Jmol^{-1}K^{-1}}}(3.0\times10^{8}\,\mathrm{Pa} \quad .$$

$$-1.0\times10^{5}\,\mathrm{Pa}) + 950\,\mathrm{K} = \underline{826\ \mathrm{K}}.$$

In order to calculate the phase transformation boundary kyanite/sillimanite we make use of the fact that the three curves intersect at the so-called triple point. At the pressure and temperature conditions of the triple point all three phase coexist and, therefore, the following relationship holds:

$$\frac{51.52\times10^{-6}\,\text{m}^3\text{mol}^{-1} - 44.15\times10^{-6}\,\text{m}^3\text{mol}^{-1}}{91.4\ \text{Jmol}^{-1}\text{K}^{-1} - 82.8\ \text{Jmol}^{-1}\text{K}^{-1}}(P_{tp}\text{Pa} - 1.0\times10^5\,\text{Pa}) + 454\text{K}$$

$$= \frac{49.86\times10^{-6}\,\text{m}^3\text{mol}^{-1} - 51.52\times10^{-6}\,\text{m}^3\text{mol}^{-1}}{95.4\ \text{Jmol}^{-1}\text{K}^{-1} - 91.4\ \text{Jmol}^{-1}\text{K}^{-1}}(P_{tp}[\text{Pa}] - 1.0\times10^5\,\text{Pa}) + 950\text{K}\,,$$

where P_{tp} is the pressure of the triple point.

Solving the above equation yields:

$$P_{tp} = \underline{0.39\ \text{GPa}}.$$

The temperature of the triple point, T_{tp}, can be calculated by inserting P_{tp} into any of the two equations that were used to calculate the phase transition boundaries. We take the equation giving the phase transition boundary andalusite/sillimanite and obtain:

$$T_{tp} = \frac{49.86\times10^{-6}\,\text{m}^3\text{mol}^{-1} - 51.52\times10^{-6}\,\text{m}^3\text{mol}^{-1}}{95.4\ \text{Jmol}^{-1}\text{K}^{-1} - 91.4\ \text{Jmol}^{-1}\text{K}^{-1}}(0.39\times10^9\,\text{Pa}$$

$$- 1.0\times10^5\,\text{Pa}) + 950\text{K} = \underline{788\ \text{K}}.$$

The pressure and temperature of the triple point are now used to determine the phase transition boundary kyanite/sillimanite. A pressure of 0.6 GPa is chosen as the second transition point and the corresponding temperature is calculated as follows:

$$T^{ky/sill}_{0.6GPa} = \frac{49.86\times10^{-6}\,\text{m}^3\text{mol}^{-1} - 44.15\times10^{-6}\,\text{m}^3\text{mol}^{-1}}{95.4\ \text{Jmol}^{-1}\text{K}^{-1} - 82.8\ \text{Jmol}^{-1}\text{K}^{-1}}(0.6\times10^9\,\text{Pa}$$

$$- 0.39\times10^9\,\text{Pa}) + 788\text{K} = \underline{883\ \text{K}}.$$

The results of the calculation are presented in Fig. 6.5, where the calculated temperatures are designated by open squares. The filled squares give the transition temperatures for kyanite/andalusite and andalusite/sillimanite at 0.1 MPa. These are the transition points that were used to undertake the calculations.

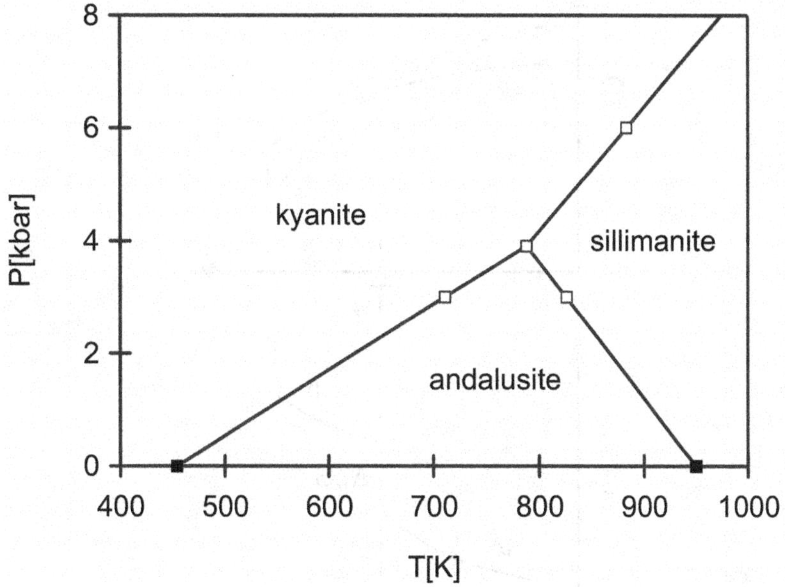

Fig. 6.5 Phase transition boundaries in the one-component system Al_2SiO_5 that were calculated using the Clausius-Clapeyron equation. Filled squares: given transition points and open squares: calculated transitions points.

6.1.2 Classification of phase transformations

According to the work of Ehrenfest, phase transitions are classified according to the behavior of a state function at the transition point. A *first-order transition* is one in which the Gibbs free energy as a function of state variables *(P,T)* is continuous, while the first derivative of the Gibbs free energy with respect to any variable of state is discontinuous. Considering the relationships that are given in Eqs. (5.16) and (5.17), the functions of state entropy, enthalpy and volume are discontinues at the transition point. Examples of first-order phase transitions are melting and transformations between isostructural phases (Toledano and Toledano 1987). Variation in the Gibbs free energy, entropy and enthalpy with temperature are illustrated schematically in Fig. 6.6.

One important feature of first-order transitions are definite values for the entropy, enthalpy and volume of transformation. This is the result of the fact that the values of these functions change stepwise at the point of transition.

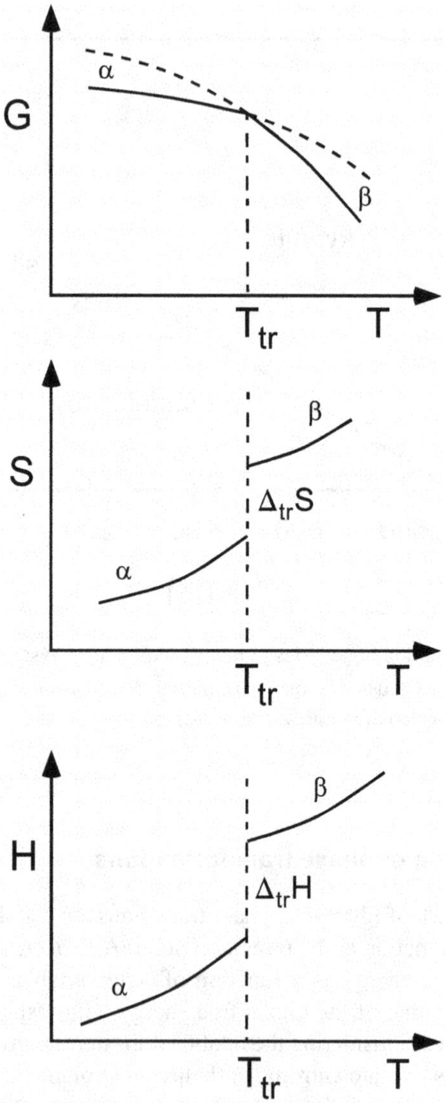

Fig. 6.6 The variation of the Gibbs free energy, entropy and enthalpy as a function of temperature at the transition point in the case of a first-order transition. $\Delta_{tr}S$ and $\Delta_{tr}H$ are the entropy and enthalpy of transformation from α to β, respectively.

The behavior of the Gibbs free energy and volume as a function of pressure is shown schematically in Fig. 6.7.

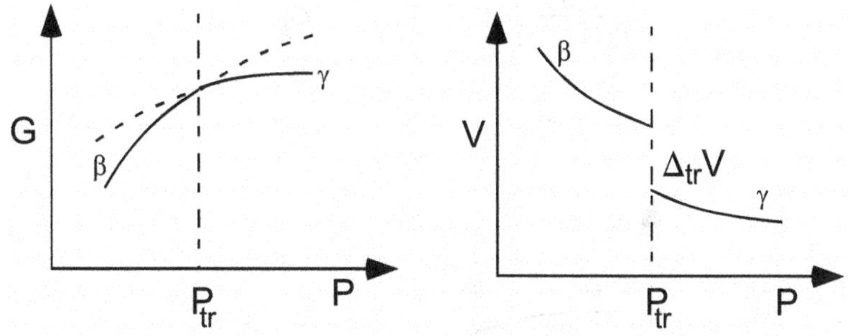

Fig. 6.7 Variation of the Gibbs free energy and volume with pressure in the case of a first-order transformation. $\Delta_{tr}V$ gives the volume of transition from β to γ.

After Ehrenfest, a *second-order transition* is one in which the Gibbs free energy function and the first derivatives are continuous while the second derivatives are discontinuous. Thus, the functions S, H, and V should show no discontinuity. However, the heat capacity, C_p, the thermal expansion coefficient, α, and the compressibility coefficient, β, should be discontinuous at the transition point, because they are the second derivatives of the Gibbs free energy with respect to temperature and pressure, respectively. Some of these relations are illustrated by the following equations:

Eq. (6.23) shows, for example, the correlation between the second derivative of the Gibbs free energy with respect to temperature, the first derivative of the entropy with respect to temperature and C_p:

$$\left(\frac{\partial^2 G}{\partial T^2}\right)_P = -\left(\frac{\partial S}{\partial T}\right)_P = -\frac{C_P}{T}. \tag{6.23}$$

Eq. (6.24) gives the relationship between the second derivative of the Gibbs free energy with respect to pressure, the first derivative of volume with respect to pressure and the compressibility, namely:

$$\left(\frac{\partial^2 G}{\partial P^2}\right)_T = \left(\frac{\partial V}{\partial P}\right)_T = -\beta V \tag{6.24}$$

and Eq. (6.25), finally, demonstrates the correlation between the second derivative of the Gibbs free energy divided by temperature, the first derivative of the enthalpy and the heat capacity at constant pressure, i.e.:

$$\frac{\partial}{\partial T}\left[\frac{\partial(G/T)}{\partial(1/T)}\right]_P = \left(\frac{\partial H}{\partial T}\right)_P = C_p. \tag{6.25}$$

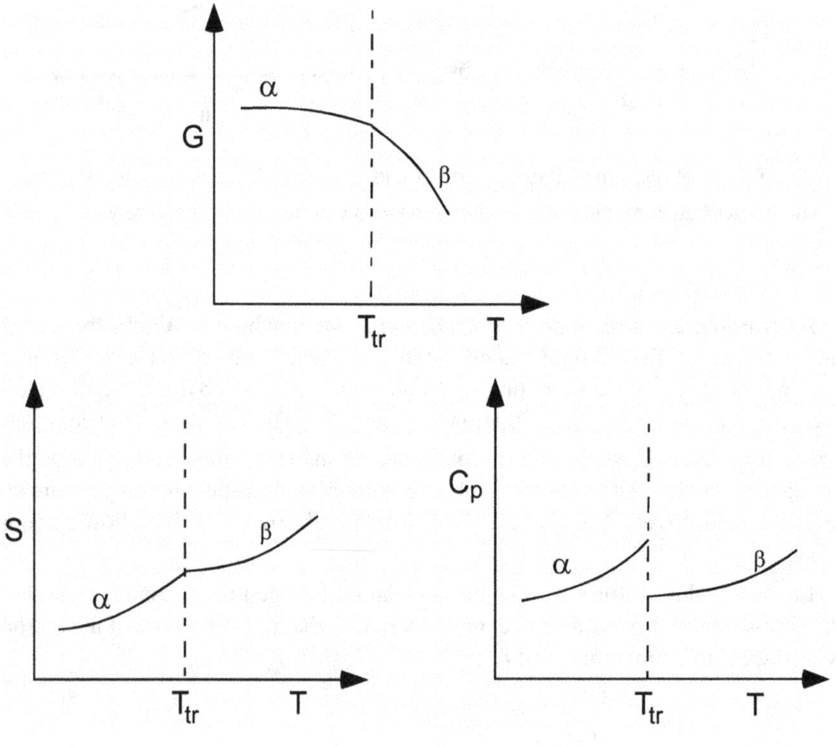

Fig. 6.8 Variation of the Gibbs free energy, *G*, entropy, *S*, and heat capacity, C_p, with temperature in the case of the second-order transition. T_{tr} gives the transition temperature.

Fig. 6.8 shows schematically the variation of the Gibbs energy, entropy and heat capacity with temperature. The Gibbs free energy and its first derivative, the entropy, do not show a discontinuity at the temperature of transition. The heat capacity, which is the second derivative of the Gibbs free energy with respect to temperature, however, is a step function. Directly at the transition temperature, there is a change from one definite value to another definite value. Ehrenfest's classification requires, therefore, that $\Delta_{tr}C_p$ is finite. This, however is normally not the case. For most second order transitions there is experimental evidence that C_p, and thus $\Delta_{tr}C_p$, becomes infinite at the transition temperature. The shape of the C_p curve near the transition temperature then resembles the greek letter λ, and thus this type of transition is often referred to as *lambda transition*. An example of a lambda transition is shown in Fig. 6.9.

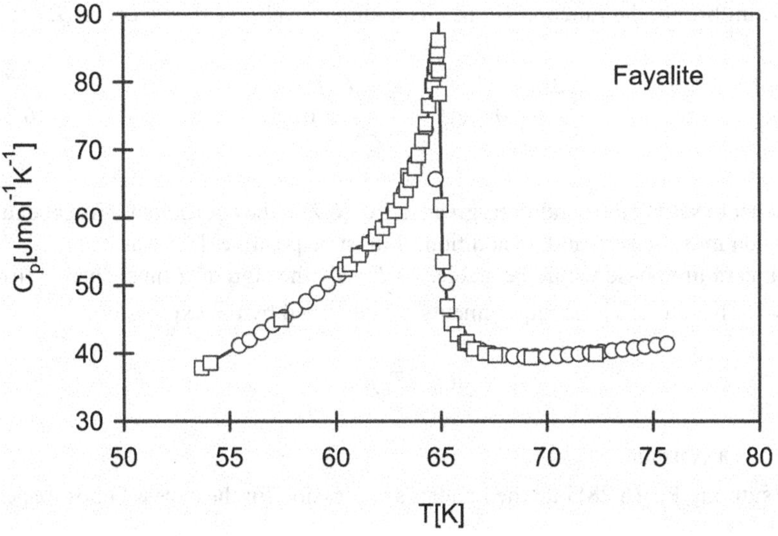

Fig. 6.9 Heat capacity of fayalite, Fe_2SiO_4, as a function of temperature showing a Lambda transition at 64.9 K.

Another way to determine the change in the Gibbs free energy associated with a phase transition is given by *Landau theory*. Central to this theory is the concept of an *order parameter*, *Q*, which is related to the change of some macroscopic property through the phase transition. Such macroscopic properties can be, for example, optical birefringence, spontaneous strain, etc. The proportionality between the order parameter, *Q*, and the measured macroscopic properties is usually either linear or quadratic. Although changes in macroscopic properties are clearly related to some microscopic interactions, a knowledge of these interactions is not necessary for Landau theory, where the Gibbs free energy associated with a phase transition is given as a polynomial expansion of the order parameter. It describes the energy changes that are superimposed on those caused by "normal" temperature variations and is therefore called the *excess Gibbs free energy* of transition.

Landau's expression for the excess Gibbs free energy associated with a phase transition reads:

$$G^L = \alpha Q + \frac{1}{2}AQ^2 + \frac{1}{3}BQ^3 + \frac{1}{4}CQ^4 \ldots, \tag{6.26}$$

where α, A, B and C are empirical coefficients. The value of the order parameter, Q, vary from 1, at absolute zero, to zero at some critical temperature, T_c, where the high temperature phase becomes stable.

At equilibrium, the function G^L has its minimum value with respect to Q, i.e:

$$\frac{\partial G^L}{\partial Q} = 0 \quad \text{and} \quad \frac{\partial^2 G^L}{\partial Q^2} > 0. \tag{6.27}$$

In order to satisfy the conditions given in Eq. (6.27), the coefficient α in Landau's expression must be zero and, in addition, A must be positive. If A was negative, the low temperature phase would be stable. At $T = T_c$, the sign of A must change from positive to negative. These requirements are governed by the expression:

$$A = a(T - T_c), \tag{6.28}$$

where a is a constant.

Substituting Eq. (6.28) into the Landau's expression for the excess Gibbs free energy yields:

$$G^L = \frac{1}{2}a(T - T_c)Q^2 + \frac{1}{3}CQ^3 + \frac{1}{4}DQ^4 \ldots. \tag{6.29}$$

In the case of a second-order transition, Q must vary continuously from 1 to 0 with temperature. This is only the case if all the odd order terms in the polynomial expansion are zero. Considering this and also the fact that two or three terms of the expansion are normally sufficient to describe the excess Gibbs free energy changes, the Landau's equation can be rewritten as:

$$G^L = \frac{1}{2}a(T - T_c)Q^2 + \frac{1}{4}CQ^4 + \frac{1}{6}DQ^6. \tag{6.30}$$

Expression Eq. (6.30) allows Q to change between 1 and 0, but it does not preclude the possibility of a discontinuity in Q. Depending upon the value of the coefficients C and D, three types of transformation can be defined, namely:

a) C is positive and the sixth-order term is negligibly small. In this case Q varies continuously between 1 and 0 as a function of temperature and the phase transition is second-order.

b) C is negative and D is positive. Here, at some given temperature, which is designated as T_{tr} and where $T_{tr} > T_c$, Q is discontinuous as a function of temperature. It jumps from a definite value $Q = Q_o$ to $Q = 0$. At this temperature, the low temperature phase with $Q = Q_o$ coexists with the high temperature phase with $Q = 0$.The phase transition is first-order and the height of the 'step' corresponds to the change in the thermodynamic functions associated with the phase transition.

c) C is zero and D is positive and the sixth-order term is present and is positive. Such an expansion describes a phase transition that is somewhere between second and first-order in nature. It is called *tricritical phase transition.*

Fig. 6.10 shows the order parameter Q as a function of temperature for a second-order and a tricritical phase transition.

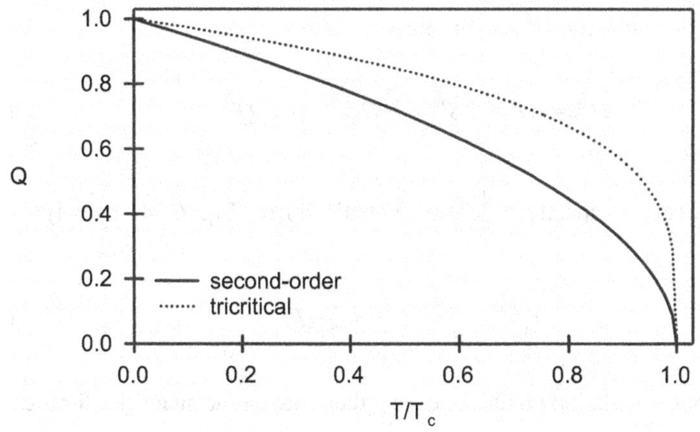

Fig. 6.10 Q as a function of temperature in the case of a second-order phase transition (solid line) and a tricritical phase transition (dotted).

In the case of a second-order phase transition, Q varies with temperature as:

$$Q = \left(\frac{T_c - T}{T_c}\right)^{1/2}, \qquad (6.31)$$

and in the case of a tricritical phase transition as:

$$Q = \left(\frac{T_c - T}{T_c}\right)^{1/4}. \tag{6.32}$$

In order to obtain the excess Gibbs free energy due to a second-order phase transition, Q in Eq. (6.30) has to be substituted by the expression given in Eq. (6.31). Considering the fact that in the case of a second-order phase transition the sixth-order term in negligibly small, the Landau expansion reads:

$$G^L = -\frac{a^2}{4C}(T - T_c)^2. \tag{6.33}$$

The other excess thermodynamic functions can be derived from Eq. (6.33). The excess entropy, S^L, is

$$S^L = -\left(\frac{\partial G^L}{\partial T}\right)_P = \frac{a^2}{2C}(T - T_c) = -\frac{1}{2}aQ^2 \tag{6.34}$$

and the excess enthalpy, H^L, accordingly

$$H^L = G^L + TS^L = \frac{1}{4}CQ^4 - \frac{1}{2}aT_cQ^2. \tag{6.35}$$

The excess heat capacity, C_p^L, can be obtained from Eq. (6.34), namely

$$T\left(\frac{\partial S^L}{\partial T}\right)_P = C_p^L = \frac{a^2}{2C}T. \tag{6.36}$$

Analogous calculation yields the excess thermodynamic quantities for the tricritical phase transition:

$$S^L = -\frac{1}{2}aQ^2, \tag{6.37}$$

$$H^L = \frac{1}{6}DQ^6 - \frac{1}{2}aT_cQ^2 \tag{6.38}$$

and

$$C_p^L = \frac{aT}{4\sqrt{T_c}}(T_c - T)^{1/2}. \tag{6.39}$$

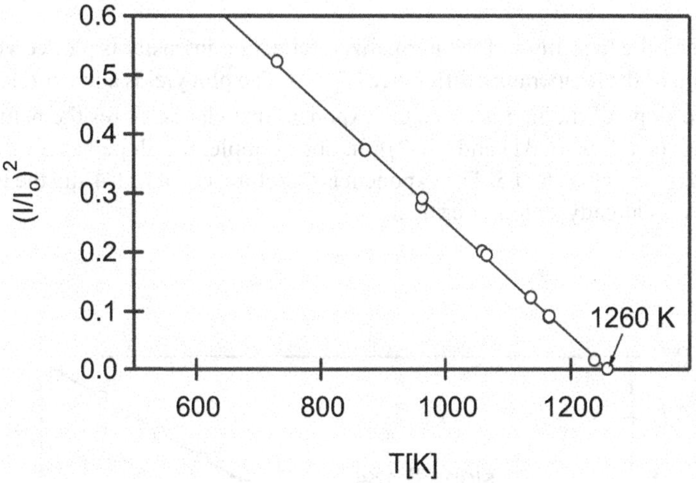

Fig. 6.11 Square of the intensity of the superlattice reflection $(11\bar{2}3)$ for $CaCO_3$ as a function of temperature. The intersection of the regression line with the abscissa gives the critical temperature $T_c = 1260$ K. All reflection intensities are normalized to 1 by dividing them by the strongest reflection (Data of Dove and Powell 1989).

Example: The high temperature phase transition in calcite, $CaCO_3$, involves the orientational disordering of CO_3 groups. In the low temperature modification, the planar CO_3 groups are ordered in alternate layers where they point in opposite direction. The ordering is associated with extra reflections in x-ray diffraction patterns, giving so-called *superlattice* reflections. In the high temperature phase, the CO_3 groups are free to rotate and, thus, are crystallographically equivalent. Therefore, the intensity of the super lattice reflections decreases on heating and they become zero at the temperature of transformation, T_c. If the square of the reflection intensity is plotted versus temperature, a linear relationship is obtained (see Fig. 6.11). This is, according to Eq. (6.32), an indication of a tricritical transition. Symmetry considerations, which cannot be discussed here, require a quadratic proportionality between the intensity of the superlattice reflections and the order parameter, Q. Thus, if the reflection intensity is expressed to the second power, the following relationship holds:

$$I^2 \propto Q^4 = (T_c - T)/T_c = 1 - \frac{T}{T_c}. \tag{6.40}$$

Fig. 6.11 shows the square of the intensity of the superlattice reflection $11\bar{2}3$ for $CaCO_3$ as a function of temperature. The best-fit line to the data points intersects the abscissa at 1260 K, which, according to Eq. (6.40), is the temperature of transformation, T_c.

In Fig. 6.12 the logarithm of the normalized reflection intensity is plotted versus the logarithm of the temperature difference (T_c - T). The plot yields a linear relationship and the slope of the line is twice the exponent that characterizes the nature of the transition (see Eqs. (6.31) and (6.32)). In our example, the slope has a value of 0.48, which is very close to 0.5. The exponent is therefore, equal to 1/4 and the transformation is, as already stated, tricritical.

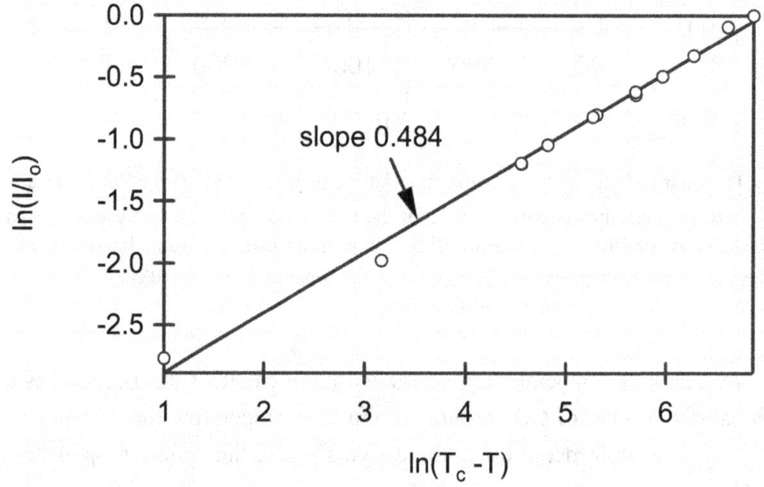

Fig. 6.12 Logarithm of the normalized reflection intensity, ln (I/I_o) vs. ln (T_c - T) for calcite, $CaCO_3$. According to Eqs. (6.31) and (6.32) the slope of the line is twice the exponent that characterizes the nature of the phase transformation.

Following Eqs. (6.37) through (6.39), the values of the Landau constants a and D are needed to calculate the excess enthalpy, H^L, excess heat capacity, C_p^L and excess entropy, S^L, associated with the tricritical phase transition in calcite. Redfern et al. 1989 determined experimentally the required constants using the calorimetry.

They measured the change in the enthalpy associated with the disordering in calcite as a function of temperature and determined the following values:

$a = 24\ \mathrm{Jmol^{-1}K^{-1}}$

$D = 30\ \mathrm{kJmol^{-1}}$.

The temperature dependence of the excess entropy arising from the disordering is calculated according to Eq. (6.37):

$$S^L = \Delta_{ord}S = -\frac{24\mathrm{Jmol^{-1}K^{-1}}}{2}\left(\frac{1260\mathrm{K} - T[\mathrm{K}]}{1260\mathrm{K}}\right)^{1/2}.$$

At a temperature of $T = 0$ K the zero point entropy of ordering, $\Delta_{ord}S_{(T=0)}$, is obtained, namely:

$$\Delta_{ord}S_{(T=0)} = -\frac{24\mathrm{Jmol^{-1}K^{-1}}}{2} = \underline{-12\ \mathrm{Jmol^{-1}K^{-1}}}.$$

In order to calculate the enthalpy change associated with the tricritical phase transition Eq. (6.38), is used:

$$H^L = \Delta_{ord}H = \frac{30000\mathrm{Jmol^{-1}}}{6}\left(\frac{1260\mathrm{K} - T[\mathrm{K}]}{1260\mathrm{K}}\right)^{3/2} - \frac{24\mathrm{Jmol^{-1}K^{-1}} \times 1260\mathrm{K}}{2}\left(\frac{1260\mathrm{K} - T[\mathrm{K}]}{1260\mathrm{K}}\right)^{1/2}.$$

The enthalpy of ordering at absolute zero is then:

$$\Delta_{ord}H_{(T=0)} = 5000\mathrm{Jmol^{-1}} - 15120\mathrm{Jmol^{-1}} = \underline{-10120\ \mathrm{Jmol^{-1}}}.$$

The excess heat capacity due to ordering is obtained using Eq. (6.39):

$$C_p^L = \Delta_{ord}C_p = \frac{24\mathrm{Jmol^{-1}K^{-1}}\,T[\mathrm{K}]}{4\sqrt{1260\mathrm{K}}}(1260\mathrm{K} - T[\mathrm{K}])^{-1/2}.$$

6.2 Stability conditions for solutions

The stability of a solid solution depends not only on temperature and pressure but also on its composition. This means a stable solid solution must not decompose spontaneously in two or more phases of different compositions if temperature and pressure are held constant.

Consider a hypothetical binary system *A-B*. According to, the Gibbs free energy of this system is

$$\overline{G} = (1 - x_B)\mu_A + x_B\mu_B. \tag{6.41}$$

Dividing the chemical potentials into a standard potential and a composition dependent part yields:

$$\overline{G} = (1 - x_B)[\mu_A^o + RT\ln a_A] + x_B[\mu_B^o + RT\ln a_B]. \tag{6.42}$$

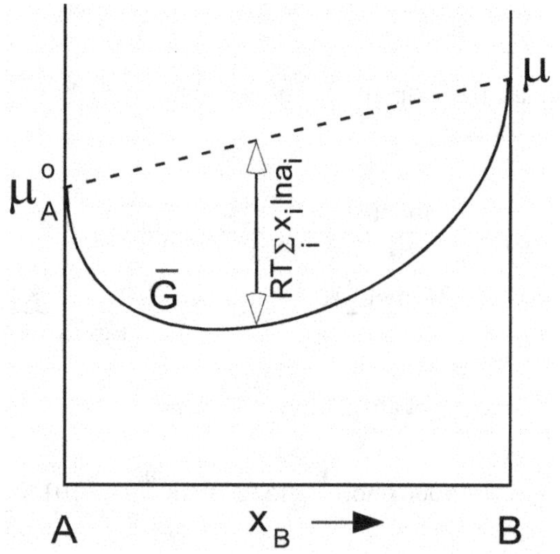

Fig. 6.13 The Gibbs free energy of a binary solution *A,B* as a function of composition at constant pressure and temperature. The double arrow designates the deviation of the Gibbs free energy of the solution from that of a mechanical mixture for the given mole fraction of x_B.

Fig. 6.13 shows schematically the variation of the Gibbs free energy as a function of the mole fraction x_B in the binary solution, *(A,B)*, at constant pressure and temperature. At $x_B = 0$ and at $x_B = 1$, the value of $\overline{G}$ equals μ_A^o and μ_B^o, respectively.

The $\overline{G}(x)$-curve, therefore, starts and ends at the values of the chemical potentials of the pure components. The dashed line gives the contribution of the standard potentials to the Gibbs free energy of the solution. Because the activities of the components A and B vary between 0 and 1, their contributions to the Gibbs free energy are negative over the whole compositional region. The $\overline{G}(x)$-curve must, therefore lie underneath the dashed line that represents the Gibbs free energy for the mechanical mixture of the pure end-members. For any x_B, the distance between the dashed line and $\overline{G}(x)$-curve corresponds to the term $RT\Sigma x_i \ln a_i$. The tangent to $\overline{G}(x)$-curve is given by

$$\left(\frac{\partial \overline{G}}{\partial x_B}\right)_{P,T} = \mu_B - \mu_A = \mu_B^{\circ} + RT\ln x_B + RT\ln\gamma_B \tag{6.43}$$

$$-\mu_A^{\circ} - RT\ln(1-x_B) - RT\ln\gamma_A$$

$$= \mu_B^{\circ} - \mu_A^{\circ} + RT\ln\frac{x_B}{1-x_B} + RT\ln\frac{\gamma_B}{\gamma_A}.$$

The limiting values of γ_B and γ_A at $x_B = 0$ and $x_B = 1$ are finite and zero, respectively. The slope of the $\overline{G}(x)$-curve at $x_B = 0$ is, therefore, $-\infty$ and at $x_B = 1$ $+\infty$. In the intermediate compositional region two different cases are to be distinguished:

Case #1: The Gibbs free energy curve as a function of composition is convex with respect to the concentrational join, as demonstrated in Fig. 6.14. Consider a solution that contains x_B moles of the component B and assume that it decomposes into the phases α and β. The Gibbs free energy of a mechanical mixture is then:

$$\overline{G}^{mm} = \frac{x_B^{\beta} - x_B}{x_B^{\beta} - x_B^{\alpha}} G^{\alpha} + \frac{x_B - x_B^{\alpha}}{x_B^{\beta} - x_B^{\alpha}} G^{\beta}, \tag{6.44}$$

where G^{α} and G^{β} are the Gibbs free energies of the phases α and β, respectively. As shown in Fig. 6.14, the Gibbs free energy of a mechanical mixture ($\overline{G}^{mm}$) is greater than that of a homogenous solution ($\overline{G}^{ss}$). The homogenous solution is, therefore, stable under the given conditions. Because the Gibbs free energy curve is convex with respect to the compositional join between $x_B = 0$ and $x_B = 1$, a solution is always more stable than any mechanical mixture of two phases of the same bulk composition. This means that there is complete miscibility between the components A and B. In this place it is important to mention that a complete miscibility is only possible if the end-members, as well as the solid solution, have the same crystal structure.

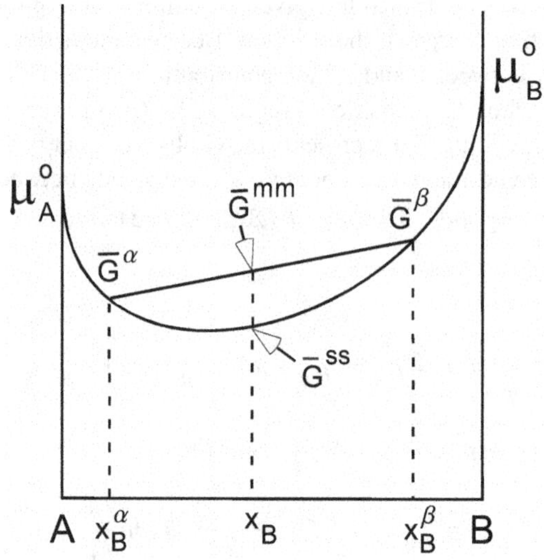

Fig. 6.14 The Gibbs free energy as a function of composition in the case of complete miscibility between the components A and B. x_B designates the mole fraction oft the component B in the solution A,B, x_B^{α} and x_B^{β} are the mole fractions of the component B in the phases α and β, respectively. $\overline{G}^{mm}$ gives the Gibbs free energy for a mechanical mixture consisting of phases α and β. $\overline{G}^{ss}$ is the Gibbs free energy of the homogenous solution. $\overline{G}^{\alpha}$ and $\overline{G}^{\beta}$ give the Gibbs free energies of the phases α and β.

Case #2: The Gibbs free energy curve as a function of composition has convex as well as concave segments (see Fig. 6.15). In the region between S^{α} and S^{β}, small separation of composition, due to the fluctuations, leads to a lowering of the energy. Any solution inside this compositional region is, therefore, unstable and decomposes spontaneously into two phases. The compositions of the *stable* phases are given by the points of tangency to the Gibbs free energy curve at Q^{α} and Q^{β}. They are referred to as *binodes*, and because they lie on the common tangent, the chemical potentials of the components A and B are equal in both phases, i.e.:

$$\mu_A^{\alpha} = \mu_A^{\beta} \quad \text{and} \quad \mu_B^{\alpha} = \mu_B^{\beta}. \tag{6.45}$$

Using the relationships that are given in Eqs. (5.49) and (5.50), one can write:

$$\overline{G}^{\alpha} - x_B^{\alpha}\left(\frac{\partial \overline{G}^{\alpha}}{\partial x_B^{\alpha}}\right)_{P,T} = \overline{G}^{\beta} - x_B^{\beta}\left(\frac{\partial \overline{G}^{\beta}}{\partial x_B^{\beta}}\right)_{P,T} \tag{6.46}$$

for the chemical potentials of the component A in the phases α and β and

$$\overline{G}^{\alpha} + (1 - x_B^{\alpha})\left(\frac{\partial \overline{G}^{\alpha}}{\partial x_B^{\alpha}}\right)_{P,T} = \overline{G}^{\beta} + (1 - x_B^{\beta})\left(\frac{\partial \overline{G}^{\beta}}{\partial x_B^{\beta}}\right)_{P,T} \tag{6.47}$$

for the chemical potentials of the component B in the coexisting phases.

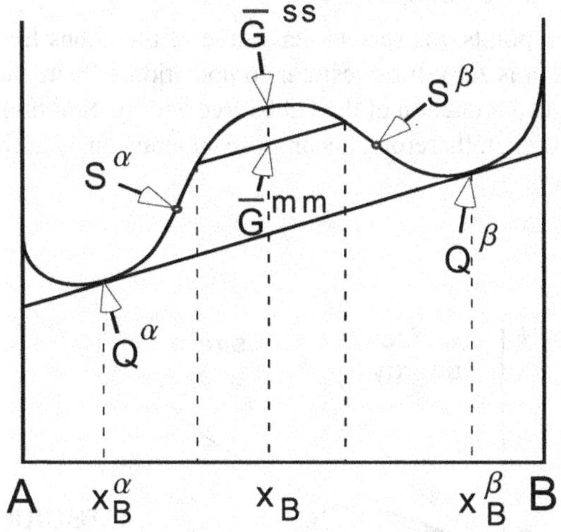

Fig. 6.15 Variation of the Gibbs free energy as a function of the mole fraction x_B in a system in which spontaneous decomposition takes place. $\overline{G}^{ss}$ and $\overline{G}^{mm}$ give the Gibbs free energy of the solution and the corresponding mechanical mixture consisting of two separate phases, respectively. S^{α} and S^{β} designate the spinodes and Q^{α} and Q^{β} the binodes.

Subtracting Eq. (6.46) from Eq. (6.47) yields:

$$\left(\frac{\partial \overline{G}^{\alpha}}{\partial x_B^{\alpha}}\right)_{P,T} = \left(\frac{\partial \overline{G}^{\beta}}{\partial x_B^{\beta}}\right)_{P,T} . \tag{6.48}$$

Eq. (6.48) requires that the slope of the tangent to the $\overline{G}(x)$ curve is the same at both points of tangency, i.e. at Q^{α} and Q^{β}, respectively.

It is important to note that the binodes do not the necessarily coincide with the minima of the Gibbs energy versus mole fraction curve.

S^{α} and S^{β} are referred to as *spinodes* and designate inflection points on the $\overline{G}(x)$ curve. In the region inside these points, the second derivative of the Gibbs free energy with respect to the mole fraction, x_B, is positive for all compositions. Such solutions are unstable, because any compositional fluctuations gives rise to a decrease in the total Gibbs free energy. The process of separation continues until two different phases attain spinodal compositions. This process of spontaneous unmixing is referred to as *spinodal decomposition*. Here, it is important to note that spinodal decomposition takes place inside a homogenous phase. On a local scale the concentration of some component in the phase increases and as a result a depletion of the same component occurs in the surrounding region.

At the inflection points, the second derivative of the Gibbs free energy with respect to composition is zero. Infinitesimal compositional fluctuations around these points do not lead to a reduction of the Gibbs free energy. Solution having the spinodal composition do not, therefore, decompose spontaneously, although the system is not at equilibrium.

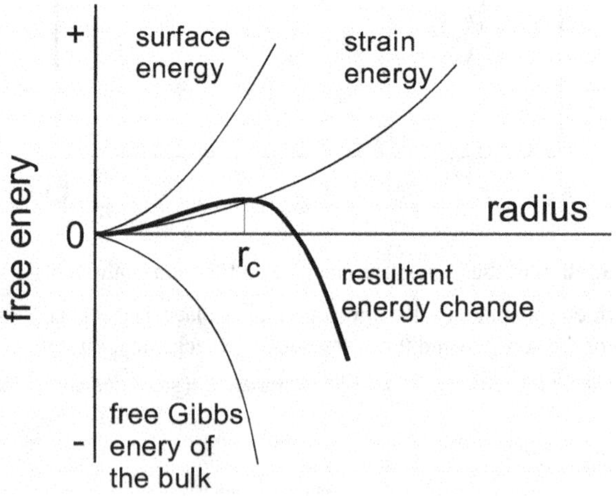

Fig. 6.16 Energy relationships associated with the formation of a spherical nucleus. Because the positive energy contributions of the surface energy and the strain energy dominate over the Gibbs free energy of the bulk, the resultant Gibbs free energy increases until it reaches a maximum at the critical radius r_c. With further growth of the nucleus, the resultant free Gibbs energy decreases.

Solutions with compositions lying between the spinodes and binodes do not decompose spontaneously but are *metastable*. The process of spontaneous decomposition can not take place in these two regions, because, as can be seen in Fig. 6.15, small fluctuations in composition lead to an increase in the Gibbs free energy. Decomposition only occurs if the degree of undercooling is large enough to overcome the energetic barrier associated with the energy of nucleation. Thereby a nucleus with composition x_B^{β} can form. Initially, the decrease in the Gibbs free energy of the bulk system resulting from the undercooling is outweighed by two positive energy contributions, which are the *strain* and the *surface energy* (see Fig. 6.16). A positive strain energy arises because the compositions and, thus, the molar volumes of the parent and the exsolving phase are different. Consequently, the exsolving phase requires more or less space than the original phase. The positive surface energy is due to the surface tension that always exists between two different phases. After the nucleus reaches some critical size (r_c), the negative Gibbs free energy of the bulk outweighs the positive Gibbs free energies originating from the strain and the surface tension. Thereby the new phase can grow. The diffusion of one of the components into the new phase causes the original solution to change in composition until the equilibrium composition, x_B^{α}, is reached.

In the early stage of phase separation, the process of spinodal decomposition differs significantly from that of nucleation and growth. In the case of spinodal decomposition, the separation of phases occurs by so-called *uphill diffusion* (Cahn 1968). This means that the component, that accumulates in the exsolving phase diffuses continuously from lower to higher concentration, a behavior that is somewhat contradictory to that which might be expected. The driving force for diffusion is a decrease in the total Gibbs free energy that is associated with the decomposition of the original solution into two phases (see Fig. 6.15). Initial separation occurs inside a grain and the result is a *wavy micro structure* with sinusoidal variation in composition about a mean without sharp interfaces between the compositional crests and troughs (see Fig. 6.17b). With time, the concentrations of the diffusing components in the new phases increase in wavelength as well as in amplitude. In the final stage, the compositional profiles square up and are not distinguishable from those that are formed by nucleation and growth (Fig. 6.17d).

The nucleation and growth process leads to nuclei having definite (binodal) compositions with sharp interfacial boundaries (see Fig. 6.17c). In the immediate local region around the nucleus, the component that accumulates in the new phase is depleted. Subsequent growth of the new phase requires, therefore, a continuous supply of this component from the host phase. Thus, the component diffuses from a region of higher to the a region of lower concentration. Hence, the new phase grows by so-called *downhill diffusion*.

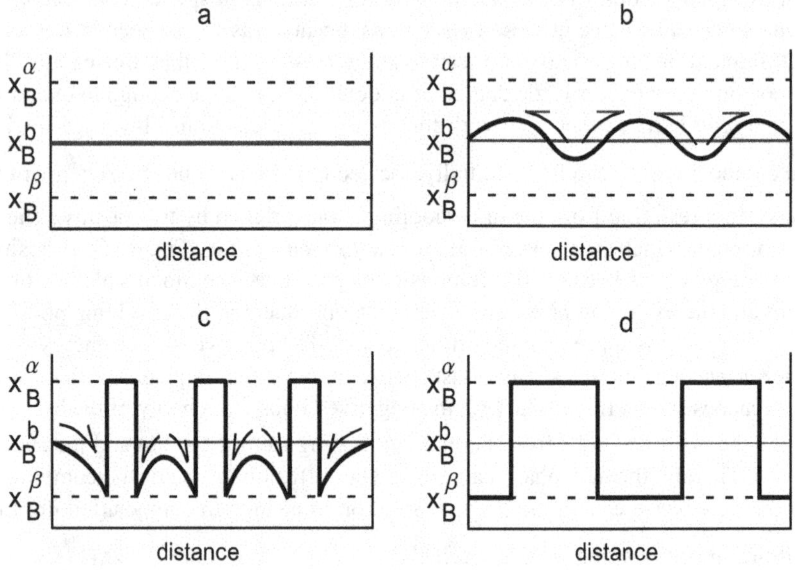

Fig. 6.17 A comparison between spinodal decomposition and nucleation and growth mechanisms. *a)* initial stage, *b)* sinusoidal fluctuation in composition due to spinodal decomposition, *c)* formation of exsolution lamellae in the case of nucleation and growth, and *d)* final state for either process. The arrows show the direction in which the component B diffuses. x_B^b is the mole fraction of the component B in the homogeneous solution before decomposition. x_B^α and x_B^β give the equilibrium mole fractions for the component B in the phases α and β, respectively.

6.2.1 Coherent exsolution

An exsolving phase and its host can have a close structural relationship with one another. In the case that their crystal structures are similar, they can be oriented such that their planes match across the interface. The degree of this lattice match is expressed in terms of *coherency*. In the case where the lattice spacings of the exsolved phase and the host match exactly, the planes of the host continue across the interface into the new phase. The interface is then called *coherent*. If the difference between the crystal structures of the parent and exsolving phase are to large to form coherent boundaries then a *semi-coherent* interface is formed. In this case some planes of the host continue into the exsolved phase and some terminate at the interface. If there is no structural relationship between the parent and the exsolved phase, a *incoherent* interface is obtained. Fig. 6.18 shows schematically the cases for coherent, semi-coherent and incoherent interfaces between host and exsolved phases.

The adjustment of the lattice at and near the interface creates strain that generates positive Gibbs free energy. This energy is termed the *Cahn energy*, Φ, and, according to Robin (1974), can be calculated as follows:

$$\Phi = \bar{G} + k(x_i - x_i^{\text{o}})^2, \tag{6.49}$$

where k is a constant that is derived from the elastic constants of the phase under consideration. x and x^{o} give mole the fractions of the component i for the exsolved phase and for the bulk crystal, respectively. $\bar{G}$ is the strainfree Gibbs free energy of the system.

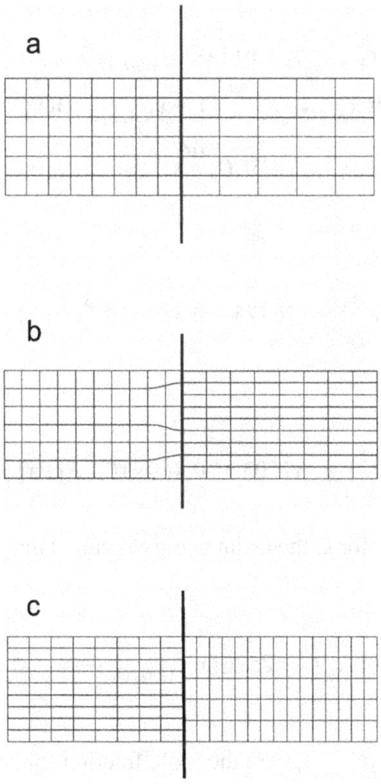

Fig. 6.18 Coherent (*a*), semi-coherent (*b*) and incoherent (*c*) interfaces (schematically)

Example: A well known case for coherent exsolution is shown by crypthopertitic alkali feldspars. Using two different data sets of elastic constants for feldspar, Robin (1974) calculated two different values for k, namely: 2525.5 Jmol^{-1} and 2948 Jmol^{-1}. In our calculations, we will use the smaller one rounded up to a value of 2526 Jmol^{-1}.

For the sake of convenience, the Gibbs free energy of mixing, $\Delta_m\overline{G}(x)$, instead of the total Gibbs free energy, $\overline{G}(x)$, will be considered here. The two functions show the same dependence on composition because, the free Gibbs energy of mixing, $\Delta_m\overline{G}(x)$, is obtained by subtracting the sum $\sum x_i\mu_i^{\text{o}}$ from the total Gibbs free energy, $\overline{G}(x)$. The same is true with respect to the relationship between $\Phi(x)$ and $\Delta\Phi(x)$. In order to calculate $\Delta_m\overline{G}(x)$ we follow Robin (1974) and use the exchange parameters, W^G_{Na-K} and W^G_{K-Na}, as determined by Thompson and Waldbaum (1969) and obtain:

$$\begin{aligned}\Delta_m\overline{G} = {} & RT[(1-x_{KAlSi_3O_8})\ln(1-x_{KAlSi_3O_8}) \\ & + x_{KAlSi_3O_8}\ln x_{KAlSi_3O_8}] + (1-x_{KAlSi_3O_8})(x_{KAlSi_3O_8})^2 W^G_{Na-K} \\ & + (1-x_{KAlSi_3O_8})^2 x_{KAlSi_3O_8} W^G_{K-Na},\end{aligned}$$

where

$$W^G_{Na-K}[\text{Jmol}^{-1}] = 26475 + 0.389\times10^{-5}P[\text{Pa}] - 19.383\,T[\text{K}]$$

and

$$W^G_{K-Na}[\text{Jmol}^{-1}] = 32103 + 0.469\times10^{-5}P[\text{Pa}] - 16.139\,T[\text{K}].$$

Using the value 2526 for k, the Cahn energy is calculated according to Eq. (6.49), as follows:

$$\Delta\Phi = \Delta_m\overline{G} + 2526(x_{KAlSi_3O_8} - x^{\text{o}}_{KAlSi_3O_8})^2,$$

where $x^{\text{o}}_{KAlSi_3O_8}$ and $x_{KAlSi_3O_8}$ are the mole fraction for sanidine in the bulk crystal and in the exsolved phase, respectively. Two mole fractions were chosen for the bulk crystal, namely: 0.3 and 0.4. The results of the calculation are presented graphically in Fig. 6.19.

As can be seen in Fig. 6.19, the slope of the common tangent to the $\Delta\Phi(x)$ curves depends on the bulk composition of the solution. The points of tangency, which determine the composition of the coexisting coherent phases, however, remain invariable. The inflection points of the $\Delta\Phi(x)$ curve give the *coherent spinodes*. They delimit the compositional region where a spontaneous coherent decomposition takes place. Coherent spinodes lay inside the *strainfree* or *chemical* spinodes. Similarly, the coherent binodes lay inside the *chemical* or *strainfree binodes*.

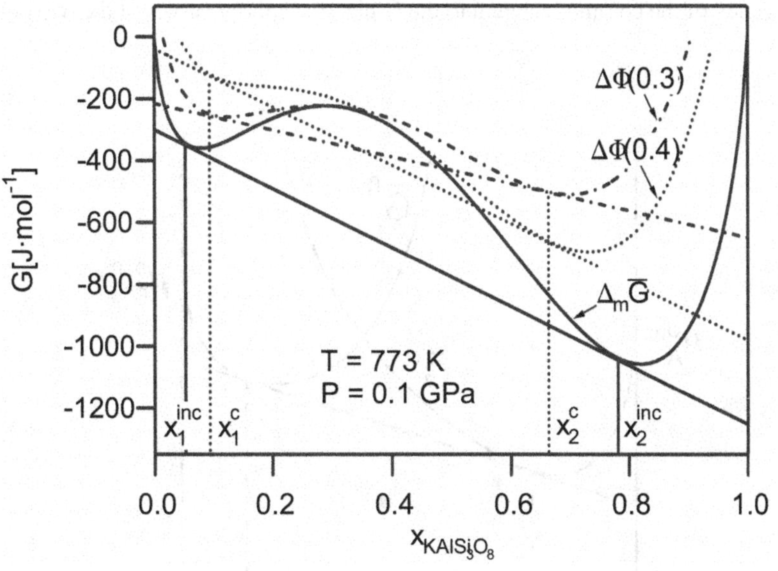

Fig. 6.19 Variation of the Gibbs free energy with composition for the system high albite - sanidine. The solid curve gives the Gibbs free energy of mixing, $\Delta_m\overline{G}(x)$. The dashed and the dotted curves represent the Cahn energy calculated using the bulk compositions $x^{o}_{KAlSi_3O_8} = 0.4$ and 0.3, respectively. $x^{inc}_{1,2}$ marks the composition of the incoherent and $x^{c}_{1,2}$ that of coherent binodes. While the slope of the common tangent to the $\Delta\Phi$ curve depends on the bulk composition, $x^{o}_{KAlSi_3O_8}$, the location of the two binodes does not. The value of the constant $k = 2526$ $\mathrm{Jmol^{-1}}$.

A $\overline{G}(x)$ curve extends over the entire compositional region only if the end-members, as well as the solid solution phases, crystallize in the same crystal structure. If this is not the case, different $\overline{G}(x)$ curves exist for the different phases. The phase with the lowest Gibbs energy is the stable one and if two $\overline{G}(x)$ curves intersect, two phases coexist. In order to obtain the composition of the coexisting phases, a common tangent to the curves must be drawn. The points of tangency give the mole fractions for the components for the coexisting phases. Fig. 6.20 shows $\overline{G}(x)$ curves for the case where the end-member phases crystallize with different crystal structures. The end-member A crystallizes as the phase α and the end-member B as phase β. The $\overline{G}(x)$ curve of the phase α does, therefore, not intersect the ordinate at $x_B = 1$. In the same way the Gibbs free energy curve of the phase β does not intersect the ordinate at $x_B = 0$. At equilibrium, the solution $(A,B)^\alpha$ coexists with the solution $(A,B)^\beta$. The concentration of B in the coexisting phases is determined by the points of tangency of the common tangent to the Gibbs free energy curves of the two phases.

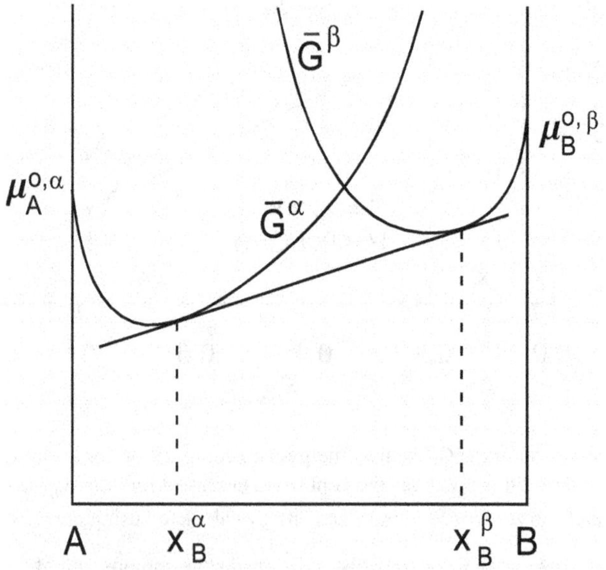

Fig. 6.20 $\overline{G}(x)$ as a function of composition for the case where the end-member A crystallizes as phase α and the end-member B as phase β. x_B^α and x_B^β are the mole fractions of the component B in the coexisting solid solutions α and β, respectively. $\mu_A^{o,\beta}$ is the standard potential of the component A in phase α and $\mu_B^{o,\beta}$ is the standard potential of the component B in phase β.

6.3 Gibbs free energy and phase diagrams of binary systems

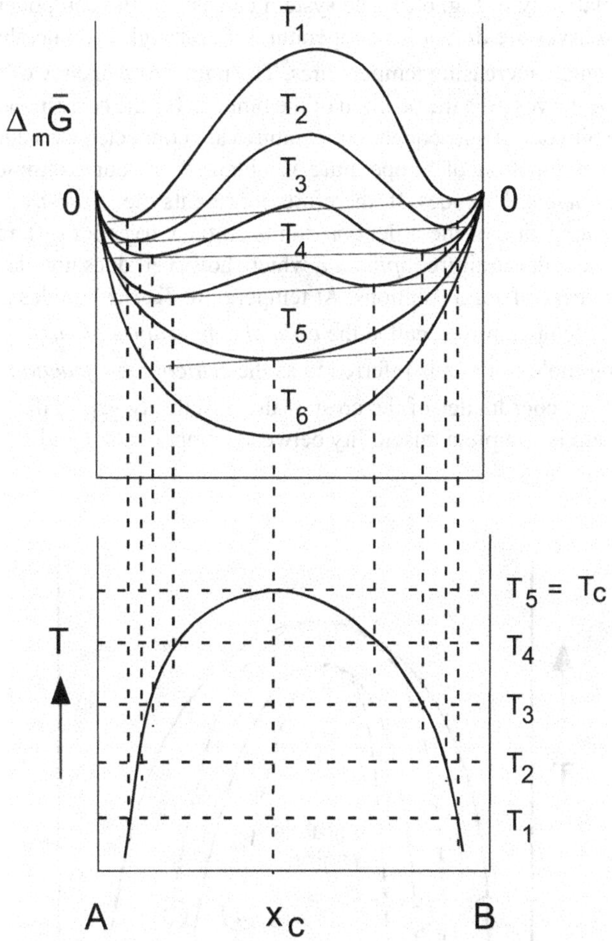

Fig. 6.21 Relationship between the Gibbs free energy of mixing and a *T*-*x* diagram in the binary system *A*-*B* containing a miscibility gap. T_c = critical temperature of mixing, and x_c = critical mole fraction.

The nature of the $\Delta_m\overline{G}(x)$ curve depends on the activities of the components in the solution. The latter depend on pressure, temperature and composition and, therefore, the Gibbs free energy of mixing depends on these variables. If at constant pressure,

$\Delta_m\overline{G}(x)$ curves are known for a series of different temperatures, a corresponding *T-x* diagram can be constructed. Similarly, a *P-x* diagram can be drawn if the $\Delta_m\overline{G}(x)$ curves at different pressures and constant temperature are known.

The relationship between the Gibbs free energy of mixing and a *T-x* diagram is shown schematically in Fig. 6.21.The system consists of the components *A* and *B*. The $\Delta_m\overline{G}(x)$ curves are drawn for temperatures T_1 through T_6, whereby increasing numbers designate increasing temperatures. The points of tangency of the common tangents to the curves give the position of the binodes for the corresponding temperatures. If the binodes of subsequent temperatures are connected by a curve, a miscibility gap as a function of temperature is obtained. In our example, the curve connects the *chemical* binodes. It, therefore, represents the *strain-free or chemical solvus*. The connection of the inflection points on the curves for different temperatures would yield the strainfree *spinodal*, which, however, does not play any practical role in the case of solid solutions. At temperature T_5, the binodes and spinodes coincide. This temperature is called the *critical temperature of mixing* (T_c) and the corresponding mole fraction is referred to as the *critical mole fraction*, x_c. The two variables are the coordinates of the crest of the miscibility gap in the *T-x* diagram. Above T_c, there is complete miscibility between components *A* and *B*.

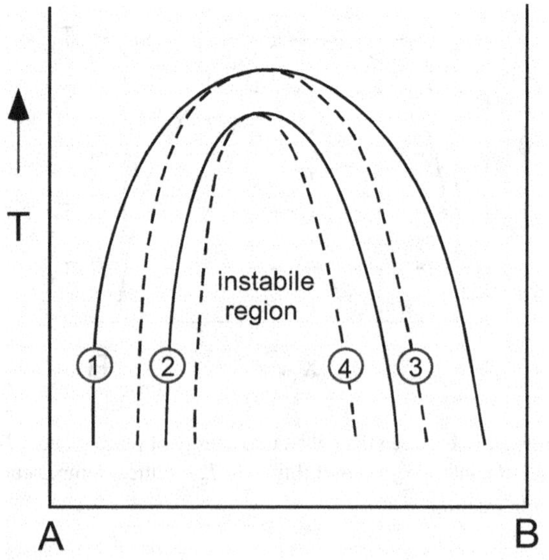

Fig. 6.22 Schematic *T-x* diagram for the system *A-B*. 1 = strain-free or chemical solvus, 2 = coherent solvus, 3 = strain-free or chemical spinodal, 4 = coherent spinodal. Inside the coherent spinodal is the region of spontaneous decomposition.

The stabile binodal compositions can only be achieved after the coherency is removed. This happens, for example, by the formation of discrete grains. As long as the coherency strain is maintained the Cahn energy determines the position of the spinodes and binodes. The *coherent spinodal* and *coherent solvus* can be constructed in the same way as in the case of incoherency. The points of tangency from the common tangent to the Cahn energy curves yield the coherent binodes and the points of inflection give the coherent spinodes. The temperature and the mole fraction at which the binodes and spinodes coincide represent the *critical conditions* for the coherent solvus. The concentration region inside the coherent spinodals is the region of spontaneous decomposition and is frequently referred to as the *instable area.*

Fig. 6.22 shows a schematic *T-x* diagram with a strain free solvus, incoherent spinodal, coherent solvus and coherent spinodal. The coherent solvus occurs inside the strain free one. The consequence is that further decomposition takes place as soon as coherency is lifted. Moreover, the critical temperature of the coherent solvus is lower than that of the strain free one.

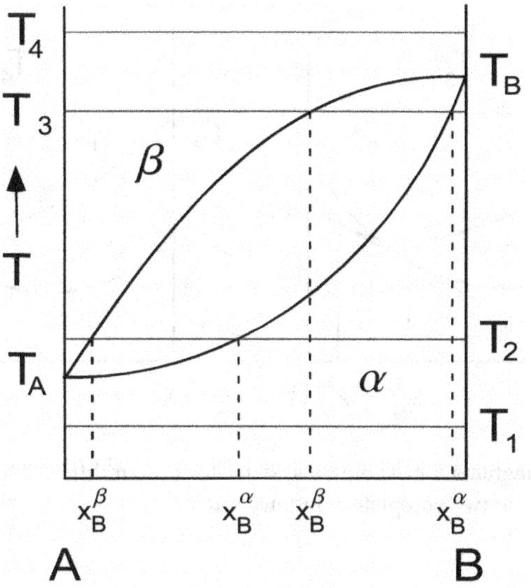

Fig. 6.23 *T-x* diagram for the system *A-B* with complete miscibility of the components in the phases α and β. x_B^α and x_B^β give the concentration of the component *B* in the coexisting phases α and β, respectively.

Fig. 6.23 shows a *T-x* diagram for a system where the end-members *A* and *B* form two complete solutions (α and β). The phase transformation α to β for the end-member A takes place at temperature T_A and that for the end-member B at T_B. The relationship between this diagram and the Gibbs free energy as a function of temperature and composition is shown in Fig. 6.24.

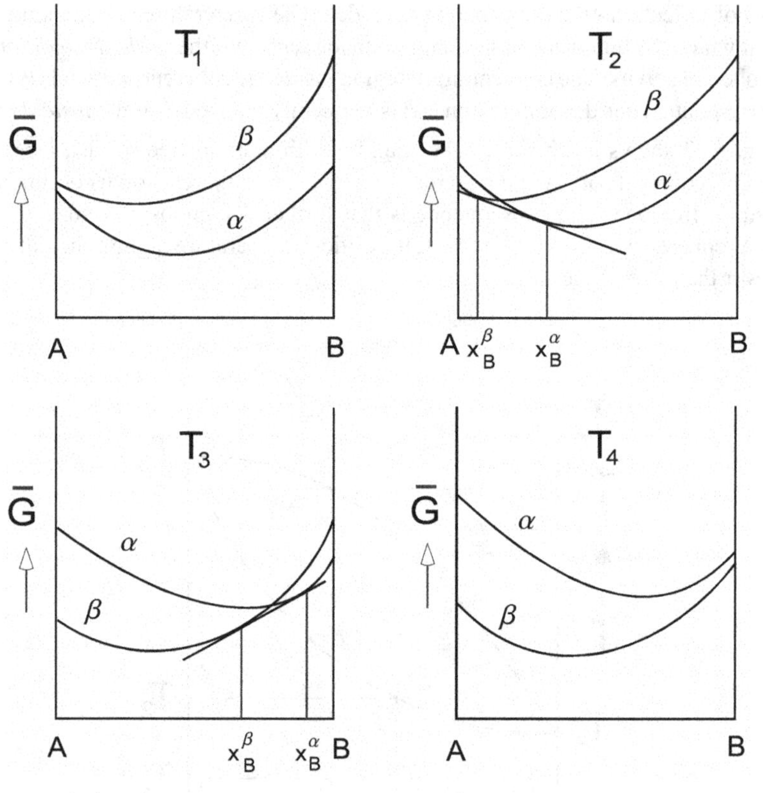

Fig. 6.24 $\overline{G}$-x diagrams for the binary system A-B at four different temperatures. The end-members form two complete solutions α and β.

At T_1, which is lower than T_A, the $\overline{G}(x)$ curve for the phase α is below that for the phase β over the entire compositional region. Hence, phase α is the only stable modification at this temperature. At T_2, the $\overline{G}(x)$ curves of the phases α and β intersect. The points of tangency of the common tangent to the curves give the composition of the coexisting phases α and β at the temperature T_2. In the compositional region between $x_B = 0$ and $x_B = x_B^\beta$, the $\overline{G}(x)$ curve of phase β is below that of phase α. Phase β is, therefore, the stabile phase over this compositional region at temperature

T_2. In the region $x_B^{\alpha} \le x_B \le 1$ the Gibbs free energy curve of phase α is lower than that of phase β. Phase α is, therefore, at this temperature the stabile modification. At T_3 a similar arrangement of curves is observed. The only difference is that their intersection is now closer to the pure end-member component B. Lastly, at T_4, the $\overline{G}(x)$ curve of phase β occurs under that of α over the entire compositional region. At this temperature phase β is, therefore, the only stable modification.

6.3.1 Calculation of phase diagrams from thermodynamic data

Critical mixing conditions, binodal, spinodal

At the critical mixing conditions the binodal and spinodal curves coincide. The spinodes are the inflection points of the $\overline{G}(x)$ curve and mark the transition between the concave and convex segments of the curve. Therefore, the following relationship holds:

$$\left(\frac{\partial^2 \overline{G}}{\partial x^2}\right)_{P,T} = 0. \tag{6.50}$$

At the spinode located on left side of the hump of the Gibbs free energy curve, the third derivative is negative because the second derivative changes from positive to negative. The spinode on the right side, however, is the point where the second derivative changes from negative to positive. Here, the third derivative is, therefore, positive. At the critical conditions the two spinodes coincide and the third derivative must be zero in order to fulfill the conditions for both spinodes simultaneously, i.e.

$$\left(\frac{\partial^3 \overline{G}}{\partial x^3}\right)_{P,T} = 0. \tag{6.51}$$

In order to formulate the critical condition using the chemical potentials and the activities, one has to start with $\overline{G}(x)$ as a function of composition, which reads:

$$\begin{aligned}\overline{G}(x) = {} & (1-x_B)\mu_A^{\mathrm{o}} + x_B\mu_B^{\mathrm{o}} + RT[(1-x_B)\ln(1-x_B) + x_B\ln x_B \\ & + (1-x_B)\ln\gamma_A + x_B\ln\gamma_B].\end{aligned} \tag{6.52}$$

Substituting the excess Gibbs free energy of mixing, $\Delta_m\overline{G}^{ex}$, for the last two terms in Eq. (6.52) yields:

$$\overline{G}(x) = (1-x_B)\mu_A^{\circ} + x_B\mu_B^{\circ} + RT[(1-x_B)\ln(1-x_B) + x_B\ln x_B] + \Delta_m\overline{G}^{ex}. \quad (6.53)$$

Consecutive differentiation of $\overline{G}(x)$ with respect to the mole fraction x_B gives:

$$\frac{\partial\overline{G}}{\partial x_B} = \frac{\partial(\Delta_m\overline{G}^{ex})}{\partial x_B} - \mu_A^{\circ} + \mu_B^{\circ} + RT\ln\frac{x_B}{1-x_B}, \quad (6.54)$$

$$\frac{\partial^2\overline{G}}{\partial x_B^2} = \frac{\partial^2(\Delta_m\overline{G}^{ex})}{\partial x_B^2} + \frac{RT}{x_B(1-x_B)}, \quad (6.55)$$

and

$$\frac{\partial^3\overline{G}^{ex}}{\partial x_B^3} = \frac{\partial^3(\Delta_m\overline{G}^{ex})}{\partial x_B^3} + \frac{RT(2x_B-1)}{x_B^2(1-x_B)^2}. \quad (6.56)$$

From Eqs. (6.50) and (6.51) it follows that

$$\frac{\partial^2(\Delta_m\overline{G}^{ex})}{\partial x_B^2} = -\frac{RT}{x_B(1-x_B)} \quad (6.57)$$

and

$$\frac{\partial^3(\Delta_m\overline{G}^{ex})}{\partial x_B^3} = -\frac{RT(2x_B-1)}{x_B^2(1-x_B)^2}. \quad (6.58)$$

In the case of *simple solutions,* the excess Gibbs free energy can be expressed, according to Eq. (5.134), using the interaction parameter W^G as:

$$\Delta_m\overline{G}^{ex} = (1-x_B)x_B W^G. \quad (6.59)$$

The second derivative of the excess Gibbs free energy is then:

$$\frac{\partial^2(\Delta_m \overline{G}^{ex})}{\partial x_B^2} = -2W^G, \tag{6.60}$$

and the third derivative is:

$$\frac{\partial^3(\Delta_m \overline{G}^{ex})}{\partial x_B^3} = 0. \tag{6.61}$$

Combining Eqs. (6.60) and (6.61) with (6.57) and (6.58) yields:

$$\frac{RT}{x_B(1-x_B)} = 2W^G \tag{6.62}$$

and

$$\frac{RT(2x_B-1)}{x_B^2(1-x_B)^2} = 0. \tag{6.63}$$

From Eq. (6.63) the critical mole fraction $x_B^c = 0.5$ is obtained and inserting this value in Eq. (6.62) gives the critical temperature, T_c as:

$$T_c = \frac{W^G}{2R}. \tag{6.64}$$

A reordering of terms in Eq. (6.64) gives:

$$\frac{W^G}{RT_c} = 2. \tag{6.65}$$

The meaning of Eq. (6.65) can be demonstrated using the activity/mole fraction relationship.

In the case of a simple binary solution, the activity of component B as a function of mole fraction is, according to Eqs. (5.82) and (5.127), given by

$$a_B = x_B \exp\left\{\frac{W^G}{RT}(1-x_B)^2\right\}, \tag{6.66}$$

where K_o in Eq. (5.127) is replaced by W^G.

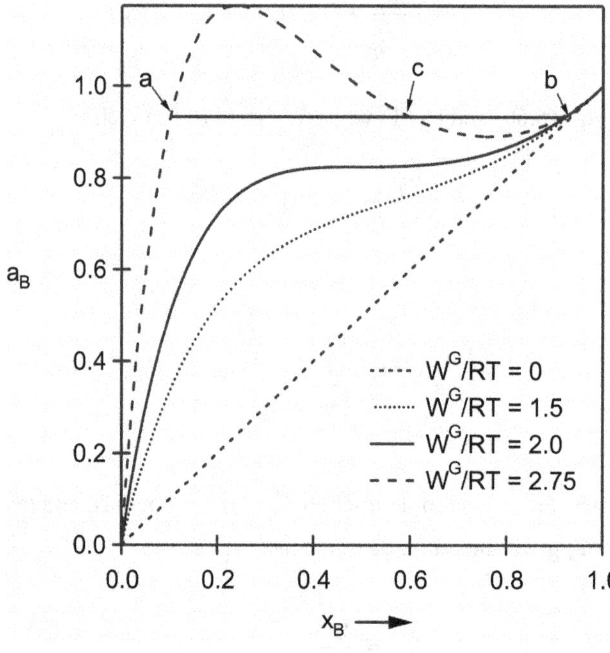

Fig. 6.25 Activity as a function of mole fraction for a simple binary solution with different values of W^G/RT. The solid curve gives the activity of the component B at the critical temperature T_c. The composition c is metastable. A solution with this composition decomposes into two phases (a and b).

Fig. 6.25 shows a plot of the thermodynamic activity, a_B, vs. mole fraction, x_B, for different values of W^G/RT. At temperatures higher than the critical temperature, T_c, W^G/RT is smaller than 2 and the activity increases monotonously with increasing mole fraction. Here, the miscibility between the end members is complete. At the critical temperature W^G/RT is equal to 2 and the $a_B(x)$-curve has an inflection point with a horizontal tangent. At temperatures below the critical temperature, W^G/RT is larger than 2 and the activity curve has a S like form. Initially, the activity increases with increasing mole fraction of B until it reaches a maximum value and then it decreases and after reaching a minimum value it slightly increases again. As a result, three different compositions can have the same activity. Two of them give the compositions of the coexisting stable phases into which a solution decomposes (a and b in Fig. 6.25). The composition c is metastable. For the case $W^G/RT = 0$, the activity is equal to the mole fraction over the entire compositional region, i.e., the activity coefficient equals 1 and consequently, the excess Gibbs free energy of mixing is zero. The solution is ideal and ideal mixtures can not have a miscibility gap.

The binodal compositions can be calculated using the conditions of chemical equilibrium, that require the chemical potentials of the components to be equal in the coexisting phases, i.e:

$$\mu_A^\alpha = \mu_A^\beta \tag{6.67}$$

and

$$\mu_B^\alpha = \mu_B^\beta. \tag{6.68}$$

In the case of a simple solution, Eqs. (6.67) and (6.68) can be written as follows:

$$\mu_A^{\mathrm{o}} + RT\ln(1 - x_B^\alpha) + (x_B^\alpha)^2 W^G = \mu_A^{\mathrm{o}} + RT\ln(1 - x_B^\beta) + (x_B^\beta)^2 W^G \tag{6.69}$$

and

$$\mu_B^{\mathrm{o}} + RT\ln x_B^\alpha + (1 - x_B^\alpha)^2 W^G = \mu_B^{\mathrm{o}} + RT\ln x_B^\beta + (1 - x_B^\beta)^2 W^G. \tag{6.70}$$

For the coexisting phases, the chemical potentials of the components A and B refer to the same standard state (pure end-member). Therefore, Eqs. (6.69) and (6.70) can be rewritten as:

$$RT\ln(1 - x_B^\alpha) + (x_B^\alpha)^2 W^G = RT\ln(1 - x_B^\beta) + (x_B^\beta)^2 W^G \tag{6.71}$$

and

$$RT\ln x_B^\alpha + (1 - x_B^\alpha)^2 W^G = RT\ln x_B^\beta + (1 - x_B^\beta)^2 W^G. \tag{6.72}$$

In the case of a simple symmetrical solution it holds that

$$x_B^\alpha + x_B^\beta = 1, \tag{6.73}$$

so that Eqs. (6.71) and (6.72) simplify to:

$$RT\ln x_B^\alpha + (1 - x_B^\alpha)^2 W^G = RT\ln(1 - x_B^\alpha) + (x_B^\alpha)^2 W^G. \tag{6.74}$$

Reordering of the terms in Eq. (6.74) and solving for T yields:

$$T = \frac{W^G}{R}\frac{(2x_B^\alpha - 1)}{\ln\frac{x_B^\alpha}{1 - x_B^\alpha}}. \tag{6.75}$$

Considering the relationship given in Eq. (6.64), Eq. (6.75) can also be written in terms of the critical temperature, T_c, i.e.:

$$T = 2T_c\frac{(2x_B^\alpha - 1)}{\ln\frac{x_B^\alpha}{1 - x_B^\alpha}}. \tag{6.76}$$

From Eq. (6.62) follows that:

$$x_B^{\alpha,\beta} = \frac{1 \pm \sqrt{1 - 2RT/W^G}}{2} \tag{6.77}$$

or

$$x_B^{\alpha,\beta} = \frac{1 \pm \sqrt{1 - T/T_c}}{2}, \tag{6.78}$$

if the term $2R/W^G$ is replaced by $1/T_c$.

Fig. 6.26 shows calculated binode and spinode for a hypothetical system *A*-*B* with a critical temperature of mixing of $T_c = 973$ K. A simple mixture model is assumed to describe the mixing behavior and the value of the interaction parameter W^G is 16.18 kJmol^{-1}.

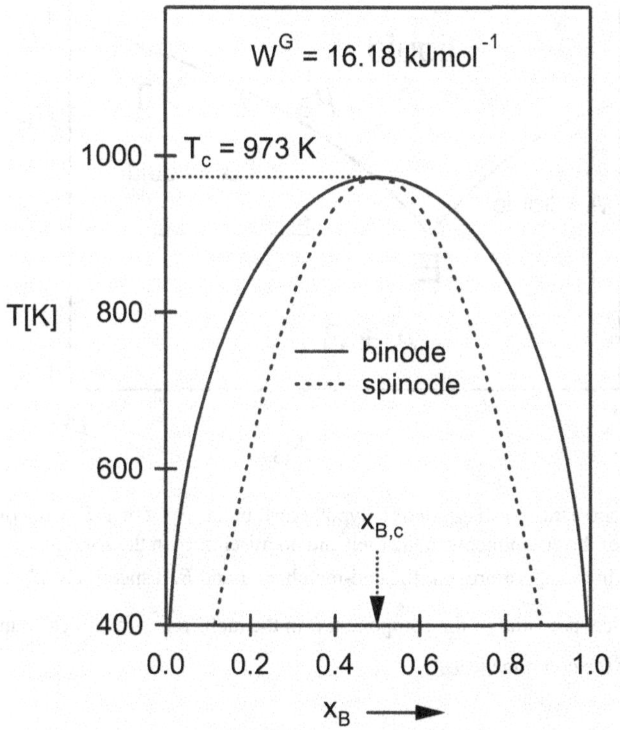

Fig. 6.26 Calculated binode and spinode in a hypothetical system *A-B*, where the components *A* and *B* form a simple mixture. The interaction parameter, W^G, is assumed to have a value of 16.18 $kJmol^{-1}$. The critical temperature of mixing is 973 K. $x_{B,c}$ = critical mole fraction of the component *B*.

Simple eutectic system

Consider the *T-x* diagram for a hypothetical binary system *A-B* as shown in Fig. 6.27. In this system, the components *A* and *B* are completely soluble in the liquid phase but are immiscible in the solid phases. The two boundaries, that separate the stability field of the melt from the areas with two coexisting phases (melt and end-member *A* and melt and end-member *B*, respectively) are called the *liquidus*. The point, where the two curves intersect designates the lowest melting temperature in the system. It is referred to as the *eutectic temperature*. Consequently, the corresponding composition is termed the *eutectic composition*.

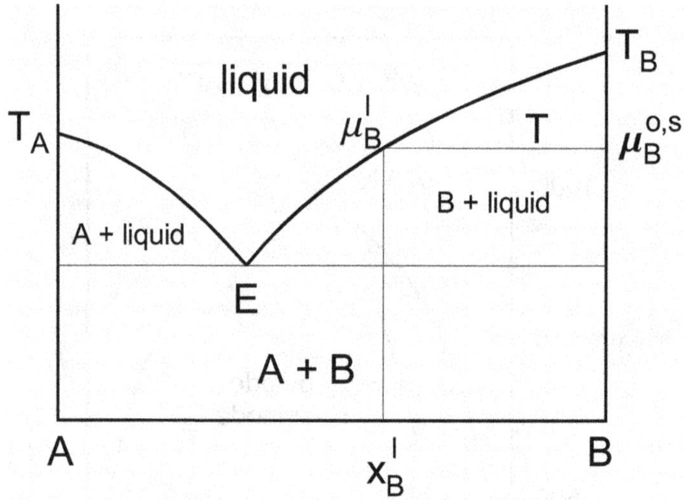

Fig. 6.27 Schematic *T-x* diagram of a hypothetical binary system *A-B* with complete miscibility of the components in the melt and no miscibility in the solid phase. T_A and T_B are the melting temperatures for the end-members *A* and *B*, respectively. *E* = eutectic. μ_B^l = chemical potential of the component *B* in the melt, $\mu_B^{o,s}$ = chemical standard potential of the component *B*.

At temperature *T* the pure solid end-member *B* and a melt with composition x_B^l coexist. The condition for the thermodynamic equilibrium requires that the chemical potentials of the component *B* are equal in both liquid and solid phase, i.e.:

$$\mu_B^l = \mu_B^{o,s} \tag{6.79}$$

or

$$\mu_B^{o,l} + RT\ln a_B^l = \mu_B^{o,s}, \tag{6.80}$$

where $\mu_B^{o,l}$ and a_B^l designate the standard chemical potential of component *B* and its activity in the melt, respectively. Assuming that the melt is an ideal solution, the activity of component *B* in Eq. (6.80) can be replaced by the mole fraction, x_B^l, and the equilibrium condition reads:

$$\mu_B^{o,l} + RT\ln x_B^l = \mu_B^{o,s}. \tag{6.81}$$

In order to maintain thermodynamic equilibrium after infinitesimal changes in temperature and pressure, changes in the chemical potentials of the component B must be equal in both phases, so that:

$$d\mu_B^l = d\mu_B^{o,s}. \tag{6.82}$$

A T-x diagram shows the phase relationships as a function of temperature at constant pressure. Therefore, in order to calculate such a phase diagram using thermodynamic data, the variation in the chemical potential with temperature must be known. According to Eqs. (5.16) and (5.17) the derivatives of the chemical potential with respect to temperature yield the entropy or, if the chemical potential is divided by T, the enthalpy. For the sake of convenience, the enthalpy is preferred over the entropy and the relationship given in Eq. (5.99) is used. Dividing Eq. (6.81) by T and differentiating with respect to temperature yields:

$$-\frac{\boldsymbol{H}_B^{o,l}}{T^2}dT + R d\ln x_B^l = -\frac{\boldsymbol{H}_B^{o,s}}{T^2}dT \tag{6.83}$$

or

$$d\ln x_B^l = \frac{\boldsymbol{H}_B^{o,l} - \boldsymbol{H}_B^{o,s}}{RT^2}dT. \tag{6.84}$$

To calculate the liquidus curve, Eq. (6.84) has to be integrated over the temperature range between the melting temperature of component B, T_B, and the temperature of interest, T:

$$\int_{x_B(T_B)}^{x_B(T)} d\ln x_B^l = \int_{T_B}^{T} \frac{\boldsymbol{H}_B^{o,l} - \boldsymbol{H}_B^{o,s}}{RT^2}dT. \tag{6.85}$$

Considering the fact that x_B equals 1 at T_B and assuming that the difference $\boldsymbol{H}_B^{o,l} - \boldsymbol{H}_B^{o,s}$ is independent of temperature, integration of Eq. (6.85) gives:

$$\ln x_B^l = -\frac{H_B^{o,l} - H_B^{o,s}}{R}\left(\frac{1}{T} - \frac{1}{T_B}\right) = -\frac{\Delta_{fus}H_B}{R}\left(\frac{T_B - T}{TT_B}\right), \tag{6.86}$$

where $\Delta_{fus}H_B$ is the enthalpy of melting for the pure component B.

The liquidus curve on the left side of the eutectic (see Fig. 6.27) can be calculated in a similar way. In this case, the equilibrium condition at temperature T is described by the following expression:

$$\mu_A^{o,l} + RT\ln(1 - x_B^l) = \mu_A^{o,s}, \tag{6.87}$$

where $\mu_A^{o,l}$ and $\mu_A^{o,s}$ designate the standard chemical potentials of component A in the melt and in the solid end-member A, respectively. Dividing Eq. (6.87) by T and differentiating with respect to temperature gives:

$$-\frac{H_A^{o,l}}{T^2}dT + Rd\ln(1 - x_B^l) = -\frac{H_A^{o,s}}{T^2}dT. \tag{6.88}$$

Integrating Eq. (6.88) over the temperature range between T and T_A leads to:

$$\ln(1 - x_B^l) = -\frac{\Delta_{fus}H_A}{R}\left(\frac{T_A - T}{TT_A}\right), \tag{6.89}$$

where $\Delta_{fus}H_A$ designates the enthalpy of melting for the pure component A.

At the temperature of the eutectic, T_E, the value of the mole fraction x_B^l is the same in Eq. (6.86) and Eq. (6.89). Hence, the two equations can be combined to give:

$$\exp\left\{-\frac{\Delta_{fus}H_A}{R}\left(\frac{T_A - T_E}{TT_A}\right)\right\} + \exp\left\{-\frac{\Delta_{fus}H_B}{R}\left(\frac{T_B - T_E}{TT_B}\right)\right\} = 1. \tag{6.90}$$

Eq. (6.90) cannot be solved analytically. In order to calculate the temperature of the eutectic, the trial-and-error or Newton's method has to be applied.

Example: Consider the binary system $CaMgSi_2O_6$ - $CaAl_2Si_2O_8$. At 0.1 MPa, the melting temperatures of diopside and anorthite are 1668 K and 1830 K, respectively. The enthalpy of melting for diopside is 137.7 $kJmol^{-1}$ and that one for anorthite is 133.0 kJmol^{-1}. (All data are taken from Robie and Hemingway 1995).

The mole fraction of $CaAl_2Si_2O_8$ in the melt that coexists with pure diopside, $CaMgSi_2O_6$, is calculated using the exponential form of Eq. (6.89), i.e.

$$x^{l}_{CaAl_2Si_2O_8} = 1 - \exp\left\{-\frac{137700\text{Jmol}^{-1}}{8.3144\text{Jmol}^{-1}\text{K}^{-1}}\left(\frac{1668\text{K} - T[\text{K}]}{1668\text{K} \times T[\text{K}]}\right)\right\}.$$

Eq. (6.86) is used to calculate the mole fraction of $CaAl_2Si_2O_8$ in the melt, which is at thermodynamic equilibrium with pure anorthite, namely:

$$x^{l}_{CaAl_2Si_2O_8} = \exp\left\{-\frac{133000\text{Jmol}^{-1}}{8.3144\text{Jmol}^{-1}\text{K}^{-1}}\left(\frac{1830\text{K} - T[\text{K}]}{1830\text{K} \times T[\text{K}]}\right)\right\}.$$

The results of the calculations are shown in Fig. 6.28.

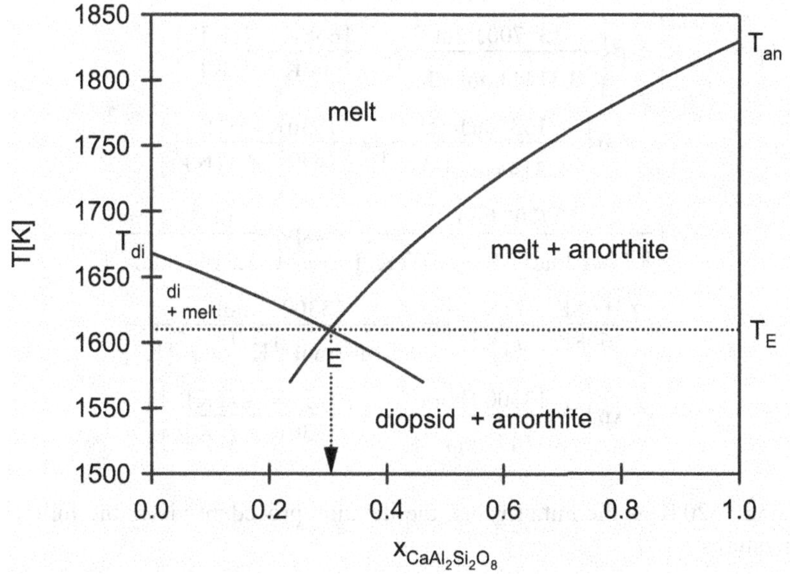

Fig. 6.28 T - x diagram for the system $CaMgSi_2O_6$ - $CaAl_2Si_2O_8$ calculated using thermodynamic data of Robie and Hemingway (1995). T_{di} = melting temperature of diopside, T_{an} = melting temperature of anorthite, E = eutectic, T_E = eutectic temperature, arrow marks the eutectic composition (30.22 mol% $CaAl_2Si_2O_8$).

The eutectic temperature is calculated using the Newton's method, where the solution to an equation is found by an iteration procedure according to the formula:

$$x_1 = x_o - \frac{f(x)}{f'(x)}, \tag{6.91}$$

where x_o is the initial value and x_1 is the result of the first cycle that is then used as the input value for the second cycle etc. The calculation is repeated until the difference between two consecutive values of x_i is close to zero. Applying this method to our example gives:

$$T_1 = T_o - \frac{\exp\left\{-\frac{\Delta_{fus}\boldsymbol{H}_A}{R}\left(\frac{T_A - T}{TT_A}\right)\right\} + \exp\left\{-\frac{\Delta_{fus}\boldsymbol{H}_B}{R}\left(\frac{T_B - T}{TT_B}\right)\right\} - 1}{\frac{\Delta_{fus}\boldsymbol{H}_A}{RT^2}\exp\left\{-\frac{\Delta_{fus}\boldsymbol{H}_A}{R}\left(\frac{T_A - T}{TT_A}\right)\right\} + \frac{\Delta_{fus}\boldsymbol{H}_B}{RT^2}\exp\left\{-\frac{\Delta_{fus}\boldsymbol{H}_B}{R}\left(\frac{T_B - T}{TT_B}\right)\right\}}$$

or

$$T_1 = T_o - \left[\exp\left\{-\frac{137700\,\mathrm{Jmol^{-1}}}{8.3144\,\mathrm{Jmol^{-1}K^{-1}}} \times \left(\frac{1668\mathrm{K} - T[\mathrm{K}]}{1668\mathrm{K} \times T[\mathrm{K}]}\right)\right\}\right.$$
$$\left. + \exp\left\{-\frac{133000\,\mathrm{Jmol^{-1}}}{8.3144\,\mathrm{Jmol^{-1}K^{-1}}} \times \left(\frac{1830\mathrm{K} - T[\mathrm{K}]}{1830\mathrm{K} \times T[\mathrm{K}]}\right)\right\} - 1\right] /$$
$$\left[\frac{137700\,\mathrm{Jmol^{-1}}}{8.3144\,\mathrm{Jmol^{-1}K^{-1}} \times T^2[\mathrm{K^2}]} \times \exp\left\{-\frac{137700\,\mathrm{Jmol^{-1}}}{8.3144\,\mathrm{Jmol^{-1}K^{-1}}}\right.\right.$$
$$\left. \times \left(\frac{1668\mathrm{K} - T[\mathrm{K}]}{1668\mathrm{K} \times T[\mathrm{K}]}\right)\right\} + \frac{133000\,\mathrm{Jmol^{-1}}}{8.3144\,\mathrm{Jmol^{-1}K^{-1}} \times T^2[\mathrm{K^2}]}$$
$$\left. \times \exp\left\{-\frac{133000\,\mathrm{Jmol^{-1}}}{8.3144\,\mathrm{Jmol^{-1}K^{-1}}} \times \left(\frac{1830\mathrm{K} - T[\mathrm{K}]}{1830\mathrm{K} \times T[\mathrm{K}]}\right)\right\}\right].$$

Using 1620 K as the initial value, the iteration procedure yields the following temperatures:

$T_o = 1620$ K (initial value),
$T_1 = 1610$ K,
$T_2 = 1609$ K and
$T_3 = 1609$ K.

As can be seen, T_2 and T_3 are the same. Hence, 1609 K is the temperature of the eutectic.

Binary system with complete miscibility of the components in two phases

Fig. 6.29 shows a schematic *T-x* diagram of a binary system in which the two components *A* and *B* form two complete solutions termed α and β. T_A and T_B mark the transition temperatures for the end-members *A* and *B*, respectively.

At some temperature, *T*, lying between the melting points of the end-members, the phases α and β are in equilibrium with one another, if the chemical potentials of the respective component are equal in both phases, i.e.

$$\mu_A^{\alpha} = \mu_A^{\beta} \tag{6.92}$$

and

$$\mu_B^{\alpha} = \mu_B^{\beta}. \tag{6.93}$$

Phase α, as well as phase β, are solutions and, therefore, Eqs. (6.92) and (6.93) can be extended to:

$$\mu_A^{o,\alpha} + RT\ln a_B^{\alpha} = \mu_A^{o,\beta} + RT\ln a_B^{\beta} \tag{6.94}$$

and

$$\mu_B^{o,\alpha} + RT\ln a_B^{\alpha} = \mu_B^{o,\beta} + RT\ln a_B^{\beta}, \tag{6.95}$$

respectively.

In the case that the solutions α and β are ideal, the activities can be replaced by the mole fractions and one obtains:

$$\mu_A^{o,\alpha} + RT\ln(1 - x_B^{\alpha}) = \mu_A^{o,\beta} + RT\ln(1 - x_B^{\beta}) \tag{6.96}$$

and

$$\mu_B^{o,\alpha} + RT\ln x_B^{\alpha} = \mu_B^{o,\beta} + RT\ln x_B^{\beta}. \tag{6.97}$$

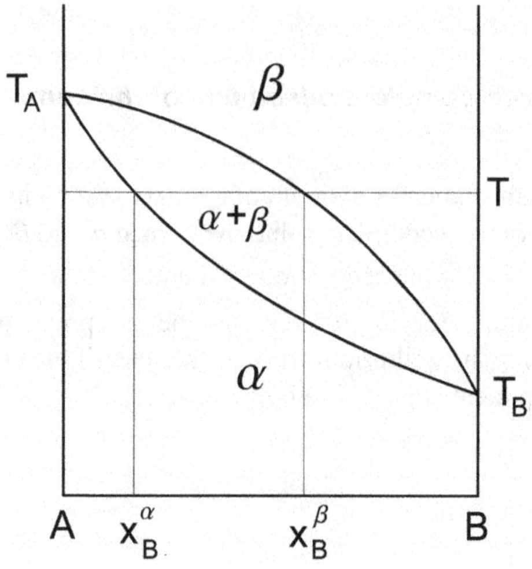

Fig. 6.29 A hypothetical binary system *A-B* with complete miscibility of the components in the two phases α and *β*. T_A marks the melting temperature of component *A* and T_B that of component *B*. The two 'single phase' stability fields of phases α and β are separated by the 'two phase' field where the phases *α* and *β* coexist. x_B^{α} and x_B^{β} are the mole fractions of the coexisting phases at temperature *T*.

Should thermodynamic equilibrium remain when temperature and pressure change, the changes in chemical potentials of the components in the coexisting phases must be equal, i.e.,

$$d\mu_A^{\alpha} = d\mu_A^{\beta} \tag{6.98}$$

and

$$d\mu_B^{\alpha} = d\mu_B^{\beta}. \tag{6.99}$$

In order to calculate a *T-x* diagram, the derivatives of the chemical potentials with respect to temperature are required. Because the enthalpy is a more convenient function for the calculation of a *T-x* diagram than the entropy, the derivative of the quotient μ/T with respect to temperature is taken again, so that:

$$-\frac{\boldsymbol{H}_A^{\text{o},\alpha}}{T^2}dT + Rd\ln(1-x_B^{\alpha}) = -\frac{\boldsymbol{H}_A^{\text{o},\beta}}{T^2}dT + Rd\ln(1-x_B^{\beta}) \tag{6.100}$$

and

$$-\frac{\boldsymbol{H}_B^{\text{o},\alpha}}{T^2}dT + Rd\ln x_B^{\alpha} = -\frac{\boldsymbol{H}_B^{\text{o},\beta}}{T^2}dT + Rd\ln x_B^{\beta} \tag{6.101}$$

is obtained.

Rearranging Eqs. (6.100) and (6.101) and solving them for the natural logarithm of the composition yields:

$$d\ln\frac{(1-x_B^{\alpha})}{(1-x_B^{\beta})} = -\frac{\boldsymbol{H}_A^{\text{o},\beta} - \boldsymbol{H}_A^{\text{o},\alpha}}{RT^2}dT = -\frac{\Delta_{tr}\boldsymbol{H}_A^{\alpha/\beta}}{RT^2}dT \tag{6.102}$$

and

$$d\ln\frac{x_B^{\alpha}}{x_B^{\beta}} = -\frac{\boldsymbol{H}_B^{\text{o},\beta} - \boldsymbol{H}_B^{\text{o},\alpha}}{RT^2}dT = -\frac{\Delta_{tr}\boldsymbol{H}_B^{\alpha/\beta}}{RT^2}dT. \tag{6.103}$$

In Eqs. (6.102) and (6.103) the terms $\Delta_{tr}\boldsymbol{H}_A^{\alpha/\beta}$ and $\Delta_{tr}\boldsymbol{H}_B^{\alpha/\beta}$ designate the enthalpy of the phase transformation of α to β for the end-members A and B, respectively.

In order to calculate the mole fractions of the components A and B in the phases α and β as a function of temperature, Eqs.(6.102) and (6.103) must be integrated. The limits of integration are given by the transformation temperature of the end-member and the temperature of interest. Considering that the mole fraction of the component B, x_B, is zero at temperature T_A and one at temperature T_B and assuming that the temperature dependence of the enthalpies of transformation is small and can, therefore, be neglected, the integration yields:

$$\ln\frac{(1-x_B^{\alpha})}{(1-x_B^{\beta})} = \frac{\Delta_{tr}\boldsymbol{H}_A^{\alpha/\beta}}{R}\left(\frac{T_A - T}{TT_A}\right) \tag{6.104}$$

for the component A and

$$\ln\frac{x_B^{\alpha}}{x_B^{\beta}} = \frac{\Delta_{tr}\boldsymbol{H}_B^{\alpha/\beta}}{R}\left(\frac{T_B - T}{TT_B}\right) \tag{6.105}$$

for the component B.

The exponential forms of the Eqs. (6.104) and (6.105) read:

$$\frac{(1-x_B^{\alpha})}{(1-x_B^{\beta})} = \exp\left\{\frac{\Delta_{tr}\boldsymbol{H}_A^{\alpha/\beta}}{R}\left(\frac{T_A - T}{TT_A}\right)\right\} \tag{6.106}$$

and

$$\frac{x_B^{\alpha}}{x_B^{\beta}} = \exp\left\{\frac{\Delta_{tr}\boldsymbol{H}_B^{\alpha/\beta}}{R}\left(\frac{T_B - T}{TT_B}\right)\right\}. \tag{6.107}$$

A reordering of Eq. (6.107) leads to the expression for the mole fraction x_B^{α}, namely:

$$x_B^{\alpha} = x_B^{\beta}\exp\left\{\frac{\Delta_{tr}\boldsymbol{H}_B^{\alpha/\beta}}{R}\left(\frac{T_B - T}{TT_B}\right)\right\}. \tag{6.108}$$

Substituting the expression in Eq. (6.108) for the mole fraction x_B^{α} in Eq. (6.106) gives:

$$1 - x_B^{\beta}\exp\left\{\frac{\Delta_{tr}\boldsymbol{H}_B^{\alpha/\beta}}{R}\left(\frac{T_B - T}{TT_B}\right)\right\} = (1 - x_B^{\beta})\exp\left\{\frac{\Delta_{tr}\boldsymbol{H}_A^{\alpha/\beta}}{R}\left(\frac{T_A - T}{TT_A}\right)\right\}. \tag{6.109}$$

Solving Eq. (6.109) for x_B^{β} yields:

$$x_B^{\beta} = \frac{\exp\left\{\frac{\Delta_{tr}\boldsymbol{H}_A^{\alpha/\beta}}{R}\left(\frac{T_A - T}{TT_A}\right)\right\} - 1}{\exp\left\{\frac{\Delta_{tr}\boldsymbol{H}_A^{\alpha/\beta}}{R}\left(\frac{T_A - T}{TT_A}\right)\right\} - \exp\left\{\frac{\Delta_{tr}\boldsymbol{H}_B^{\alpha/\beta}}{R}\left(\frac{T_B - T}{TT_B}\right)\right\}}. \tag{6.110}$$

Eq. (6.110) gives the mole fraction of component B vs. temperature curve for the phase β. The corresponding curve for the phase α is calculated according to Eq. (6.108) using the mole fractions x_B^β obtained from Eq. (6.110).

Example: Forsterite, Mg_2SiO_4, and fayalite, Fe_2SiO_4, are completely soluble in the liquid as well as in the solid phase. At ambient pressure, forsterite melts at 2174 K (Richet et al. 1993) and fayalite at 1490 K (Stebbins and Carmichael 1984). The enthalpies of melting are 142 $kJmol^{-1}$ (Richet et al. 1993) and 89.3 $kJmol^{-1}$ (Stebbins and Carmichael 1984) for forsterite and fayalite, respectively. The calculation of the liquidus curve as a function of temperature is carried out by applying Eq. (6.110) as follows:

$$x_{Fe_2SiO_4}^{l} = \left[\exp\left\{\frac{142000\,\mathrm{Jmol}^{-1}}{2\times 8.3144\,\mathrm{Jmol}^{-1}\mathrm{K}^{-1}}\times\left(\frac{2174\mathrm{K}-T[\mathrm{K}]}{2174\mathrm{K}\times T[\mathrm{K}]}\right)\right\}-1\right]\Big/$$
$$\left[\exp\left\{\frac{142000\,\mathrm{Jmol}^{-1}}{2\times 8.3144\,\mathrm{Jmol}^{-1}\mathrm{K}^{-1}}\times\left(\frac{2174\mathrm{K}-T[\mathrm{K}]}{2174\mathrm{K}\times T[\mathrm{K}]}\right)\right\}\right.$$
$$\left.-\exp\left\{\frac{89300\,\mathrm{Jmol}^{-1}}{2\times 8.3144\,\mathrm{Jmol}^{-1}\mathrm{K}^{-1}}\times\left(\frac{1490\mathrm{K}-T[\mathrm{K}]}{1490\mathrm{K}\times T[\mathrm{K}]}\right)\right\}\right].$$

In the expression given above, the enthalpies of melting are divided by 2 because a *one cation* basis is used in the calculation. This means that $(Mg,Fe)Si_{0.5}O_2$ is taken as the chemical formula of olivine and not $(Mg,Fe)_2SiO_4$ as normally done. The enthalpy of melting given in the thermodynamic tables refers to the full formula unit and, therefore, half of these value has to be used in the calculation of the phase boundaries.

The solidus curve as a function of temperature is calculated using Eq. (6.108). Inserting the numerical data in this equation, the calculation reads:

$$x_{Fe_2SiO_4}^{s} = x_{Fe_2SiO_4}^{l}\exp\left\{\frac{89300\,\mathrm{Jmol}^{-1}}{2\times 8.3144\,\mathrm{Jmol}^{-1}\mathrm{K}^{-1}}\left(\frac{1490\mathrm{K}-T[\mathrm{K}]}{1490\mathrm{K}\times T[\mathrm{K}]}\right)\right\}.$$

The results of the calculation are shown graphically in Fig. 6.30.

In order to calculate a P-x diagram for a hypothetical binary system A-B, the derivatives of the chemical potential with respect to pressure are required. In the case of an infinitesimally small changes in pressure at constant temperature the equilibrium between two ideal solid solutions (e.g. α and β) is maintained if the conditions:

$$V_A^\alpha dP + RTd\ln(1-x_B^\alpha) = V_A^\beta dP + RTd\ln(1-x_B^\beta) \qquad (6.111)$$

and

$$V_B^{\alpha} dP + RT d\ln x_B^{\alpha} = V_B^{\beta} dP + RT d\ln x_B^{\beta} \tag{6.112}$$

are fulfilled.

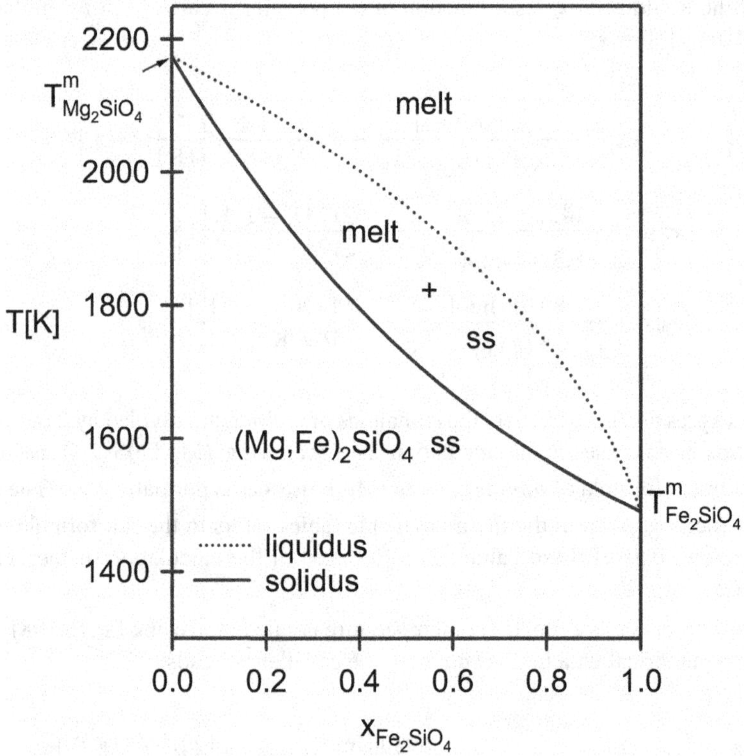

Fig. 6.30 *T-x* diagram of the binary system Mg_2SiO_4 - Fe_2SiO_4 at 0.1 MPa calculated using the heats of fusion of the end-members. $T^m_{Mg_2SiO_4}$ = melting temperature of forsterite, $T^m_{Fe_2SiO_4}$ = melting temperature of fayalite.

Rearranging terms in Eqs. (6.111) and (6.112) leads to

$$d\ln\frac{(1-x_B^{\alpha})}{(1-x_B^{\beta})} = \frac{V_A^{\beta}-V_A^{\alpha}}{RT}dP \tag{6.113}$$

and

$$d\ln\frac{x_B^{\alpha}}{x_B^{\beta}} = \frac{V_B^{\beta}-V_B^{\alpha}}{RT}dP, \tag{6.114}$$

where the terms $V_A^{\beta}-V_A^{\alpha}$ and $V_B^{\beta}-V_B^{\alpha}$ give the changes in volume of the pure components A and B associated with the phase transformation from α to β, respectively.

In order to obtain the compositions of the coexisting phases as a function of pressure Eqs. (6.113) and (6.114) must be integrated. The limits of integration are the transformation pressure of the respective end-member component and the pressure of interest. At P_A and P_B the mole fraction of either component is zero and one, respectively.The integration therefore yields:

$$\ln\frac{(1-x_B^{\alpha})}{(1-x_B^{\beta})} = \frac{V_A^{\beta}-V_A^{\alpha}}{RT}(P-P_A) \tag{6.115}$$

and

$$\ln\frac{x_B^{\alpha}}{x_B^{\beta}} = \frac{V_B^{\beta}-V_B^{\alpha}}{RT}(P-P_B), \tag{6.116}$$

where P_A is the transformation pressure of the pure end-member component A and P_B that one of the pure end-member B.

In the exponential form Eqs. (6.115) and (6.116) read:

$$\frac{(1-x_B^{\alpha})}{(1-x_B^{\beta})} = \exp\left\{\frac{\Delta_{tr}V_A^{\alpha/\beta}}{RT}(P-P_A)\right\} \tag{6.117}$$

and

$$\frac{x_B^\alpha}{x_B^\beta} = \exp\left\{\frac{\Delta_{tr}V_B^{\alpha/\beta}}{RT}(P - P_B)\right\}, \tag{6.118}$$

where $\Delta_{tr}V_A^{\alpha/\beta}$ and $\Delta_{tr}V_B^{\alpha/\beta}$ designate the volumes of transition of the components A and B, respectively.

Rearranging Eq. (6.118) gives:

$$x_B^\alpha = x_B^\beta \exp\left\{\frac{\Delta_{tr}V_B^{\alpha/\beta}}{RT}(P - P_B)\right\}. \tag{6.119}$$

Substituting Eq. (6.119) for x_B^α and rearranging the terms yields:

$$x_B^\beta = \frac{\exp\left\{\frac{\Delta_{tr}V_A^{\alpha/\beta}}{RT}(P - P_A)\right\} - 1}{\exp\left\{\frac{\Delta_{tr}V_A^{\alpha/\beta}}{RT}(P - P_A)\right\} - \exp\left\{\frac{\Delta_{tr}V_B^{\alpha/\beta}}{RT}(P - P_B)\right\}}. \tag{6.120}$$

Using Eq. (6.120) the mole fraction of component B in the phase β as a function of pressure can be calculated. The results can then be used to calculate the composition of the coexisting phase α according to Eq. (6.119).

6.4 Gibbs phase rule

The Gibbs phase rule gives the number of variables that can be changed independently of one another without changing the number of phases in a system. It defines the variance at equilibrium in the case where the number of components and the number of coexisting phases are known. The number of free variables is referred to as the *number of the degrees of freedom*. It is equal to the difference between the number of state variables and the number of equations relating these variables. The Gibbs phase rule can be derived as follows:

Consider a system containing C different components that occur in ϕ different phases. The thermodynamic state of each phase is determined by pressure, temperature and composition, where the latter is generally expressed in mole fraction. If all components occur in all phases, there are $\phi \times C$ mole fractions in total. In each phase, however, the mole fractions sum to 1 and, therefore, only

$$\phi(C-1) = \phi C - \phi \tag{6.121}$$

mole fractions are required to describe the composition of a system. One mole fraction of each phase is determined by the rest of the (C-1) mole fractions. In addition to the composition, i.e. mole fractions, pressure and temperature must be known in order to define the system at equilibrium. The total number of variables that are required to specify the thermodynamic state of a system is, therefore, given by:

$$\phi C - \phi + 2. \tag{6.122}$$

The variables are related to each other by the chemical potentials of the components, which, in the case of thermodynamic equilibrium, are equal in all phases. For a component i that is contained in the phases 1 through ϕ, for example, it holds that:

$$\overset{1}{\mu_i} = \overset{2}{\mu_i} = \overset{3}{\mu_i} = \ldots \overset{\phi}{\mu_i}.$$

In the case that a system contains ϕ different phases, (ϕ -1) independent pairs of equal chemical potential can be formulated. For example, if the component i occurs in three phases, only the pairs:

$$\overset{1}{\mu_i} = \overset{2}{\mu_i} \text{ and } \overset{2}{\mu_i} = \overset{3}{\mu_i}$$

are independent, while the equality

$$\overset{1}{\mu_i} = \overset{3}{\mu_i}$$

results automatically from the two relations that are given above. Generally, it holds that in a system containing C different components, there are in total $C(\phi$-1) independent equations interrelating the variables. In order to obtain the number of the degrees of freedom or variance, the number of independent equations has to be subtracted from the number of independent variables, i.e:

$$F = (\phi C - \phi + 2) - C(\phi - 1) \tag{6.123}$$

or

$$F = C + 2 - \phi, \tag{6.124}$$

which is the usual form of the Gibbs phase rule.

Depending on the number of degrees of freedom, equilibria are designated as *in-*

variant ($F = 0$), *univariant* ($F = 1$), or *divariant* ($F = 2$), etc. The largest number of coexisting phases is equal to $C + 2$.

Example: The one-component system Al_2SiO_5 is trimorphic. The three phases are kyanite, andalusite and sillimanite. If all three phases coexist, the number of the degrees of freedom is, according to the Gibbs rule, zero, namely

$$F = 1 + 2 - 3.$$

Hence, the equilibrium is invariant. The coexistence of kyanite, andalusite and sillimanite is only possible at one definite pressure and one definite temperature. In a *P-T*-diagram these two variables define a point, the so-called *triple point.* A change in either variable causes the loss of equilibrium and at least one phase disappears.

Although the Gibbs phase rule appears to be simple, its application to natural systems is in many cases problematical. A violation of the phase rule clearly indicates disequilibrium, but if it is obeyed, this does not prove a state of equilibrium.

The number of possible invariant, univariant, divariant, etc. equilibria can be calculated using a combinatorial formula, that determines in how many ways N elements can be arranged in groups of k elements. The formula reads:

$$\Gamma_{N,k} = \binom{N}{k} = \frac{N!}{k!(N-k)!}. \tag{6.125}$$

In our case, the total number of elements, N, corresponds to the largest number of coexisting phases in a system containing C components. As can be determined from the Gibbs phase rule, this number is equal to $C + 2$. Accordingly, the number of elements, k, contained in the group corresponds to the number of coexisting phases. Thus, one obtains:

$$\Gamma_{C+2,\phi} = \frac{(C+2)!}{\phi!(C+2-\phi)!} = \frac{(C+2)!}{\phi!F!}. \tag{6.126}$$

Example: In the one-component system Al_2SiO_5 the number of invariant triple points is:

$$\Gamma_{3,3} = \frac{3!}{3!0!} = 1.$$

The number of univariant curves, where two phases coexist is:

$$\Gamma_{3,2} = \frac{3!}{2!1!} = 3$$

and the number of divariant fields containing one phase is:

$$\Gamma_{3,1} = \frac{3!}{1!2!} = 3.$$

6.5 Problems

1. According to Akimoto et al. (1977), fayalite, Fe_2SiO_4, undergoes a phase transition from α-Fe_2SiO_4 (olivine) to γ-Fe_2SiO_4 (spinel) at 700°C and 52.1 kbar.

- Calculate the pressure of the phase transition at 1000°C using the Clausius-Clapeyron equation. Assume that entropy of transition and the volume of transition are temperature and pressure independent.

$$^{a}V_{ol}^{Fe_2SiO_4} = 46.31\ \text{cm}^3\text{mol}^{-1},$$

$$^{b}V_{sp}^{Fe_2SiO_4} = 42.04\ \text{cm}^3\text{mol}^{-1},$$

$$^{a}S_{ol}^{Fe_2SiO_4} = 151.00\ \text{Jmol}^{-1}\text{K}^{-1} \quad \text{and}$$

$$^{c}S_{sp}^{Fe_2SiO_4} = 131.03\ \text{Jmol}^{-1}\text{K}^{-1}.$$

[a]Holland and Powell (1998), [b]Marumo et al. (1977), [c]Navrotsky et al. (1979)

2. The interaction parameter of a hypothetical regular binary solution A-B has a value of 20 kJmol^{-1}.

- Calculate the critical temperature of mixing.
- Calculate the spinodal compositions at 700°C.

3. Calculate the melting temperature of a mechanical mixture consisting of 10 mol% sphene, $CaTiSiO_5$, and of 90 mol% anorthite, $CaAl_2Si_2O_8$. Assume that the melt behaves ideally, that the melting enthalpy of anorthite is independent of temperature and that no $CaTiSiO_5$ is insoluble in solid anorhtite and vice versa.

$$\Delta_{fus}H^{an}_{CaAl_2Si_2O_8} = 133.0\ \text{kJmol}^{-1}, \quad T_{fus,\ CaAl_2Si_2O_8} = 1830\ \text{K}.$$

- Calculate the activity coefficient for anorthite in the case that the same mixture melts at 1818 K.

4. For the binary system Mg_2SiO_4 - Fe_2SiO_4, the pressure of the phase transition olivine → spinel depends strongly upon the composition of the transforming phase. An increasing content of Mg_2SiO_4 increases transformation pressures. Pure forsterite, Mg_2SiO_4, does not transform directly to spinel type structure, but to a phase that is termed β-Mg_2SiO_4. At very high pressures β-Mg_2SiO_4 finally undergoes the transition to the spinel phase.

   Using the thermodynamic data of Navrotsky et al. (1979) a pressure of 10.7 GPa is obtained for the metastable phase transformation olivine → spinel for pure forsterite at 700°C. The volume of transition equals - 4.02 cm^3mol^{-1}. Pure fayalite undergoes this phase transition at 5.21 GPa (Akimoto et al. 1977). The change in volume is equal to - 4.27 cm^3mol^{-1}.

- Calculate the mole fraction of fayalite for coexisting olivine and spinel at 700°C and 6.5 GPa assuming that olivine and spinel behave ideally and that the volumes of transition are pressure independent.

  Recall that there are two thermodynamically equivalent crystallographic sites per formula unit in olivine!

5. Consider the ternary system MgO-Al_2O_3-SiO_2 containing the phases: forsterite (fo), enstatite (en), kyanite (ky), cordierite (crd) and spinel (sp).

- Draw the Gibbs triangle for this system and plot the compositions of the phases given above.
- Show the univariant and divariant assemblages.

Chapter 7 Chemical reactions

7.1 Phase equilibria in reacting systems

A system is, according to Eq. (6.7), at equilibrium if its Gibbs free energy has a minimum value. This means that the total differential of the Gibbs free energy must be equal to zero, i e.

$$dG = 0.$$

Applying this condition to reacting systems (see Eqs. (5.178) and (5.179)) yields:

$$\Delta_r G = \sum_i \nu_i \mu_i = 0, \tag{7.1}$$

provided the reaction takes place at constant pressure and temperature. If the chemical potentials, μ_i, in Eq. (7.1) are written in the extended form as the sum of the standard potentials and the logarithm of activities, the following expression results:

$$\Delta_r G = \sum_i \nu_i \mu_i^{\mathrm{o}} + RT \sum_i \nu_i \ln a_i = 0 \tag{7.2}$$

or

$$\Delta_r G = \sum_i \nu_i \mu_i^{\mathrm{o}} + RT \sum_i \nu_i \ln x_i + RT \sum_i \nu_i \ln \gamma_i = 0 \tag{7.3}$$

if the definition of activity is considered.

According to Eq. (5.182), the stoichiometric sum of the standard potentials of the components coefficients gives the standard Gibbs free energy of reaction, $\Delta_r \boldsymbol{G}^{\mathrm{o}}$. Thus, Eq. (7.3) can be rewritten as:

$$\Delta_r G = \Delta_r \boldsymbol{G}^{\mathrm{o}} + RT \sum_i \nu_i \ln x_i + RT \sum_i \nu_i \ln \gamma_i = 0. \tag{7.4}$$

Further, it holds that

$$\sum_i \nu_i \ln a_i = \ln K_{P,T}, \qquad (7.5)$$

where

$$K_{P,T} = \Pi[a_i]^{\nu_i}. \qquad (7.6)$$

The value $K_{P,T}$ in Eq. (7.6) is referred to as the *thermodynamic equilibrium constant.*

Considering Eqs (7.2) through (7.5), the condition for thermodynamic equilibrium in a reacting system reads:

$$\Delta_r \boldsymbol{G}^{\mathrm{o}} + RT \ln K_{P,T} = 0 \qquad (7.7)$$

or

$$\Delta_r \boldsymbol{G}^{\mathrm{o}} = -RT \ln K_{P,T}. \qquad (7.8)$$

7.1.1 Reactions in systems containing pure solid phases

In the case of a chemical reaction between pure phases, the activities of all components equal 1. Consequently, the logarithm of the thermodynamic constant, $K_{P,T}$, is equal to zero and the standard Gibbs free energy of reaction, $\Delta_r \boldsymbol{G}^{\mathrm{o}}$, also equals zero. In order to describe thermodynamic equilibrium in a reacting system as a function of temperature and pressure, the standard Gibbs free energy as a function of these two variables must be known.

The temperature dependence of the Gibbs free energy of reaction is given by the Gibbs-Helmholtz equation, according to which it holds that

$$\Delta_r \boldsymbol{G}_T^{\mathrm{o}} = \Delta_r \boldsymbol{H} - T\Delta_r \boldsymbol{S}. \qquad (7.9)$$

If, additionally, the temperature dependence of the enthalpy and entropy of reaction is considered, Eq. (7.9) can be written as:

$$\Delta_r G_T^{\mathrm{o}} = \Delta_r H_{298} + \int_{298}^{T} \Delta_r C_p dT - T\left[\Delta_r S_{298} + \int_{298}^{T} \frac{\Delta_r C_p}{T} dT\right]. \tag{7.10}$$

In Eq. (7.10) $\Delta_r H_{298}$ designates the standard enthalpy of reaction and, thus, represents the stoichiometric sum of the standard enthalpies of formation of the participating phases. $\Delta_r S_{298}$ is the standard entropy of reaction and is calculated using third-law entropies. Eq. (7.10) holds for the standard pressure of 0.1 MPa.

The pressure dependence of the standard Gibbs energy of reaction can be derived using the cross-differentiation identity, which holds for an exact differential. According to this rule, one can write:

$$\frac{\partial}{\partial P}\left(\frac{\partial G}{\partial \xi}\right)_T = \frac{\partial}{\partial \xi}\left(\frac{\partial G}{\partial P}\right)_T \tag{7.11}$$

or

$$\left(\frac{\partial \Delta_r G}{\partial P}\right)_T = \left(\frac{\partial V}{\partial \xi}\right)_T = \Delta_r V, \tag{7.12}$$

where $\Delta_r V$ in Eq. (7.12) is the volume of reaction. It corresponds to the change in volume of a system per unit reaction progress variable, ξ.

Using the relationship given in Eq. (7.12), the standard Gibbs free energy of reaction at pressure P and temperature T reads:

$$\Delta_r G_{P,T}^{\mathrm{o}} = \Delta_r G_T^{\mathrm{o}} + \int_{P_o}^{P} \Delta_r V(T) dP, \tag{7.13}$$

if the reactants as well as the products are pure end-member phases. In Eq. (7.13) $\Delta_r V(T)$ indicates the volume of reaction at temperature T. If the volume of reaction is considered to be independent of pressure and temperature, the integration of the last term in Eq. (7.13) within the limits of P_{o} and P yields:

$$\Delta_r G_{P,T}^{\mathrm{o}} = \Delta_r G_T^{\mathrm{o}} + \Delta_r V_{298}(P - P_o). \tag{7.14}$$

Substituting the expression present in Eq. (7.10) for $\Delta_r G_T^{\mathrm{o}}$ in Eq. (7.14) gives:

$$\Delta_r G^{\circ}_{P,T} = \Delta_r \boldsymbol{H}_{298} + \int_{298}^{T} \Delta_r \boldsymbol{C}_p dT - T\left[\Delta_r \boldsymbol{S}_{298} + \int_{298}^{T} \frac{\Delta_r \boldsymbol{C}_p}{T} dT\right] \tag{7.15}$$

$$+ \Delta_r \boldsymbol{V}_{298}(P - P_o).$$

If additionally, the enthalpy and entropy of reaction are assumed independent of temperature, Eq. (7.15) simplifies to:

$$\Delta_r G^{\circ}_{P,T} = \Delta_r \boldsymbol{H}_{298} - T\Delta_r \boldsymbol{S}_{298} + \Delta_r \boldsymbol{V}_{298}(P - P_o). \tag{7.16}$$

In the case that the pressure and temperature dependence of volume is taken into consideration, the integral in Eq. (7.13) obtains the form:

$$\int_{P_o}^{P} \Delta_r \boldsymbol{V} dP = \int_{P_o}^{P} [\Delta_r \boldsymbol{V}_{298} + \Delta_r(\alpha_i \boldsymbol{V}_{i,298})(T - 298) - \Delta_r(\beta_i \boldsymbol{V}_{i,298})P]dP$$

$$= \left[\Delta_r \boldsymbol{V}_{298} + \Delta_r(\alpha_i \boldsymbol{V}_{i,298})(T - 298) - \frac{\Delta_r(\beta_i \boldsymbol{V}_{i,298})P}{2}\right](P - P_o). \tag{7.17}$$

In Eq. (7.17), the term $(\Delta(\beta_i \boldsymbol{V}_{i,298})P_o)/2$ is neglected. Its contribution to the result is negligibly small, because $P_o \ll P$. α_i and β_i are the thermal expansion and compressibility coefficient of the component i, respectively. The term $\Delta(\alpha_i \boldsymbol{V}_{i,298})$ represents the stoichiometric sum of the products of the molar volumes times the corresponding thermal expansion coefficients, i.e.

$$\Delta(\alpha_i \boldsymbol{V}_{i,298}) = \sum_i \nu_i \alpha_i \boldsymbol{V}_{i,298}. \tag{7.18}$$

Analogously, the term $\Delta(\beta_i \boldsymbol{V}_{i,298})$ gives the stoichiometric sum of the products of the compressibility coefficients times the molar volumes, so that

$$\Delta(\beta_i \boldsymbol{V}_{i,298}) = \sum_i \nu_i \beta_i \boldsymbol{V}_{i,298}. \tag{7.19}$$

The volumes in Eqs. (7.18) and (7.19) refer to the ambient pressure and temperature. The bold letter is used to indicate the pure phases.

Combining Eqs. (7.10) and (7.17) yields the standard Gibbs free energy of reaction for the case where the temperature and pressure dependence of the state functions is considered, namely:

$$\Delta_r G^o_{P,T} = \Delta_r H_{298} + \int_{298}^{T} \Delta_r C_p dT - T\left[\Delta_r S_{298} + \int_{298}^{T} \frac{\Delta_r C_p}{T} dT\right]$$
$$+ \left[\Delta_r V_{298} + \Delta_r(\alpha_i V_{i,298})(T-298) - \frac{\Delta_r(\beta_i V_{i,298})P}{2}\right](P-P_o). \quad (7.20)$$

or

$$\Delta_r G^o_{P,T} = \Delta_r H_{298} + \Delta_r a(T-298) + \frac{\Delta_r b}{2}(T^2 - 298^2) \quad (7.21)$$
$$+ \Delta_r c\left(\frac{1}{T} - \frac{1}{298}\right) - T\left[\Delta_r S_{298} + (\Delta_r a)\ln\frac{T}{298}\right.$$
$$\left. + \Delta_r b(T-298) + \frac{\Delta_r c}{2}\left(\frac{1}{T^2} - \frac{1}{298^2}\right)\right] + \left[\Delta_r V_{298}\right.$$
$$\left. + \Delta_r(\alpha_i V_{i,298})(T-298) - \frac{\Delta_r(\beta_i V_{i,298})P}{2}\right](P-P_o)$$

when the polynomials of the molar heat capacities of the phases involved in the reaction have the form: $C_p = a + bT - cT^{-2}$.

For a system in thermodynamic equilibrium, $\Delta_r G_{P,T}$ equals zero and Eq. (7.20) reads:

$$\Delta_r H_{298} + \int_{298}^{T} \Delta_r C_p dT - T\left[\Delta_r S_{298} + \int_{298}^{T} \frac{\Delta_r C_p}{T} dT\right] \quad (7.22)$$
$$+ \left[\Delta_r V_{298} + \Delta_r(\alpha_i V_{i,298})(T-298) - \frac{\Delta_r(\beta_i V_{i,298})P}{2}\right](P-P_o) = 0.$$

Eq. (7.22) can be used to calculate the equilibrium conditions for reacting systems when all reactants are solid phases and when a high precision is required. In many cases, however, the pressure and temperature dependence of the state functions can be neglected and Eq. (7.22) simplifies to:

$$\Delta_r H_{298} - T\Delta_r S_{298} + \Delta_r V_{298}(P - P_o) = 0. \quad (7.23)$$

Example 1: Consider the reaction:

$$Ca_3Al_2Si_3O_{12} + SiO_2 \rightarrow CaAl_2Si_2O_8 + 2CaSiO_3.$$

Assuming that all changes in thermodynamic functions associated with the reaction are independent of temperature and pressure, the equilibrium conditions can be calculated using, Eq. (7.23).

Table 7.1 Thermodynamic data for grossular, quartz, anorthite and wollastonite (Robie and Hemingway 1995)

Phase	$\Delta_f H_{298}$ [kJmol^{-1}]	S_{298} [Jmol^{-1}K^{-1}]	V_{298} [cm^{3}mol^{-1}]
Grossular	-6640.0	260.1	125.28
Quartz	-910.7	41.5	22.69
Anorthite	-4234.0	199.3	100.79
Wollastonite	-1634.8	81.7	39.90

First, the enthalpy, entropy and the volume of reaction are to be calculated. In order to calculate the enthalpy of reaction the standard enthalpies of formation are used and the calculation reads:

$$\Delta_r H_{298} = -\Delta_f H_{298}^{grt} - \Delta_f H_{298}^{qtz} + \Delta_f H_{298}^{an} + 2\Delta_f H_{298}^{wo}.$$

The entropy of reaction represents the stoichiometric sum of the conventional standard entropies of the phases involved, namely:

$$\Delta_r S_{298} = -S_{298}^{grt} - S_{298}^{qtz} + S_{298}^{an} + 2S_{298}^{wo}$$

and the volume of reaction is obtained using the molar volumes of the phases at standard conditions as follows:

$$\Delta_r V_{298} = -V_{298}^{grt} - V_{298}^{qtz} + V_{298}^{an} + 2V_{298}^{wo}.$$

The required thermodynamic data are given Tab. 7.1. Using these data one obtains:

$$\Delta_r H_{298} = 6640.0\ \text{kJ} + 910.7\ \text{kJ} - 4234.0\ \text{kJ} - 2 \times 1634.8\ \text{kJ} = \underline{47.1\ \text{kJ}}$$

for the enthalpy of reaction,

$$\Delta_r S_{298} = -260.1\ \text{JK}^{-1} - 41.5\ \text{JK}^{-1} + 199.3\ \text{JK}^{-1} + 2 \times 81.7\ \text{JK}^{-1} = \underline{61.1\ \text{JK}^{-1}}$$

for the entropy of reaction and

$$\Delta_r V_{298} = -125.28\text{cm}^3 - 22.69\text{cm}^3 + 100.79\text{cm}^3 + 2 \times 39.90\text{cm}^3 = \underline{32.62\ \text{cm}^3}.$$

for the volume of reaction.

In order to calculate the equilibrium temperature at any given pressure, the terms in Eq. (7.23) must be rearranged, namely:

$$T = \frac{\Delta_r H_{298} + \Delta_r V(P - P_o)}{\Delta_r S_{298}}. \qquad (7.24)$$

Inserting the values for the enthalpy of reaction, entropy of reaction and volume of reaction into Eq. (7.24) gives:

$$T = \frac{47100\ \text{J}}{61.1\ \text{JK}^{-1}} = \underline{771\ \text{K}}$$

at ambient pressure and

$$T = \frac{47100\ \text{J} + 32.62 \times 10^{-6}\text{m}^3(5 \times 10^8\text{Pa} - 10^5\text{Pa})}{61.1\ \text{JK}^{-1}} = \underline{1038\ \text{K}}$$

at 0.5 GPa. The two temperatures determine the curve where grossular and quartz react to anorthite and wollastonite. All four phases (grossular, quartz, anorthite and wollastonite) coexist along this line. The area on the left side of the line represents the stability field of the assemblage grossular and quartz and on the right side the stability field of anorthite and wollastonite. Fig. 7.1 shows the calculated reaction curve together with the experimental results of Boettcher (1970) and Huckenholz et al. (1975).

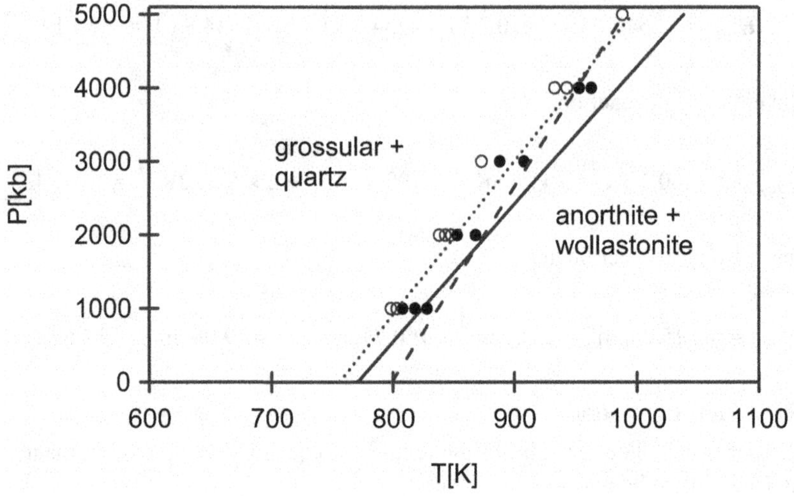

Fig. 7.1 Pressure-temperature diagram for the reaction grossular + quartz = anorthite + wollastonite. Solid line = calculated using thermodynamic data, circles and dotted line = experimental results of Huckenholz et al. (1975), dashed line = experimental results of Boettcher (1970). Open circles = growth of grossular and quartz, closed circles = growth of anorthite and wollastonite. The calculations are based on the thermodynamic data of Robie and Hemingway (1995).

Example 2: In section 6.1.1 the phase diagram for the one-component system Al_2SiO_5 was calculated using the Clausius-Clapeyron equation. The calculation was rather intricate. The knowledge of two transition points was necessary in order to determine the phase boundaries of the stability fields of the three different modifications: kyanite, andalusite and sillimanite. This phase diagram can much easier be calculated by treating the phase transitions as chemical reactions.

The phase transition kyanite to andalusite can be described by the chemical reaction:

$$(Al_2SiO_5)^{ky} = (Al_2SiO_5)^{and}$$

and the reaction:

$$(Al_2SiO_5)^{and} = (Al_2SiO_5)^{sill}$$

is equivalent to the phase transition andalusite to sillimanite.

The phase transition kyanite to sillimanite gives the reaction

$$(Al_2SiO_5)^{ky} = (Al_2SiO_5)^{sill}.$$

The heat of transformation, the entropy of transformation and the volume of transformation correspond to the enthalpy of reaction, the entropy of reaction and the volume of reaction. Assuming that all changes in the thermodynamic functions associated with the phase transition are pressure and temperature independent, the equation for equilibrium reads:

$$\Delta_{tr}\boldsymbol{H}_{298} - T\Delta_{tr}\boldsymbol{S}_{298} + \Delta_{tr}\boldsymbol{V}_{298}(P - P_o) = 0. \tag{7.25}$$

Rearranging Eq. (7.25) and solving for T yields:

$$T = \frac{\Delta_{tr}\boldsymbol{H}_{298} + \Delta_{tr}\boldsymbol{V}_{298}(P - P_o)}{\Delta_{tr}\boldsymbol{S}_{298}}. \tag{7.26}$$

The thermodynamic tables for minerals of Robie and Hemingway (1995) give the following standard enthalpies of formation: -2589.9 $kJmol^{-1}$ for andalusite, -2586.1 $kJmol^{-1}$ for sillimanite and -2593.8 $kJmol^{-1}$ for kyanite. Using these data and the data from Tab. 6.1, the changes in the thermodynamic functions associated with the transitions can be determined, namely:

$$\Delta_{tr}\boldsymbol{H}_{298}^{ky/and} = 2593.8\ \text{kJmol}^{-1} - 2589.9\ \text{kJmol}^{-1} = \underline{3.9\ \text{kJmol}^{-1}},$$

$$\Delta_{tr}\boldsymbol{H}_{298}^{ky/sill} = 2593.8\ \text{kJmol}^{-1} - 2586.1\ \text{kJmol}^{-1} = \underline{7.7\ \text{kJmol}^{-1}},$$

$$\Delta_{tr}\boldsymbol{H}_{298}^{and/sill} = 2589.9\ \text{kJmol}^{-1} - 2586.1\ \text{kJmol}^{-1} = \underline{3.8\ \text{kJmol}^{-1}},$$

$$\Delta_{tr}\boldsymbol{S}_{298}^{ky/and} = -82.8\ \text{Jmol}^{-1}\text{K}^{-1} + 91.4\ \text{Jmol}^{-1}\text{K}^{-1} = \underline{8.6\ \text{Jmol}^{-1}\text{K}^{-1}},$$

$$\Delta_{tr}\boldsymbol{S}_{298}^{ky/sill} = -82.8\ \text{Jmol}^{-1}\text{K}^{-1} + 95.4\ \text{Jmol}^{-1}\text{K}^{-1} = \underline{12.6\ \text{Jmol}^{-1}\text{K}^{-1}},$$

$$\Delta_{tr}\boldsymbol{S}_{298}^{and/sill} = -91.4\ \text{Jmol}^{-1}\text{K}^{-1} + 95.4\ \text{Jmol}^{-1}\text{K}^{-1} = \underline{4.0\ \text{Jmol}^{-1}\text{K}^{-1}},$$

$$\Delta_{tr}\boldsymbol{V}_{298}^{ky/and} = -44.15\text{cm}^3\text{mol}^{-1} + 51.52\text{cm}^3\text{mol}^{-1} = \underline{7.37\ \text{cm}^3\text{mol}^{-1}},$$

$$\Delta_{tr}\boldsymbol{V}_{298}^{ky/sill} = -44.15\text{cm}^3\text{mol}^{-1} + 49.86\text{cm}^3\text{mol}^{-1} = \underline{5.71\ \text{cm}^3\text{mol}^{-1}}$$

and

$$\Delta_{tr} V_{298}^{and/sill} = -51.52\text{cm}^3\text{mol}^{-1} + 49.86\text{cm}^3\text{mol}^{-1} = \underline{-1.66\ \text{cm}^3\text{mol}^{-1}}.$$

The temperature for the transition of kyanite to andalusite at ambient pressure is then:

$$T_{tr}^{ky/and} = \frac{3900\ \text{Jmol}^{-1}}{8.6\ \text{Jmol}^{-1}\text{K}^{-1}} = \underline{453\ \text{K}}$$

and at 0.45 GPa:

$$T_{tr}^{ky/and} = \frac{3900\ \text{Jmol}^{-1} + 7.37\times 10^{-6}\text{m}^3\text{mol}^{-1}(0.45\times 10^9\text{Pa} - 10^5\text{Pa})}{8.6\ \text{Jmol}^{-1}\text{K}^{-1}} = \underline{839\ \text{K}}.$$

The temperature for the transition of andalusite to sillimanite at 0.1 MPa is:

$$T_{tr}^{and/sill} = \frac{3800\ \text{Jmol}^{-1}}{4.0\ \text{Jmol}^{-1}\text{K}^{-1}} = \underline{950\ \text{K}}$$

and at 0.45 GPa:

$$T_{tr}^{and/sill} = \frac{3800\ \text{Jmol}^{-1} - 1.66\times 10^{-6}\text{m}^3\text{mol}^{-1}(0.45\times 10^9\text{Pa} - 10^5\text{Pa})}{4.0\ \text{Jmol}^{-1}\text{K}^{-1}} = \underline{763\ \text{K}}.$$

The two lines drawn through the data points, calculated above, intersect at the triple point, where all three phases (kyanite, andalusite and sillimanite) coexist. At the triple point, it therefore holds that:

$$\frac{\Delta_{tr}\boldsymbol{H}_{298}^{ky/and} + \Delta_{tr}\boldsymbol{V}_{298}^{ky/and}(P_{tp} - 10^5)}{\Delta_{tr}\boldsymbol{S}_{298}^{ky/and}} = \frac{\Delta_{tr}\boldsymbol{H}_{298}^{and/sill} + \Delta_{tr}\boldsymbol{V}_{298}^{and/sill}(P_{tp} - 10^5)}{\Delta_{tr}\boldsymbol{S}_{298}^{and/sill}}, \tag{7.27}$$

where P_{tp} designates the pressure of the triple point.

Rearranging Eq. (7.27) and solving for P_{tp} gives:

$$P_{tp} = \frac{\Delta_{tr}S_{298}^{ky/and} \times \Delta_{tr}H_{298}^{and/sill} - \Delta_{tr}S_{298}^{and/sill} \times \Delta_{tr}H_{298}^{ky/and}}{\Delta_{tr}S_{298}^{and/sill} \times \Delta_{tr}V_{298}^{ky/and} - \Delta_{tr}S_{298}^{ky/and} \times \Delta_{tr}V_{298}^{and/sill}} + 10^5 . \qquad (7.28)$$

Inserting the numeric values into Eq. (7.28) yields:

$$\begin{aligned} P_{tp} &= (8.6\ \text{Jmol}^{-1}\text{K}^{-1} \times 3800\ \text{Jmol}^{-1} - 4.0\ \text{Jmol}^{-1}\text{K}^{-1} \\ &\quad \times 3900\ \text{Jmol}^{-1})/(4.0\ \text{Jmol}^{-1}\text{K}^{-1} \times 7.37\times10^{-6}\text{m}^3\text{mol}^{-1} \\ &\quad + 8.6\ \text{Jmol}^{-1}\text{K}^{-1} \times 1.66\times10^{-6}\text{m}^3\text{mol}^{-1}) + 10^5\text{Pa} = \underline{0.39\ \text{GPa}}. \end{aligned}$$

In order to determine the temperature of the triple point, the pressure of the triple point (0.39 GPa) has to be inserted into one of the two equations that were used to calculate the transition temperature as a function of pressure. We take the one, that was used to calculate the phase transition andalusite to sillimanite and obtain:

$$T_{tp} = \frac{3900\ \text{Jmol}^{-1} + 7.37\times10^{-6}\text{m}^3\text{mol}^{-1}(0.39\times10^9\text{Pa} - 10^5\text{Pa})}{8.6\ \text{Jmol}^{-1}\text{K}^{-1}} = \underline{788\ \text{K}}.$$

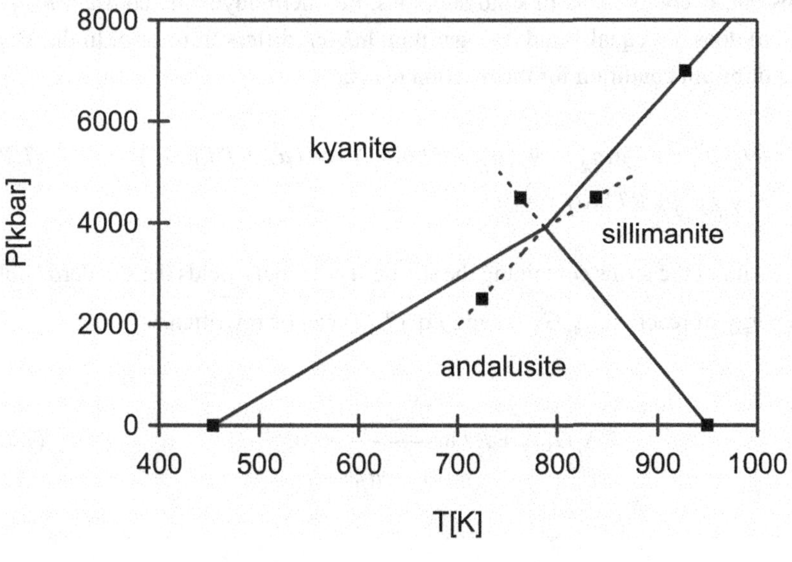

Fig. 7.2 P-T-diagram of the Al_2SiO_5 system constructed using thermodynamic data of Robie and Hemingway (1995). Filled squares designate the calculated transition points.

In addition, only one temperature for the phase transition kyanite to sillimanite at some pressure above 0.39 GPa is required in order to construct the complete *P*-*T* diagram for the Al_2SiO_5 system. We take a pressure of 0.7 GPa and calculate:

$$T_{tr}^{ky/sill} = \frac{7700\ \text{Jmol}^{-1} + 5.71\times10^{-6}\text{m}^3\text{mol}^{-1}(0.7\times10^9\text{Pa} - 10^5\text{Pa})}{12.6\ \text{Jmol}^{-1}\text{K}^{-1}} = \underline{928\ \text{K}}.$$

The resulting phase diagram is shown in Fig. 7.2. The extensions of the transition lines (dashed) beyond the triple point give the temperatures and pressures for the metastable phase transitions. Hence, the transition of kyanite to sillimanite can occur metastably within the stability field of andalusite, that of andalusite to sillimanite in the stability field of kyanite and that of kyanite to andalusite in the stability field of sillimanite. The squares mark the transition points that were calculated above.

7.1.2 Reactions in systems containing solid solutions

Consider a general reaction

$$\nu_a A + \nu_b B = \nu_c C + \nu_d D$$

that takes place at some *P* and *T*. If the reactants and products do not occur as pure phases but as components in solid solutions, the thermodynamic constant $K_{P,T}$ in Eq. (7.6) does not equal 1 and its logarithm, $\ln K_{P,T}$, differs from zero. In this case, the equilibrium condition for the reaction reads:

$$-\nu_a(\mu_A^o + RT\ln a_A) - \nu_b(\mu_B^o + RT\ln a_B) + \nu_c(\mu_C^o + RT\ln a_C) + \nu_d(\mu_D^o + RT\ln a_D) = 0. \tag{7.29}$$

The sum of the terms containing the standard potentials yields the standard Gibbs free energy of reaction, $\Delta_r G_{P,T}^o$, and Eq. (7.29) can be rewritten as:

$$\Delta_r G_{P,T}^o + RT\ln\frac{a_C^{\nu_c} \cdot a_D^{\nu_d}}{a_A^{\nu_a} \cdot a_B^{\nu_b}} = 0, \tag{7.30}$$

or

$$\Delta_r G^o_{P,T} = -RT\ln\frac{a_C^{\nu_c} \cdot a_D^{\nu_d}}{a_A^{\nu_a} \cdot a_B^{\nu_b}}. \tag{7.31}$$

If $\Delta_r G^o_{P,T}$ is replaced by $(\Delta_r H_{298} - T\Delta_r S_{298} + \Delta_r V_{298}(P - P_o))$, an expression that allows the calculation of the equilibrium pressure and temperature is obtained, namely:

$$\Delta_r H_{298} - T\Delta_r S_{298} + \Delta_r V_{298}(P-P_o) = -RT\ln\frac{a_C^{\nu_c} \cdot a_D^{\nu_d}}{a_A^{\nu_a} \cdot a_B^{\nu_b}} \tag{7.32}$$

for the case of non-ideal solid solutions, and

$$\Delta_r H_{298} - T\Delta_r S_{298} + \Delta_r V_{298}(P-P_o) = -RT\ln\frac{x_C^{\nu_c} \cdot x_D^{\nu_d}}{x_A^{\nu_a} \cdot x_B^{\nu_b}} \tag{7.33}$$

in the case that the solutions are ideal and the activities can be replaced by the mole fractions of the components. Eqs. (7.32) and (7.33), of course, do not account for the temperature and pressure dependence of the changes in the thermodynamic functions associated with the reaction.

Example: Consider the anorthite breakdown reaction:

$$3CaAl_2Si_2O_8 \rightarrow Ca_3Al_2Si_3O_{12} + 2Al_2SiO_5 + SiO_2$$

and assume that anorthite, $CaAl_2Si_2O_8$, kyanite, Al_2SiO_5, and quartz, SiO_2, are pure phases and grossular, $Ca_3Al_2Si_3O_{12}$, is a component of a pyrope-grossular solid solution. Assume further that garnet is an ideal solution and the mole fraction of grossular equals 0.22.

For the system at equilibrium, the following condition holds according to Eq. (7.29):

$$\begin{aligned} &-3(\mu^{o,pl}_{CaAl_2Si_2O_8} + RT\ln x^{pl}_{CaAl_2Si_2O_8}) + (\mu^{grt}_{Ca_3Al_2Si_3O_{12}} + 3RT\ln x^{grt}_{Ca_3Al_2Si_3O_{12}}) \\ &+ 2(\mu^{o,ky}_{Al_2SiO_5} + RT\ln x^{ky}_{Al_2SiO_5}) + (\mu^{o,qtz}_{SiO_2} + RT\ln x^{qtz}_{SiO_2}) = 0 \end{aligned}$$

or

$$\Delta_r G^o_{P,T} + RT\ln K_{P,T} = 0,$$

where

$$K_{P,T} = \frac{(x^{grt}_{Ca_3Al_2Si_3O_{12}})^3 \cdot (x^{ky}_{Al_2SiO_5})^2 \cdot x^{qtz}_{SiO_2}}{(x^{pl}_{CaAl_2Si_2O_8})^3}.$$

If one assumes that the pressure and temperature dependence of the changes in the thermodynamic functions associated with the reaction is negligibly small, the expression given in Eq. (7.23) can be used to calculate the equilibrium conditions and one obtains:

$$\Delta_r H_{298} - T\Delta_r S_{298} + \Delta_r V_{298}(P - P_o) = -RT\ln K_{P,T}.$$

Because all but the mole fraction of grossular equal one, the thermodynamic equilibrium constant reads:

$$K_{P,T} = 0.22^3$$

and, therefore:

$$\Delta_r H_{298} - T\Delta_r S_{298} + \Delta_r V_{298}(P - P_o) = -3RT\ln(0.22).$$

At any given temperature, the equilibrium pressure is then given by:

$$P = \frac{-\Delta_r H_{298} + T\Delta_r S_{298} - 3RT\ln(0.22)}{\Delta_r V_{298}} + P_o.$$

Using the data on the enthalpies of formation, third-law entropies and molar volumes given in the literature (Holland and Powell 1990) yields:

$$\Delta_r H_{298} = -41.54 \text{ kJ}$$

$$\Delta_r S_{298} = -135.8 \text{ JK}^{-1} \text{ and}$$

$$\Delta_r V_{298} = -66.05 \text{ cm}^3.$$

The equilibrium pressure at a temperature of 1273 K is then calculated as follows:

$$P = [41540\text{J} - 1273\text{K} \times 135.8\text{JK}^{-1} - 3 \times 8.3144\ \text{JK}^{-1} \times 1273\text{K} \times \ln(0.22)]/[-66.05\times10^{-6}\text{m}^3] + 10^5\text{Pa} = \underline{1.26\ \text{GPa}}.$$

This result is to be compared with the equilibrium pressure of 1.99 GPa that is obtained for the end-member reaction at 1273 K.

In order to consider the non-ideality of the garnet solid solution, the activity coefficient, γ, has to be introduced into the equation and one obtains:

$$\Delta_r \boldsymbol{H}_{298} - T\Delta_r \boldsymbol{S}_{298} + \Delta_r \boldsymbol{V}_{298}(P - P_o) = -3RT\ln(x \cdot \gamma)^{grt}_{Ca_3Al_2Si_3O_{12}}.$$

According to the results of the calorimetric measurements in the system CaO-MgO-Al_2O_3-SiO_2 (Newton et al. 1977), the garnet pyrope solution is asymmetric and can be modelled by the subregular solution model.

Using the exchange parameters recommended by Ganguly and Saxena (1984) the logarithm of the activity coefficient for grossular reads:

$$\ln\gamma^{grt}_{Ca_3Al_2Si_3O_{12}} = (1 - x^{grt}_{Ca_3Al_2Si_3O_{12}})^2 \left\{(16933 - 6.28T) + 2[(4184 - 6.28T) - (16933 - 6.28T)](1 - x^{grt}_{Ca_3Al_2Si_3O_{12}})\right\}/RT.$$

With the equation given above a value of $\gamma^{grt}_{Ca_3Al_2Si_3O_{12}} = 0.53$ is obtained for $x^{grt}_{Ca_3Al_2Si_3O_{12}} = 0.22$. Using these two values, the equilibrium pressure for T = 1273 K is calculated as follows:

$$P = [41540\text{J} - 1273\text{K} \times 135.8\ \text{JK}^{-1} - 3 \times 8.3144\ \text{JK}^{-1} \times 1273\text{K} \times \ln(0.22 \times 0.53)]/[-66.05\times10^{-6}\text{m}^3] + 10^5\text{Pa} = \underline{0.96\ \text{GPa}}.$$

7.1.3 Reactions in systems containing solid and gas phases

The standard potentials of solid phases refer to pure phases at the temperature and pressure of interest. The standard potentials of gases, however, refer to pure ideal gases at the temperature T and the standard pressure $P_0 = 0.1$ MPa. In order to account for this difference, the Gibbs free energy of reaction is divided into three terms, namely:

$$\Delta_r G_{P,T} = \sum_i \nu_i^s \mu_i^{o,s} + \sum_j \nu_j^g \mu_j^{o,g} + RT\sum_j \nu_j^g \ln f_j^g, \tag{7.34}$$

where the superscripts s and g designate solids and gases, respectively.

In terms of the standard Gibbs free energy of reaction, Eq. (7.34) reads:

$$\Delta_r G_{P,T} = \Delta_r G_{P,T}^{o,s} + \Delta_r G_{P_o,T}^{o,g} + RT\sum_j \nu_j^g \ln f_j^g. \tag{7.35}$$

In the case where the solids as well as the gases are ideal mixtures and mixing in the solids occurs on one crystallographic site only, Eq. (7.35) obtains the form:

$$\Delta_r G_{P,T} = \Delta_r G_{P,T}^{o,s} + RT\sum_i \nu_i^s \ln x_i^s + \Delta_r G_{P_o,T}^{o,g} + RT\sum_j \nu_j^g \ln \frac{P_j}{P_o}, \tag{7.36}$$

where P_j designates the partial pressure of gas j.

According to Eq. (7.15) it holds that

$$\Delta_r G_{P,T}^{o,s} = \Delta_r H_{298}^s + \int_{298}^{T} \Delta_r C_p^s dT - T\left[\Delta_r S_{298}^s + \int_{298}^{T} \frac{\Delta_r C_p^s}{T} dT\right]$$
$$+ \left[\Delta_r V_{298}^s + \Delta_r(\alpha_i V_{i,298})^s (T-298) - \frac{\Delta_r(\beta_i V_{i,298})^s P}{2}\right](P - P_o) \tag{7.37}$$

and

$$\Delta_r G_{P_o,T}^{o,g} = \Delta_r H_{298}^g + \int_{298}^{T} \Delta_r C_p^g dT - T\left[\Delta_r S_{298}^g + \int_{298}^{T} \frac{\Delta_r C_p^g}{T} dT\right]. \tag{7.38}$$

Substituting Eqs. (7.37) and (7.38) into (7.36) yields:

$$\Delta_r G_{P,T} = \Delta_r H_{298} + \int_{298}^{T} \Delta_r C_p dT - T\left[\Delta_r S_{298} + \frac{\Delta_r C_p}{T} dT\right] \tag{7.39}$$

$$+\left[\Delta_r V_{298}^s + \Delta_r(\alpha_i V_{i,\,298})^s (T-298) - \frac{\Delta_r(\beta_i V_{i,\,298})^s P}{2}\right](P-P_o)$$

$$+RT\sum_i \nu_i^s \ln x_i^s + RT\sum_j \nu_j^g \ln\frac{P_j}{P_o}$$

if

$$\Delta_r H_{298} = \Delta_r H_{298}^s + \Delta_r H_{298}^g, \tag{7.40}$$

$$\Delta_r S_{298} = \Delta_r S_{298}^s + \Delta_r S_{298}^g \tag{7.41}$$

and

$$\Delta_r C_p = \Delta_r C_p^s + \Delta_r C_p^g. \tag{7.42}$$

In the case where the temperature and pressure dependence of the enthalpy of reaction, entropy of reaction and volume of reaction is neglected, Eq. (7.39) simplifies to:

$$\Delta_r G_{P,T} = \Delta_r H_{298} - T\Delta_r S_{298} + \Delta_r V_{298}^s (P-P_o) \tag{7.43}$$

$$+RT\sum_i \nu_i^s \ln x_i^s + RT\sum_j \nu_j^g \ln\frac{P_j}{P_o}.$$

For a system in thermodynamic equilibrium $\Delta_r G_{P,T}$ equals zero, and Eq. (7.43) reads:

$$\Delta_r H_{298} - T\Delta_r S_{298} + \Delta_r V_{298}^s (P-P_o) = -RT\sum_i \nu_i^s \ln x_i^s - RT\sum_j \nu_j^g \ln\frac{P_j}{P_o}. \tag{7.44}$$

The terms on the right side of equation Eq. (7.44) correspond to the term $-RT\ln K_{P,T}$. In the case where the reacting solids and gases both represent ideal solutions, the thermodynamic equilibrium constant has the form:

$$K_{P,T} = \Pi[x_i]^{\nu_i^s}\Pi[P_j/P_o]^{\nu_j^g}. \tag{7.45}$$

When pressure is measured in bar, then P_o equals one and instead of Eq. (7.45) one can write:

$$K_{P,T} = \Pi[x_i]^{\nu_i^s}\Pi[P_j]^{\nu_j^g}. \tag{7.46}$$

Example 1: Consider the dehydration of pure muscovite in the presence of quartz. The reaction reads:

$$KAl_2[AlSi_3O_{10}](OH)_2 + SiO_2 = KAlSi_3O_8 + Al_2SiO_5 + H_2O.$$

Assume that the enthalpy, entropy and the volume of reaction are pressure and temperature independent and that steam behaves as an ideal gas. In this case, the equilibrium conditions are given by Eq. (7.44) and the dehydration temperature at any given pressure is calculated as follows:

$$T = \frac{\Delta_r \boldsymbol{H}_{298} + \Delta_r \boldsymbol{V}_{298}^s(P - P_o)}{\Delta_r \boldsymbol{S}_{298} - R\ln(\boldsymbol{P}_{H_2O}/P_o)}.$$

Tab. 7.2 contains the standard heats of formation, third-law entropies and the volume data for the reacting phases. (Holland and Powell, 1990).

Table 7.2 Thermodynamic data for muscovite, quartz, sanidine, sillimanite and steam (Holland and Powell 1990)

Phase	$\Delta_f \boldsymbol{H}_{298}$[kJmol^{-1}]	$\boldsymbol{S}_{298}$[Jmol^{-1}K^{-1}]	$\boldsymbol{V}_{298}$[cm^{3}mol^{-1}]
Muscovite	- 5981.63	289.00	140.83
Quartz	- 910.80	41.50	22.69
Sanidine	- 3959.06	230.00	108.92
Sillimanite	- 2586.67	96.00	50.03
Steam	- 241.81	188.80	-

With the data from Tab. 7.2, the following values for the changes in thermodynamic functions associated with the dehydration reaction are obtained:

$$\Delta_r H_{298} = 104.89 \text{ kJ},$$

$$\Delta_r S_{298} = 184.3 \text{ JK}^{-1} \quad \text{and}$$

$$\Delta_r V_{298} = -4.57 \text{ cm}^3.$$

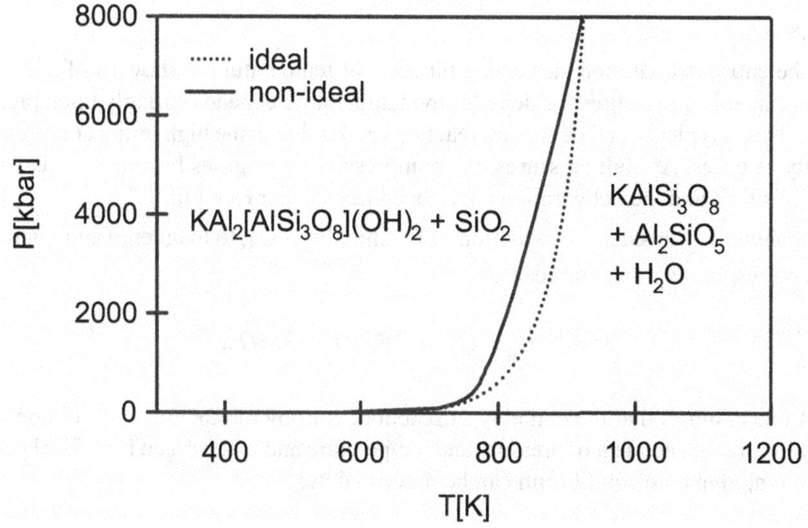

Fig. 7.3 Upper thermal stability of muscovite in the presence of quartz, calculated using the thermodynamic data of Holland and Powell (1990) and H_2O fugacity data of Grevel and Chatterjee (1992).

Assume that the total pressure and the water pressure are equal, that is:

$$P_{tot} = P_{H_2O}.$$

The dehydration temperature for muscovite in the presence of quartz, is then:

$$T = \frac{104890\,\text{J}}{184.3\,\text{JK}^{-1}} = \underline{569\text{ K}}$$

at 0.1 MPa,

$$T = \frac{104890\text{J} - 4.57\times10^{-6}\text{m}^3(0.05\times10^9\text{Pa} - 10^5\text{Pa})}{184.3\text{JK}^{-1} - 8.3144\text{JK}^{-1}\times\ln(0.05\times10^9\text{Pa}/10^5\text{Pa})} = \underline{789\text{ K}}$$

at 0.05 GPa and

$$T = \frac{104890\text{J} - 4.57\times10^{-6}\text{m}^3(0.8\times10^9\text{Pa} - 10^5\text{Pa})}{184.3\text{JK}^{-1} - 8.3144\text{JK}^{-1}\times\ln((0.8\times10^9\text{Pa})/(10^5\text{Pa}))} = \underline{924\text{ K}}$$

at 0.8 GPa.

The entire dehydration curve as a function of temperature is shown in Fig. 7.3. Note that at low pressures the dehydration temperature depends strongly upon pressure. This is typical of all degassing reactions and is due to the high initial compressibility of gases. At high pressures, the compressibility of gases becomes similar to that of solids and the dehydration curve becomes steeper (see Fig. 7.3).

Example 2: Consider the oxidation of fayalite, Fe_2SiO_4, to magnetite and quartz. The corresponding reaction reads:

$$3Fe_2SiO_4 + O_2 \rightarrow 2Fe_3O_4 + 3SiO_2.$$

If one assumes that the enthalpy of reaction, entropy of reaction and volume of reaction are independent of pressure and temperature and that oxygen is an ideal gas, the thermodynamic equilibrium can be described by:

$$\Delta_r \boldsymbol{H}_{298} - T\Delta_r \boldsymbol{S}_{298} + \Delta_r \boldsymbol{V}^S_{298}(P - P_o) = RT\ln\frac{\boldsymbol{P}_{O_2}}{P_o}. \tag{7.47}$$

The oxidation of fayalite occurs at very low oxygen pressures and, therefore, the assumption that oxygen pressure and total pressure are equal is inappropriate. It is, however, important to know the equilibrium oxygen pressure of the assemblage fayalite/magnetite/quartz at any given temperature and total pressure.

In order to calculate the equilibrium oxygen pressure as a function of temperature at constant pressure, Eq. (7.47) is divided by RT, which gives:

$$\ln\boldsymbol{P}_{O_2} = \frac{\Delta_r \boldsymbol{H}_{298} + \Delta_r \boldsymbol{V}^S_{298}(P - P_o)}{RT} - \frac{\Delta_r \boldsymbol{S}_{298}}{R} + \ln P_o. \tag{7.48}$$

As it is apparent from Eq. (7.48), a plot of the logarithm of the oxygen pressure versus reciprocal temperature gives a straight line with the slope equal to

$$\frac{\Delta_r H_{298} + \Delta_r V^s_{298}}{R}(P - P_o).$$

Thus, at constant temperature and pressure the assemblage fayalite/magnetite/quartz defines the uniquely oxygen pressure. This assemblage is, therefore used as a buffer in experiments where the oxygen fugacity must be controlled. This, for example, is the case when reactants contain transition elements such as iron, manganese, chromium, etc.

Table 7.3 Thermodynamic data for fayalite, magnetite, quartz and oxygen (Holland and Powell 1990)

Phase	$\Delta_f H_{298}[\text{kJmol}^{-1}]$	$S_{298}[\text{Jmol}^{-1}\text{K}^{-1}]$	$V_{298}[\text{cm}^3\cdot\text{mol}^{-1}]$
Fayalite	- 1478.80	151.00	46.30
Magnetite	- 1115.81	146.10	44.52
Quartz	- 910.80	41.50	22.69
Oxygen	0.00	205.20	-

Using the thermodynamic data in Tab. 7.3, the following values for the changes in the thermodynamic functions associated with the oxidation of fayalite are obtained:

$$\Delta_r H_{298} = -527.62 \text{ kJ},$$

$$\Delta_r S_{298} = -241.50 \text{ JK}^{-1} \text{ and}$$

$$\Delta_r V^s_{298} = -18.21 \text{ cm}^3.$$

The oxygen pressure at 1173 K and 0.1 MPa is then:

$$\ln P_{O_2} = \frac{-527620\,\text{J}}{8.3144\,\text{JK}^{-1} \times 1173\,\text{K}} + \frac{241.50\,\text{JK}^{-1}}{8.3144\,\text{JK}^{-1}} + \ln(10^5) = -13.54$$

$$P_{O_2} = \exp(-13.54) = \underline{1.32\times10^{-6} \text{ Pa.}}$$

Fig 7.3 shows the logarithm of the oxygen pressure as a function of reciprocal temperature at 0.1 MPa and 1.0 GPa total pressure.

Example 3: The reaction between phlogopite, $KMg_3[AlSi_3O_{10}](OH)_2$, calcite, $CaCO_3$, and quartz, SiO_2, yields tremolite, $Ca_2Mg_5Si_8O_{22}(OH)_2$, sanidine, $KAlSi_3O_8$, carbon dioxide, CO_2, and water, H_2O, and reads:

$$5KMg_3[AlSi_3O_{10}](OH)_2 + 6CaCO_3 + 24SiO_2 \rightarrow \\ 3Ca_2Mg_5Si_8O_{22}(OH)_2 + 5KAlSi_3O_8 + 6CO_2 + 2H_2O.$$

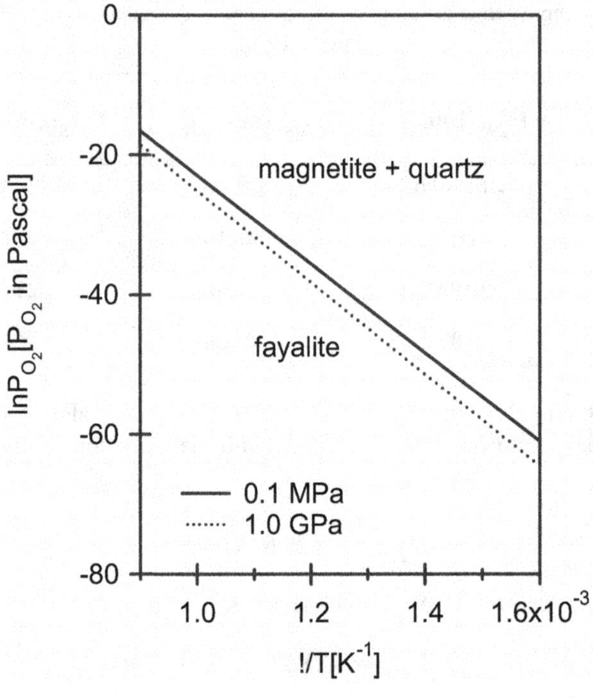

Fig. 7.4 Plot of the logarithm of the oxygen pressure as a function of the reciprocal temperature for the reaction $3Fe_2SiO_4 + O_2 = 2Fe_3O_4 + 3SiO_2$ at ambient pressure and at 1.0 GPa, calculated using the thermodynamic data of Holland and Powell (1990).

If all solids phases are pure end-members and the temperature and pressure dependence of the enthalpy, entropy and the volume of reaction is neglected, the equilibrium condition is given by:

$$\Delta_r \boldsymbol{H}_{298} - T\Delta_r \boldsymbol{S}_{298} + \Delta_r V^s_{298}(P - P_o) = -6RT\ln\frac{P_{CO_2}}{P_o} - 2RT\ln\frac{P_{H_2O}}{P_o}.$$

The partial pressures of the two gases can be calculated applying Dalton's law, according to which holds that:

$$P_i = x_i P_{tot}, \tag{7.49}$$

where P_{tot} and x_i are the total pressure and the mole fraction of gas i, respectively.

Using Eq. (7.49), the following partial pressures are obtained:

$$P_{CO_2} = \frac{6}{6+2} P_{tot} = \frac{3}{4} P_{tot}$$

and

$$P_{H_2O} = \frac{2}{6+2} P_{tot} = \frac{1}{4} P_{tot}.$$

Assuming that the total pressure is made up by the two gases and that the changes in the thermodynamical functions associated with the reaction are independent of pressure, the equilibrium temperature at any given pressure is calculated according to

$$T = \frac{\Delta_r \boldsymbol{H}_{298} + \Delta_r \boldsymbol{V}^s_{298}(P - P_o)}{\Delta_r \boldsymbol{S}_{298} - 6RT\ln(3/4 \times P/P_o) - 2RT\ln(1/4 \times P/P_o)},$$

where P and P_0 give the total and the ambient pressure, respectively.

Using the thermodynamic data for the pure end-member phases of Holland and Powell (1990) the following values for the changes in the thermodynamic functions associated with the reaction of phlogopite, calcite and quartz to tremolite, sanidine, carbon dioxide and water are obtained:

$$\Delta_r \boldsymbol{H}_{298} = 617.14 \text{ kJ},$$

$$\Delta_r \boldsymbol{S}_{298} = 1288.6 \text{ JK}^{-1} \text{ and}$$

$$\Delta_r \boldsymbol{V}^s_{298} = -153.5 \text{ cm}^3.$$

The calculation of the equilibrium temperature at the ambient pressure of 0.1 MPa yields:

$$T = \frac{617140\,\text{J}}{1288.6\,\text{JK}^{-1} - 8.3144\,\text{JK}^{-1} \times [6\ln(3/4) + 2\ln(1/4)]} = \underline{465\ \text{K}}$$

and at 0.5 GPa:

$$T = [617140\,\text{J} - 153.5\times10^{-6}\,\text{m}^3(0.5\times10^9\,\text{Pa}$$

$$-\,0.1\times10^6\,\text{Pa})]/\left\{1288.6\,\text{JK}^{-1} - 8.3144\,\text{JK}^{-1}\right.$$

$$\left.\times\left[6\ln\left(3/4\times\frac{0.5\times10^9}{0.1\times10^6}\right) + 2\ln\left(1/4\times\frac{0.5\times10^9}{0.1\times10^6}\right)\right]\right\} = \underline{711\ \text{K}}.$$

Example 3: In the foregoing example the mole fraction of carbon dioxide, x_{CO_2}, in the gas phase is 0.75. We want now to calculate the upper thermal stability of the phase assemblage phlogopite, calcite and quartz for two other gas compositions, namely for $x_{CO_2} = 0.25$ and 0.5. We begin with $x_{CO_2} = 0.25$ and calculate the equilibrium temperature at the ambient pressure of 0.1 MPa:

$$T = \frac{617140\,\text{J}}{1288.6\,\text{JK}^{-1} - 8.3144\,\text{JK}^{-1} \times [6\ln(0.25) + 2\ln(0.75)]} = \underline{453\ \text{K}},$$

and at 0.5 GPa:

$$T = \frac{617140\,\text{J} - 153.5\times10^{-6}\,\text{m}^3(0.5\times10^9\,\text{Pa} - 0.1\times10^6\,\text{Pa})}{1288.6\,\text{JK}^{-1} - 8.3144\,\text{JK}^{-1} \times \left[6\ln\left(0.25\frac{0.5\times10^9}{0.1\times10^6}\right) + 2\ln\left(0.75\frac{0.5\times10^9}{0.1\times10^6}\right)\right]}$$

$$= \underline{679\ \text{K}}.$$

Analogous calculations for $x_{CO_2} = 0.5$ yield:

$$T = \frac{617140\,\text{J}}{1288.6\,\text{JK}^{-1} - 8.3144\,\text{JK}^{-1} \times [6\ln(0.5) + 2\ln(0.5)]} = \underline{462\ \text{K}}$$

at 0.1 MPa and

$$T = \frac{617140\,\text{J} - 153.5\times10^{-6}\,\text{m}^3 \times (0.5\times10^9\,\text{Pa} - 0.1\times10^6\,\text{Pa})}{1288.6\,\text{JK}^{-1} - 8.3144\,\text{JK}^{-1} \times \left[6\ln\left(0.5\frac{0.5\times10^9}{0.1\times10^6}\right) + 2\ln\left(0.5\frac{0.5\times10^9}{0.1\times10^6}\right)\right]}$$

$$= \underline{703\ \text{K}}$$

at 0.5 GPa.

Fig. 7.5 shows the equilibrium curves as a function of temperature for all three gas compositions.

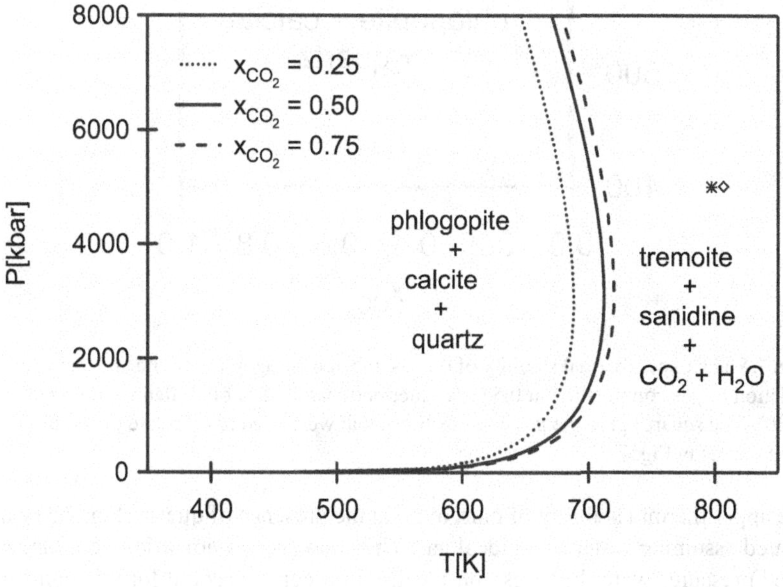

Fig. 7.5 The upper thermal stability of the assemblage phlogopite + calcite + quartz for three different gas compositions. The curves are calculated assuming ideal thermodynamic behavior of the gases. The asterisk and the diamond give the equilibrium temperatures for $x_{CO_2} = 0.75$ for the case that CO_2 and H_2O are non-ideal gases in ideal and non-ideal mixtures, respectively (see text).

Fig. 7.6 shows the upper thermal stability of the assemblage phlogopite/calcite/quartz as a function of gas composition at a constant pressure of 0.5 GPa. The three gas compositions which were used in the calculation of the equilibrium curves in Fig. 7.5 are marked by the open squares.

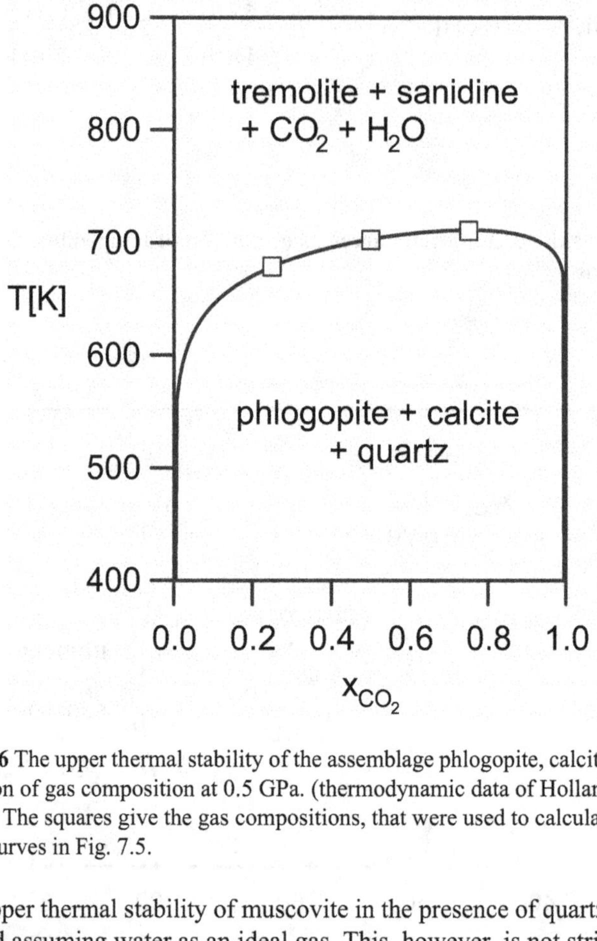

Fig. 7.6 The upper thermal stability of the assemblage phlogopite, calcite and quartz as a function of gas composition at 0.5 GPa. (thermodynamic data of Holland and Powell 1990). The squares give the gas compositions, that were used to calculate the equilibrium curves in Fig. 7.5.

The upper thermal stability of muscovite in the presence of quartz (Fig. 7.3) was calculated assuming water as an ideal gas. This, however, is not strictly correct. At elevated pressures water behaves non-ideally. In order to account for this non-ideality, the fugacity coefficient has to be introduced into the calculations. If the solid components are pure end-member phases and the changes in thermodynamic functions are considered to be independent of pressure and temperature, the equilibrium condition reads:

$$\Delta_r \boldsymbol{H}_{298} - T\Delta_r \boldsymbol{S}_{298} + \Delta_r \boldsymbol{V}^{s}_{298}(P - P_o) + RT\sum_j \nu_j^{g} \ln\left(\frac{\boldsymbol{P}_j}{P_o}\varphi_j\right) = 0, \qquad (7.50)$$

where $\boldsymbol{P}_j$ and φ_j are the pressure and the fugacity coefficient of the gas, respectively.

At any given pressure, the equilibrium temperature is calculated according to:

$$T = \frac{\Delta_r \boldsymbol{H}_{298} + \Delta_r \boldsymbol{V}^s_{298}(P - P_o)}{\Delta_r \boldsymbol{S}_{298} - RT\sum_j \mathrm{v}^g_j \ln\left(\frac{\boldsymbol{P}_j}{P_o}\varphi_j\right)}. \tag{7.51}$$

Because the fugacity coefficient is a function of temperature and pressure, Eq. (7.51) can only be solved iteratively.

Example: Consider now the reaction between muscovite and quartz once again, but this time assume that steam is a non-ideal gas. Inserting the thermodynamic data from page 295 into Eq. (7.51) yields:

$$T = \frac{104890\,\mathrm{J} - 4.57\times10^{-6}\mathrm{m}^3 \times (P[\mathrm{Pa}] - P_o[\mathrm{Pa}])}{184.3\,\mathrm{JK}^{-1} - 8.3144\,\mathrm{JK}^{-1} \times \ln\left(\frac{\boldsymbol{P}_{H_2O}}{P_o}\varphi_{H_2O}\right)}.$$

We assume, again, that the total pressure is equal to the pressure of steam and, therefore, the following relation holds $P = \boldsymbol{P}_{H_2O}$.

The equilibrium temperature at the ambient pressure is the same as in the case for an ideal gas because all terms containing pressure vanish.

For a pressure of 0.05 GPa, however, the equation can not be solved analytically, because the fugacity coefficient is a function of both pressure and temperature. An iterative calculation method is, therefore, made. To begin, the fugacity coefficient is set to 1 and the same result is obtained as in the case of an ideal gas, namely:

$$T = \frac{104890\,\mathrm{J} - 4.57\times10^{-6}\mathrm{m}^3 \times (0.05\times10^9\mathrm{Pa} - 10^5\mathrm{Pa})}{184.3\,\mathrm{JK}^{-1} - 8.3144\,\mathrm{JK}^{-1} \times \ln(0.05\times10^9/10^5 \times 1)} = \underline{789\ \mathrm{K}}.$$

At 789 K and 0.05 GPa, the fugacity coefficient for water has a value of 0.681 (Grevel and Chatterjee 1992). Using this value, the second cycle is calculated and a temperature of:

$$T = \frac{104890\,\mathrm{J} - 4.57\times10^{-6}\mathrm{m}^3 \times (0.05\times10^9\mathrm{Pa} - 10^5\mathrm{Pa})}{184.3\,\mathrm{JK}^{-1} - 8.3144\,\mathrm{JK}^{-1} \times \ln(0.05\times10^9/10^5 \times 0.681)} = \underline{771\ \mathrm{K}}$$

results. For a temperature of 771 K and at 0.05 GPa, a fugacity coefficient value of 0.649 is obtained (Grevel and Chatterjee 1992). In the next cycle, a temperature of 768 K is calculated and the corresponding value of the fugacity coefficient is 0.644. The next calculation yields a temperature of 768 K, which is the same as in the foregoing cycle and, therefore, it represents the final result. The entire dehydration curve is shown in Fig. 7.3.

If more than one gas participates in a reaction, the gaseous phase is as a mixture. Thus, the partial fugacities of the different gas components must be known to calculate the equilibrium conditions. However, if the mixing properties of the gases are unknown, the fugacities of pure gases can be used instead. That is, the solution is assumed to be ideal.

Example: In the examples where we calculated the equilibrium conditions for the reaction of phlogopite, calcite and quartz to tremolite, sanidine, carbon dioxide and water, the gaseous phase was considered to be an ideal mixture of ideal gases. In order to account for the non-ideal behavior of the gases, we will now introduce fugacity coefficients into the calculation. As a first approximation, we will assume that the gases are non-ideal, but that they mix ideally. Hence, we can use the fugacities for pure carbon dioxide and pure water. We want to calculate the equilibrium temperature at $P = 0.5$ GPa for a gas composition following the stoichiometry of the reaction, namely $x_{CO_2} = 3/4$. For this purpose, the equation that was used in the calculations where both gases were considered to be ideal has, thus, to be modified and reads then:

$$T = \frac{\Delta_r \boldsymbol{H}_{298} + \Delta_r \boldsymbol{V}_{298}^{s}(P - P_o)}{\Delta_r \boldsymbol{S}_{298} - 6RT\ln\left(3/4 \times \frac{P}{P_o}\varphi_{CO_2}\right) - 2RT\ln\left(1/4 \times \frac{P}{P_o}\varphi_{H_2O}\right)}.$$

Inserting the numeric data yields:

$$T = [617140\,\mathrm{J} - 153.5\times10^{-6}\mathrm{m}^3 \times (0.5\times10^9 - 0.1\times10^6)] / \left\{ 1288.6\,\mathrm{JK}^{-1} \right.$$

$$\left. - 8.3144\,\mathrm{JK}^{-1} \times \left[6\ln\left(3/4 \times \frac{0.5\times10^9}{0.1\times10^6}\varphi_{CO_2}\right) + 2\ln\left(1/4 \times \frac{0.5\times10^9}{0.1\times10^6}\varphi_{H_2O}\right)\right]\right\}.$$

Because the fugacity coefficients are a function of both temperature and pressure, the iteration procedure has to be applied once again. As a first step, the activity coefficients are set to one and the equilibrium temperature is calculated. Then, the values for the activity coefficients corresponding to the temperature obtained from the first step are used to calculate the next temperature, etc. This procedure is repeated until a constant value for the equilibrium temperature is obtained. The entire iteration procedure with intermediate and final results is given in Table 7.4.

Table 7.4 Calculation of the equilibrium temperature for the reaction phlogopite + calcite = tremolite + sanidine + CO_2 + H_2O at 0.5 GPa assuming an ideal mixing of non-ideal gases CO_2 and H_2O (data of Holland and Powell 1991 and Grevel and Chatterjee 1992).

	φ_{CO_2}	φ_{H_2O}	resulting temp. [K]
1. step	1.0	1.0	711
2. step	7.513	0.305	796
3. step	6.680	0.457	797
4. step	6.671	0.459	**797**

As can be seen from Tab. 7.4, the resulting value for the temperature is 797 K. This value has to be compared with 711 K, which is the equilibrium temperature in the case where the two gases are considered to be ideal.

If the gas phase is considered to be a non-ideal mixture of two non-ideal gases, according to Eq. (5.73), the fugacity coefficients for the pure gases CO_2 and H_2O must be replaced by their partial fugacity coefficients φ_i, which depend not only on temperature and pressure but also on composition of the gas mixture. Thus, the calculation reads:

$$T = [617140\,\text{J} - 153.5\times10^{-6}\text{m}^3 \times (0.5\times10^9\text{Pa} - 0.1\times10^6\text{Pa})]/\Bigg\{1288.6\,\text{JK}$$

$$-8.3144\,\text{JK}^{-1} \times \left[6\ln\left(3/4\times\frac{0.5\times10^9}{0.1\times10^6}\varphi_{CO_2}\right) + 2\ln\left(1/4\times\frac{0.5\times10^9}{0.1\times10^6}\varphi_{H_2O}\right)\right]\Bigg\}.$$

As in the foregoing examples, the equation can, once again, be solved only iteratively. The iteration steps, the intermediate and the final result are listed in Tab. 7.5.

The resulting temperature of 807 K differs by only 10 degrees from that obtained when considering CO_2 and H_2O to be non-ideal gases in an ideal gas mixture. Apparently, disregarding the effect of non-ideality of the gases introduces a much larger error than disregarding the non-ideality of the gas mixture. The two equilibrium temperatures that were obtained in the last two calculations are shown in Fig. 7.5 as an asterisk and diamond.

Table 7.5 Iteration procedure used to determine the equilibrium temperature for the assemblage phlogopite/calcite/sanidine/tremolite/CO_2/H_2O assuming the gas phase to be a non-ideal mixture of two non-ideal gases (thermodynamic data of Holland and Powell 1991; Aranovich and Newton 1999).

	φ_{CO_2}	φ_{H_2O}	resulting temp- [K]
1. step	1.0	1.0	711
2. step	7.713	0.496	808
3. step	6.732	0.734	807
4. step	6.741	0.732	807

7.1.4 Distribution coefficient, K_D

The distribution of an element between two or more coexisting phases depends upon pressure, temperature and the composition of the phases. This is one of the consequences resulting from equilibrium conditions that require equality of chemical potentials of components in coexisting phases. Because the standard chemical potential of a component is different in different phases, the term $RT\ln a$ is required to compensate for the difference.

Consider a hypothetical ternary system *A-B-C* where the component *C* is soluble in *A* and *B*, but *B* is not soluble in *A* and *A* is also not soluble in *B*.

For the state of thermodynamic equilibrium it holds that

$$\mu_C^{(A,\,C)} = \mu_C^{(B,\,C)}, \tag{7.52}$$

or if the chemical potentials of the component *C* in the two coexisting phases are expressed in terms of the standard potential and activity, $RT\ln a_i$:

$$\mu_C^{\mathrm{o},\,(A,\,C)} + RT\ln a_C^{(A,\,C)} = \mu_C^{\mathrm{o},\,(B,\,C)} + RT\ln a_C^{(B,\,C)}. \tag{7.53}$$

Rearranging terms in Eq. (7.53) and considering the definition of chemical activity gives:

$$\frac{a_C^{(A,C)}}{a_C^{(A,C)}} = \frac{x_C^{(A,C)} \cdot \gamma_C^{(A,C)}}{x_C^{(B,C)} \cdot \gamma_C^{(B,C)}} = \exp\left\{\frac{\mu_C^{o,(B,C)} - \mu_C^{o,(A,C)}}{RT}\right\} = const. \qquad (7.54)$$

The relationship between the mole fractions of the component C in the coexisting phases is referred to as the *distribution coefficient, K_D*. In ideal solutions, where the activity coefficient γ_i equals 1, the distribution coefficient depends on pressure and temperature but is independent of composition. At constant pressure and temperature, therefore, it holds:

$$K_D = \frac{x_C^{(A,C)}}{x_C^{(B,C)}}. \qquad (7.55)$$

A special case of the relationship, as given in Eq. (7.55), is the so-called *Nernst distribution law*. It holds for the partitioning of a dilute component that obeys the Henry's law between two liquid phases. In this case, the activity coefficients differ from 1 but have constant values. They can. therefore, be incorporated into the standard potentials and Eq. (7.54) obtains the following form:

$$\frac{x_C^{(A,C)}}{x_C^{(B,C)}} = \exp\left\{\frac{\mu_C^{\infty(B,C)} - \mu_C^{\infty(A,C)}}{RT}\right\} = const, \qquad (7.56)$$

where A and B are two immiscible solvents. $\mu_C^{\infty(A,C)}$ and $\mu_C^{\infty(A,C)}$ are the standard chemical potentials of the component C referring to the state of infinite dilution of C in A and B, respectively.

Example: Seck (1971) determined experimentally the phase relations in the ternary system $NaAlSi_3O_8$ - $KAlSi_3O_8$ - $CaAl_2Si_2O_8$. According to his experiments, at 650°C and 0.1 GPa, the solubility of anorthite, $CaAl_2Si_2O_8$, in an alkali feldspar solid solution, $(Na,K)AlSi_3O_8$, is very low (1.5 to 2.0 mol%) and can be neglected as a first approximation. Similarly, the solubility of potassium feldspar, $KAlSi_3O_8$, in a plagioclase solid solution, $(Ca,Na)(Al,Si)AlSi_2O_8$, at low albite concentrations is very low. Thus, two solid solution series exist (alkali feldspar and plagioclase) in which the $NaAlSi_3O_8$ component is distributed. The equilibrium condition, therefore reads:

$$\mu_{NaAlSi_3O_8}^{akf} = \mu_{NaAlSi_3O_8}^{pl}$$

or

$$\mu^{o,\,akf}_{NaAlSi_3O_8} + RT\ln a^{akf}_{NaAlSi_3O_8} = \mu^{o,\,pl}_{NaAlSi_3O_8} + RT\ln a^{pl}_{NaAlSi_3O_8},$$

if pure albite is taken as a standard state for the component $NaAlSi_3O_8$ in both solid solution series.

Considering the definition of the activity ($a_i = x_i\gamma_i$), the equation given above can be rewritten as follows:

$$RT\ln\frac{x^{akf}_{NaAlSi_3O_8}}{x^{pl}_{NaAlSi_3O_8}} = RT\ln\frac{\gamma^{pl}_{NaAlSi_3O_8}}{\gamma^{akf}_{NaAlSi_3O_8}}.$$

Saxena (1973) used this relationship to calculate the activity of albite in alkali feldspar and plagioclase based on the data of Seck (1971).

In the preceding example, there was only one component partitioning between two coexisting phases. Another possibility is that two components are distributed between two coexisting phases whose compositions can be described by the hypothetical formulas: (A,B)X and (A,B)Y. In this case, the distribution of the components *A* and *B* between the two phases is interdependent.

The formulation of the thermodynamic equilibrium conditions for this type of distribution are typically expressed in terms of *exchange reaction*. In the case of our hypothetical system it reads:

$$AX + BY = AY + BX, \tag{7.57}$$

where AX, BX, AY and BY are the end-member components.

The thermodynamic equilibrium constant for the reaction (7.57) is given by

$$K_{P,\,T} = \frac{a_{AY}\cdot a_{BX}}{a_{AX}\cdot a_{BY}} = \frac{x_{AY}\cdot x_{BX}}{x_{AX}\cdot x_{BY}}\cdot\frac{\gamma_{AY}\cdot\gamma_{BX}}{\gamma_{AX}\cdot\gamma_{BY}} \tag{7.58}$$

or

$$K_{P,\,T} = K_D\cdot K_\gamma, \tag{7.59}$$

where

$$K_D = \frac{x_{AY} \cdot x_{BX}}{x_{AX} \cdot x_{BY}} \tag{7.60}$$

and

$$K_\gamma = \frac{\gamma_{AY} \cdot \gamma_{BX}}{\gamma_{AX} \cdot \gamma_{BY}}. \tag{7.61}$$

According to Ramberg and DeVore (1951), K_D in Eq. (7.60) is also called the distribution coefficient, although it differs from that defined by Nerst.

In the case that the phases (A,B)X and (A,B)Y are ideal mixtures the distribution coefficient depends only upon temperature and pressure. However, if the mixtures are non-ideal, it depends also on total composition of the system.

Example: The partitioning of magnesium iron between garnet, $(Mg,Fe)_3Al_2Si_3O_{12}$, and biotite, $K(Mg,Fe)_3AlSi_3O_{10}(OH)_2$, can be used to determine the crystallization temperature of metamorphic rocks containing these minerals.

The correspondent exchange reaction reads:

$$\begin{aligned} &Fe_3Al_2Si_3O_{12} + KMg_3AlSi_3O_{10}(OH)_2 \\ &= Mg_3Al_2Si_3O_{12} + KFe_3AlSi_3O_{10}(OH)_2. \end{aligned} \tag{7.62}$$

If the enthalpy of reaction, entropy of reaction, and volume of reaction are independent of temperature and pressure, thermodynamic equilibrium is given by:

$$\Delta_r \boldsymbol{H}_{298} - T\Delta_r \boldsymbol{S}_{298} + \Delta_r V_{298}(P - P_o) + RT\ln K_{P,T} = 0,$$

where

$$K_{P,T} = \frac{a^{grt}_{Mg_3Al_2Si_3O_{12}} \cdot a^{bt}_{KFe_3AlSi_3O_{10}(OH)_2}}{a^{grt}_{Fe_3Al_2Si_3O_{12}} \cdot a^{bt}_{KMg_3AlSi_3O_{10}(OH)_2}} \tag{7.63}$$

or

$$K_{P,T} = \frac{(x^{grt}_{Mg_3Al_2Si_3O_{12}})^3 \cdot (x^{bt}_{KFe_3AlSi_3O_{10}(OH)_2})^3}{(x^{grt}_{Fe_3Al_2Si_3O_{12}})^3 \cdot (x^{bt}_{KMg_3AlSi_3O_{10}(OH_2)})^3} \times \frac{(\gamma^{grt}_{Mg_3Al_2Si_3O_{12}})^3 \cdot (\gamma^{bt}_{KFe_3AlSi_3O_{10}(OH)_2})^3}{(\gamma^{grt}_{Fe_3Al_2Si_3O_{12}})^3 \cdot (\gamma^{bt}_{KMg_3AlSi_3O_{10}(OH)_2})^3}. \quad (7.64)$$

The exponents in Eq. (7.64) account for the three crystallographic sites per formula unit that are available to magnesium and iron in garnet as well as in mica. If the thermodynamic constant $K_{P,T}$ is expressed in terms of K_D and K_γ, one obtains:

$$K_{P,T} = 3K_D \cdot 3K_\gamma . \quad (7.65)$$

Assuming that garnet, as well as mica, are ideal solid solutions, thermodynamic equilibrium is given by:

$$\Delta_r \boldsymbol{H}_{298} - T\Delta_r \boldsymbol{S}_{298} + \Delta_r \boldsymbol{V}_{298}(P - P_o) + 3RT\ln K_D = 0 \quad (7.66)$$

or solving for $\ln K_D$:

$$\ln K_D = -\frac{\Delta_r \boldsymbol{H}_{298}}{3RT} + \frac{\Delta_r \boldsymbol{S}_{298}}{3R} - \frac{\Delta_r \boldsymbol{V}_{298}}{3RT}(P - P_o). \quad (7.67)$$

If the negative logarithm of the distribution coefficient $\ln K_D$, is plotted versus reciprocal temperature $1/T$, a line with the slope

$$\frac{\Delta_r \boldsymbol{H}_{298} + \Delta_r \boldsymbol{V}_{298}(P - P_o)}{3R} \quad (7.68)$$

is obtained.

The term

$$\frac{\Delta_r \boldsymbol{S}_{298}}{3R} \quad (7.69)$$

gives the intersection of the $\ln K_D$ vs. $1/T$ line at $1/T = 0$.

Ferry and Spear (1978) determined experimentally the equilibrium composition of garnet and biotite as a function of temperature at 0.207 GPa. Their results are shown graphically in Fig. 7.7. In the experiments marked by open squares, the equilibrium was approached from the left side of Eq. (7.62) that means, the iron component in mica increased in these runs and that in garnet decreased. The filled squares indicate the runs in which the exchange reaction proceeded in the opposite direction, that is, the iron content in garnet increased and that in mica decreased. The dotted line was calculated using the values of the thermodynamic functions $\Delta_r \boldsymbol{H}$, $\Delta_r \boldsymbol{S}$ and $\Delta_r \boldsymbol{V}$ at standard conditions. In order to calculate the dashed line, these functions were calculated to 970 K, which is the mean value of all the temperatures that Ferry and Spear (1978) used in their experiments.

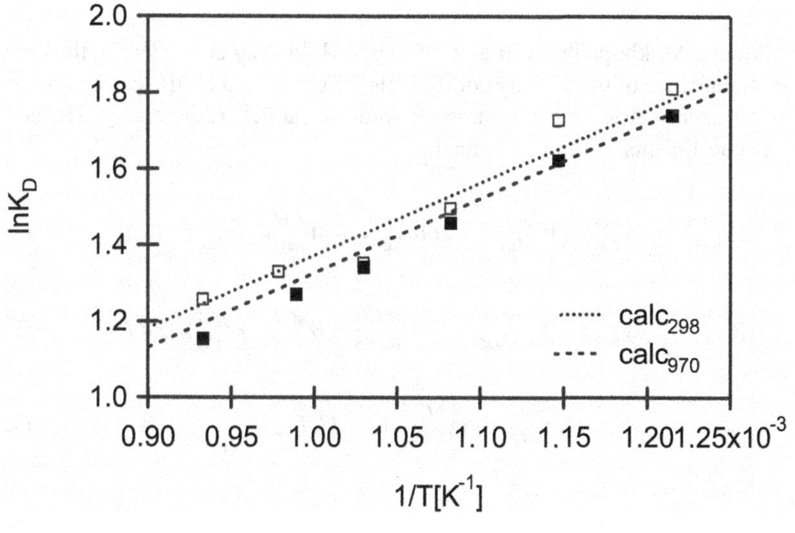

Fig. 7.7 Plot of $\ln K_D$ vs. reciprocal temperature for the exchange reaction of magnesium and iron between garnet and biotite. The squares give the experimental results of Ferry and Spear (1978). Equilibrium was approached from both sides of the reaction (filled and open symbols). The dotted and dashed lines are calculated using the thermodynamic data of Holland and Powell (1990) at 298 K and at 970 K, respectively.

As can be seen in Fig. 7.7, the experimental $\ln K_D$ values agree fairly well with the calculated lines. This agreement does not, however, necessarily mean that the assumption of ideal mixing for garnet and biotite is correct. This is actually not the case, as the experimental determinations of the activity-composition relations for biotite and garnet show (e.g. Holdaway et al. 1997). Nonetheless, if the activity coefficients for garnet and mica have similar values, the value of K_γ approaches 1 and the thermodynamic equilibrium constant, $K_{P,T}$, can be replaced by the distribution coefficient, K_D.

If the phases involved in the reaction are considered to be non-ideal solid solutions, Eq. (7.67) obtains the form:

$$\ln K_D = -\frac{\Delta_r \boldsymbol{H}_{298}}{3RT} + \frac{\Delta_r \boldsymbol{S}_{298}}{3R} - \frac{\Delta_r \boldsymbol{V}_{298}}{3RT}(P - P_o) - \ln K_\gamma \,, \tag{7.70}$$

where

$$\ln K_\gamma = \ln\left(\frac{\gamma_{Mg}^{grt}}{\gamma_{Fe}^{grt}}\right) - \ln\left(\frac{\gamma_{\mathrm{Mg}}^{bt}}{\gamma_{\mathrm{Fe}}^{bt}}\right). \tag{7.71}$$

Following Mukhopadhyay et al. (1993) and Holdaway et al. (1997), the composition dependence of the activity coefficients for garnet and biotite can be represented by an asymmetric and a symmetric solution model, respectively. Hence, the activity coefficients in Eq. (7.71) read:

$$\ln\gamma_{Mg}^{grt} = (x_{Fe}^{grt})^2 [W_{MgFe}^{grt} + 2(W_{FeMg}^{grt} - W_{MgFe}^{grt}) x_{Mg}^{grt}]/3RT, \tag{7.72}$$

$$\ln\gamma_{Fe}^{grt} = (x_{Mg}^{grt})^2 [W_{FeMg}^{grt} + 2(W_{MgFe}^{grt} - W_{FeMg}^{grt}) x_{Mg}^{grt}]/3RT, \tag{7.73}$$

$$\ln\gamma_{Mg}^{bt} = (x_{Fe}^{bt})^2 W^{bt}/3RT, \tag{7.74}$$

and

$$\ln\gamma_{Fe}^{bt} = (x_{Mg}^{bt})^2 W^{bt}/3RT, \tag{7.75}$$

where W_{FeMg}^{grt}, W_{MgFe}^{grt} and W^{bt} are the interaction parameters for the components in the asymmetric garnet and in the symmetric biotite solid solution, respectively.

Substituting Eqs. (7.72) through (7.75) into Eq. (7.71), doing some algebra and remembering that $x_{Mg} + x_{Fe} = 1.0$, leads to:

$$\ln K_\gamma = [(x_{Fe}^{grt})^2 W_{MgFe}^{grt} - (x_{Mg}^{grt})^2 W_{FeMg}^{grt} + 2x_{Mg}^{grt} x_{Fe}^{grt} (W_{FeMg}^{grt} - W_{MgFe}^{grt}) - W^{bt}(x_{Fe}^{bt} - x_{Mg}^{bt})]/3RT. \tag{7.76}$$

Replacing $\ln K_\gamma$ in Eq. (7.70) by the expression given in Eq. (7.76) yields:

$$\ln K_D = -\frac{\Delta_r \boldsymbol{H}_{298}}{3RT} + \frac{\Delta_r \boldsymbol{S}_{298}}{3R} - \frac{\Delta_r \boldsymbol{V}_{298}}{3RT}(P - P_o) \tag{7.77}$$

$$-[(x_{Fe}^{grt})^2 W_{MgFe}^{grt} - (x_{Mg}^{grt})^2 W_{FeMg}^{grt}$$

$$+ 2x_{Mg}^{grt} x_{Fe}^{grt} (W_{FeMg}^{grt} - W_{MgFe}^{grt}) - W^{bt}(x_{Fe}^{bt} - x_{Mg}^{bt})]/3RT.$$

The interaction parameters in Eq. (7.77) are, according to Holdaway et al. (1997), temperature and pressure dependent and they read:

$$W_{MgFe}^{grt} = 22265 - 12.40\mathrm{T} + 0.05 \times 10^5 P \ [\mathrm{Jmol}^{-1}],$$

$$W_{FeMg}^{grt} = -24166 + 22.09\mathrm{T} - 0.034 \times 10^5 P \ [\mathrm{Jmol}^{-1}] \text{ and}$$

$$W^{bt} = 40719 - 30T \ [\mathrm{Jmol}^{-1}],$$

where T and P are given in K and Pa, respectively.

The relationship given in Eq. (7.77) can be used to evaluate the equilibration temperature of metamorphic rocks containing garnet and mica in the case that both phases contain no other components than iron and magnesium.

Reciprocal solutions

In the foregoing examples the cation mixing occurred only over one crystallographic site. In many solid solutions, however, mixing occurs over more than one site. These solutions are of the type $(A,B...)_\alpha(X,Y...)_\beta Z$. Examples among the rock-forming minerals are: garnet, $A_3B_2Si_3O_{12}$, pyroxene, $ABSi_2O_6$, spinel, AB_2O_4, etc. In such solid solutions the chemical potentials of the various possible end-member components are not all mutually independent. This is because the number of independent components necessary to describe the chemical variability of the phases is smaller than the number of the possible end-member components.

In order to derive the energetics a the reciprocal solution we follow Wood and Nicolls (1978).

Consider a hypothetical mineral with two different crystallographic sites having the general chemical formula:

$$(A, B)_\alpha (X, Y)_\beta Z_\gamma \, . \tag{7.78}$$

There are four possible end-member components, namely: $A_\alpha X_\beta Z_\gamma$, $B_\alpha X_\beta Z_\gamma$, $A_\alpha Y_\beta Z_\gamma$ and $B_\alpha Y_\beta Z_\gamma$. The composition of the solution, however, can be completely described using only three components as demonstrated by the following example.

Assume the composition of a solid solution $(A_{0.2}B_{0.8})_\alpha(X_{0.3}Y_{0.7})_\beta Z_\gamma$ is to be described using the three components $A_\alpha X_\beta Z_\gamma$, $B_\alpha X_\beta Z_\gamma$ and $A_\alpha Y_\beta Z_\gamma$. Because the mole fraction of the component $B_\alpha Y_\beta Z_\gamma$ is set to zero, the component $A_\alpha Y_\beta Z_\gamma$ is the only one that contains the cation *Y*. Its mole fraction must, therefore, be equal to the atomic fraction x_Y, namely 0.7. On the other hand, the mole fraction of the component $B_\alpha X_\beta Z_\gamma$ must be equal to the atomic fraction x_B, because $B_\alpha X_\beta Z\gamma$ is the only component containing *B* after the component $B_\alpha Y_\beta Z_\gamma$ has been excluded form consideration. Its mole fraction, therefore, equals 0.8. Finally, the mole fraction for the component $A_\alpha X_\beta Z_\gamma$ must equal the difference x_X - x_B. In our example this is 0.3 - 0.8 = - 0.5.

Clearly the values for the mole fractions depend on the choice of the components. This is demonstrated in Tab. 7.6, where all four possible sets of the ternary components and their relationships to the atomic fractions of the cations are presented.

Table 7.6 Mole fraction of the components in the solution $(A_{0.2}B_{0.8})_\alpha(X_{0.3}Y_{0.7})_\beta Z_\gamma$ using different ternary phase combinations. x_A, x_B, x_X and x_Y designate the atomic fractions of the cations mixing on the two different sites.

AXZ	BYZ	AYZ	BXZ
0.0	$0.5 = (x_Y - x_A)$	$0.2 = x_A$	$0.3 = x_X$
$-0.5 = (x_X - x_B)$	0.0	$0.7 = x_Y$	$0.8 = x_B$
$0.2 = x_A$	$0.7 = x_Y$	0.0	$0.1 = (x_X - x_A)$
$0.3 = x_X$	$0.8 = x_B$	$-0.1 = (x_Y - x_B)$	0.0

The molar Gibbs free energy of the solid solution consists of the sum of the standard potentials of the end-member components at the temperature and pressure of interest, times the corresponding mole fractions plus the Gibbs energy of mixing, i.e:

$$\bar{G} = \sum_i x_i \mu_i^{o} + \Delta_m \bar{G}. \tag{7.79}$$

The first term on the right side of equation Eq. (7.79) gives the Gibbs free energy of a mechanical mixture, hence it holds that:

$$\sum_i x_i \mu_i^{o} = \Delta_m \bar{G}^{mech}. \tag{7.80}$$

The second term in Eq. (7.79), contains the excess Gibbs free energy of mixing, $\Delta_m\overline{G}^{ex}$, and the configurational entropy, $\Delta_m\overline{S}^{conf}$ multiplied by the temperature T.

Using the relationships between the atomic fractions for the cations and the mole fractions for the ternary components as given in Tab. 7.6, four different expressions for the free Gibbs energy of a mechanical mixture can be formulated, namely:

$$\Delta_m\overline{G}^{mech}_{(A_\alpha X_\beta Z_\gamma)} = x_A\mu^o_{A_\alpha Y_\beta Z_\gamma} + x_X\mu^o_{B_\alpha X_\beta Z_\gamma} + (x_Y - x_A)\mu^o_{B_\alpha Y_\beta Z_\gamma}, \tag{7.81}$$

$$\Delta_m\overline{G}^{mech}_{(B_\alpha Y_\beta Z_\gamma)} = x_B\mu^o_{B_\alpha X_\beta Z_\gamma} + x_Y\mu^o_{A_\alpha Y_\beta Z_\gamma} + (x_X - x_B)\mu^o_{A_\alpha X_\beta Z_\gamma}, \tag{7.82}$$

$$\Delta_m\overline{G}^{mech}_{(A_\alpha Y_\beta Z_\gamma)} = x_A\mu^o_{A_\alpha X_\beta Z_\gamma} + x_Y\mu^o_{B_\alpha Y_\beta Z_\gamma} + (x_X - x_A)\mu^o_{B_\alpha X_\beta Z_\gamma} \tag{7.83}$$

and

$$\Delta_m\overline{G}^{mech}_{(B_\alpha X_\beta Z_\gamma)} = x_B\mu^o_{B_\alpha Y_\beta Z_\gamma} + x_x\mu^o_{A_\alpha X_\beta Z_\gamma} + (x_Y - x_B)\mu^o_{A_\alpha Y_\beta Z_\gamma}. \tag{7.84}$$

In Eqs. (7.81) through (7.84) the component set to zero is given in parenthesis as a subscript.

The configurational entropy is, according to Eq. (4.82), given by

$$\Delta_m\overline{S}^{conf} = -\alpha R(x_A \ln x_A + x_B \ln x_B) - \beta R(x_X \ln x_X + x_Y \ln x_Y). \tag{7.85}$$

According to Eqs. (7.81) through (7.84) there are four different possibilities for expressing the total Gibbs free energy of a solid solution, depending on which of the four equations are chosen to give the Gibbs free energy of the mechanical mixture. For example, if one takes the first expression, where the component $A_\alpha X_\beta Z_\gamma$ is set to zero, the total Gibbs free energy of the solution reads:

$$\overline{G} = x_A\mu^o_{A_\alpha Y_\beta Z_\gamma} + x_X\mu^o_{B_\alpha X_\beta Z_\gamma} + (x_Y - x_A)\mu^o_{B_\alpha Y_\beta Z_\gamma} + \Delta_m\overline{G}^{ex} - T\Delta_m\overline{S}^{conf}. \tag{7.86}$$

The chemical potentials of the end-member components remain the same regardless of which three components are taken. Hence, the Gibbs free energy of a mechanical mixture depends on the choice of components. The total Gibbs free energy of a solid solution does, however, not depend on the choice of the components. The difference must, therefore, be compensated by the term $\Delta_m\overline{G}^{ex}$.

In order to evaluate the term $\Delta_m\overline{G}^{ex}$, a random mixing on each site with no interaction between the atoms is assumed. Under these conditions, the probabilities of occurrence of configurations AX, AY, BX and BY in the solution equals the products between the atomic fractions, i.e. x_Ax_X, x_Ax_Y, x_Bx_X and x_Bx_Y. If the interaction energies between the pairs A-X, A-Y, B-X and B-Y are additive, the Gibbs free energy of the mechanical mixture should be given by the sum of the standard chemical potentials of the end-members multiplied by the probability of occurrence of the respective configuration in the solution:

$$\Delta_m\overline{G}^{mech} = x_Ax_X\mu^o_{A_\alpha X_\beta Z_\gamma} + x_Ax_Y\mu^o_{A_\alpha Y_\beta Z_\gamma} + x_Bx_X\mu^o_{B_\alpha X_\beta Z_\gamma} + x_Bx_Y\mu^o_{B_\alpha Y_\beta Z_\gamma}. \tag{7.87}$$

For the case of ideal mixing, the total Gibbs free energy on the individual sites is therefore:

$$\overline{G}^{id} = x_Ax_X\mu^o_{A_\alpha X_\beta Z_\gamma} + x_Ax_Y\mu^o_{A_\alpha Y_\beta Z_\gamma} + x_Bx_X\mu^o_{B_\alpha X_\beta Z_\gamma} + x_Bx_Y\mu^o_{B_\alpha Y_\beta Z_\gamma} - T\Delta_m\overline{S}^{conf}. \tag{7.88}$$

Subtracting Eq. (7.88) from Eqs. Eq. (7.86) and solving for $\Delta_m\overline{G}^{ex}$ yields:

$$\Delta_m\overline{G}^{ex}_{(A_\alpha X_\beta Z_\gamma)} = x_A\mu^o_{A_\alpha Y_\beta Z_\gamma} + x_X\mu^o_{B_\alpha X_\beta Z_\gamma} + (x_Y - x_A)\mu^o_{B_\alpha Y_\beta Z_\gamma} - x_Ax_X\mu^o_{A_\alpha X_\beta Z_\gamma} - x_Ax_Y\mu^o_{A_\alpha Y_\beta Z_\gamma} - x_Bx_X\mu^o_{B_\alpha X_\beta Z_\gamma} - x_Bx_Y\mu^o_{B_\alpha Y_\beta Z_\gamma}. \tag{7.89}$$

Substituting x_B by (1 - x_A) and x_Y by (1 - x_X) and doing some algebra leads to:

$$\Delta_m\overline{G}^{ex}_{(A_\alpha X_\beta Z_\gamma)} = x_Ax_X(\mu^o_{A_\alpha X_\beta Z_\gamma} + \mu^o_{B_\alpha Y_\beta Z_\gamma} - \mu^o_{A_\alpha Y_\beta Z_\gamma} - \mu^o_{B_\alpha X_\beta Z_\gamma}). \tag{7.90}$$

The expression in the parenthesis gives the Gibbs free energy of the exchange reaction:

$$A_\alpha Y_\beta Z_\gamma + B_\alpha X_\beta Z_\gamma = A_\alpha X_\beta Z_\gamma + B_\alpha Y_\beta Z_\gamma\,. \tag{7.91}$$

Reaction (7.91) relates the three end-member components with the fourth and is generally referred to as the *reciprocal reaction*. Using the symbol $\Delta_r\boldsymbol{G}^{rec}$ for the Gibbs free energy of the reciprocal reaction, Eq. (7.90) can be written as:

$$\Delta_m \overline{G}^{ex}_{(A_\alpha X_\beta Z_\gamma)} = x_A x_X \Delta_r \boldsymbol{G}^{rec}. \tag{7.92}$$

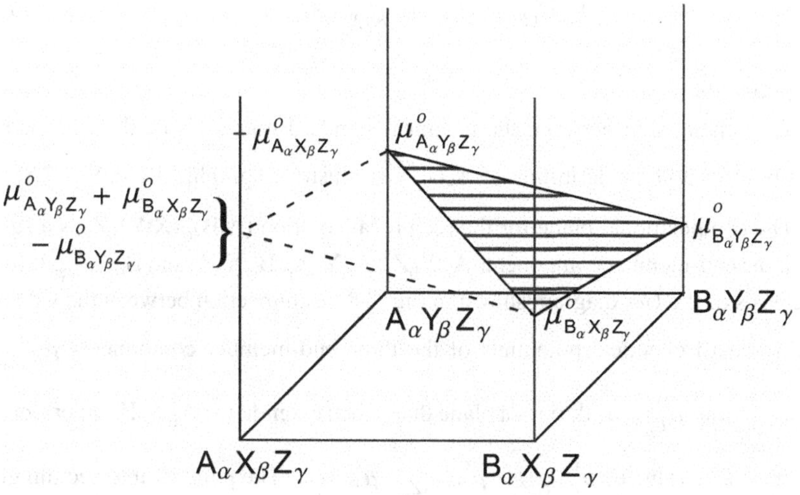

Fig. 7.8 The standard chemical potentials for the four end-member components of the reciprocal solution $(A,B)_\alpha(X,Y)_\beta Z_\gamma$. The hatched triangle is formed by the connection of the standard chemical potential values of the three end-member components $\mu^o_{A_\alpha Y_\beta Z_\gamma}$, $\mu^o_{B_\alpha X_\beta Z_\gamma}$ and $\mu^o_{B_\alpha Y_\beta Z_\gamma}$. The extension of this plane to $A_\alpha X_\beta Z_\gamma$ intersects the ordinate at $\mu^o_{A_\alpha Y_\beta Z_\gamma} + \mu^o_{B_\alpha X_\beta Z_\gamma} - \mu^o_{B_\alpha Y_\beta Z_\gamma}$, which is not the actual value of the standard chemical potential for the dependent component $A_\alpha X_\beta Z_\gamma$, $\mu^o_{A_\alpha X_\beta Z_\gamma}$. The difference between the point of intersection and $\mu^o_{A_\alpha X_\beta Z_\gamma}$ equals, therefore, $\mu^o_{A_\alpha X_\beta Z_\gamma} - (\mu^o_{A_\alpha Y_\beta Z_\gamma} + \mu^o_{B_\alpha X_\beta Z_\gamma} - \mu^o_{B_\alpha Y_\beta Z_\gamma})$ and corresponds to $\Delta_r \boldsymbol{G}^{rec}$.

Eq. (7.92) was derived using the expression for the Gibbs free energy for the mechanical mixture that was obtained by dropping the component $A_\alpha X_\beta Z_\gamma$. Using the other three expressions in Eqs. (7.82) through (7.84), one obtains:

$$\Delta_m \overline{G}^{ex}_{(B_\alpha Y_\beta Z_\gamma)} = x_B x_Y \Delta_r \boldsymbol{G}^{rec}, \tag{7.93}$$

$$\Delta_m \overline{G}^{ex}_{(A_\alpha Y_\beta Z_\gamma)} = -x_A x_Y \Delta_r \boldsymbol{G}^{rec} \tag{7.94}$$

and

$$\Delta_m \overline{G}^{ex}_{(B_\alpha X_\beta Z_\gamma)} = -x_B x_X \Delta_r \boldsymbol{G}^{rec}, \tag{7.95}$$

respectively.

The relationship between the chemical standard potentials of the end-member components and the definition of $\Delta_r \boldsymbol{G}^{rec}$ is illustrated graphically in Fig. 7.8.

The compositional plane for the reciprocal solution $(A,B)_\alpha(X,Y)_\beta Z_\gamma$ is a square with the end-member components $A_\alpha X_\beta Z_\gamma$, $A_\alpha Y_\beta Z_\gamma$, $B_\alpha X_\beta Z_\gamma$ and $B_\alpha Y_\beta Z_\gamma$. It forms the basis of the $\overline{G}$-x diagram shown in Fig. 7.8. A connection between the values of the standard chemical potentials of the three end-member components $\overset{o}{\mu}_{A_\alpha Y_\beta Z_\gamma}$, $\overset{o}{\mu}_{B_\alpha X_\beta Z_\gamma}$ and $\overset{o}{\mu}_{B_\alpha Y_\beta Z_\gamma}$ defines a plane that, when extended to $A_\alpha X_\beta Z_\gamma$, intersects the ordinate at a value of $\overset{o}{\mu}_{A_\alpha Y_\beta Z_\gamma} + \overset{o}{\mu}_{B_\alpha X_\beta Z_\gamma} - \overset{o}{\mu}_{B_\alpha Y_\beta Z_\gamma}$. The point of intersection gives the value of the standard chemical potential of component $A_\alpha X_\beta Z_\gamma$ only if the standard Gibbs free energy of the reciprocal reaction, $\Delta_r \boldsymbol{G}^{rec}$, is zero. According to Eqs. (7.90) and (7.92) it holds that

$$\Delta_r \boldsymbol{G}^{rec} = \overset{o}{\mu}_{A_\alpha X_\beta Z_\gamma} + \overset{o}{\mu}_{B_\alpha Y_\beta Z_\gamma} - \overset{o}{\mu}_{A_\alpha Y_\beta Z_\gamma} - \overset{o}{\mu}_{B_\alpha X_\beta Z_\gamma}$$

and this equation, thus, represents, thus, the difference between the actual value of the standard chemical potential for the end-member component $A_\alpha X_\beta Z_\gamma$ and the value at which the plane intersects the coordinate (see Fig. 7.8). In our example, it is assumed that the value of the standard chemical potential of the fourth component is greater than the sum $\overset{o}{\mu}_{A_\alpha Y_\beta Z_\gamma} + \overset{o}{\mu}_{B_\alpha X_\beta Z_\gamma} - \overset{o}{\mu}_{B_\alpha Y_\beta Z_\gamma}$, so that $\Delta_r \boldsymbol{G}^{rec} > 0$.

The excess chemical potential of any component in the solution is obtained by differentiating $\Delta_m \overline{G}^{ex}$ with respect to composition, i.e:

$$\mu^{ex}_{A_\alpha X_\beta Z_\gamma} = \Delta_m \overline{G}^{ex} - x_B \left(\frac{\partial \Delta_m \overline{G}^{ex}}{\partial x_B} \right)_{P,\,T,\,x_Y} - x_Y \left(\frac{\partial \Delta_m \overline{G}^{ex}}{\partial x_Y} \right)_{P,\,T,\,x_B}, \tag{7.96}$$

$$\mu^{ex}_{B_\alpha Y_\beta Z_\gamma} = \Delta_m\overline{G}^{ex} - x_A\left(\frac{\partial\Delta_m\overline{G}^{ex}}{\partial x_A}\right)_{P,\,T,\,x_X} - x_X\left(\frac{\partial\Delta_m\overline{G}^{ex}}{\partial x_X}\right)_{P,\,T,\,x_A}, \tag{7.97}$$

$$\mu^{ex}_{A_\alpha Y_\beta Z_\gamma} = \Delta_m\overline{G}^{ex} - x_B\left(\frac{\partial\Delta_m\overline{G}^{ex}}{\partial x_B}\right)_{P,\,T,\,x_X} - x_X\left(\frac{\partial\Delta_m\overline{G}^{ex}}{\partial x_X}\right)_{P,\,T,\,x_B}, \tag{7.98}$$

and

$$\mu^{ex}_{B_\alpha X_\beta Z_\gamma} = \Delta_m\overline{G}^{ex} - x_A\left(\frac{\partial\Delta_m\overline{G}^{ex}}{\partial x_A}\right)_{P,\,T,\,x_Y} - x_Y\left(\frac{\partial\Delta_m\overline{G}^{ex}}{\partial x_Y}\right)_{P,\,T,\,x_A}. \tag{7.99}$$

The differentiation procedure gives following results:

$$\mu^{ex}_{A_\alpha X_\beta Z_\gamma} = -x_B x_Y \Delta_r \boldsymbol{G}^{rec}, \tag{7.100}$$

$$\mu^{ex}_{B_\alpha Y_\beta Z_\gamma} = -x_A x_X \Delta_r \boldsymbol{G}^{rec}, \tag{7.101}$$

$$\mu^{ex}_{A_\alpha Y_\beta Z_\gamma} = x_B x_X \Delta_r \boldsymbol{G}^{rec} \tag{7.102}$$

and

$$\mu^{ex}_{B_\alpha X_\beta Z_\gamma} = x_A x_Y \Delta_r \boldsymbol{G}^{rec}. \tag{7.103}$$

In accordance with Eq. (5.114) the chemical potentials in Eqs. (7.100) through (7.103) can be replaced by the corresponding activity coefficients and one obtains:

$$RT\ln\gamma^{rec}_{A_\alpha X_\beta Z_\gamma} = -x_B x_Y \Delta_r \boldsymbol{G}^{rec}, \tag{7.104}$$

$$RT\ln\gamma^{rec}_{B_\alpha Y_\beta Z_\gamma} = -x_A x_X \Delta_r \boldsymbol{G}^{rec}, \tag{7.105}$$

$$RT\ln\gamma^{rec}_{A_\alpha Y_\beta Z_\gamma} = x_B x_X \Delta_r \boldsymbol{G}^{rec} \tag{7.106}$$

and

$$RT\ln\gamma^{rec}_{B_\alpha X_\beta Z_\gamma} = x_A x_Y \Delta_r \boldsymbol{G}^{rec}. \tag{7.107}$$

Using the relationships given in Eqs. (7.104) through (7.107), the chemical potentials for the four ternary components read:

$$\mu_{A_\alpha X_\beta Z_\gamma} = \boldsymbol{\mu}^{o}_{A_\alpha X_\beta Z_\gamma} + RT\ln(x^{\alpha}_{A} x^{\beta}_{X}) + RT\ln\gamma^{rec}_{A_\alpha X_\beta Z_\gamma} \tag{7.108}$$

$$\mu_{B_\alpha Y_\beta Z_\gamma} = \boldsymbol{\mu}^{o}_{B_\alpha Y_\beta Z_\gamma} + RT\ln(x^{\alpha}_{B} x^{\beta}_{Y}) + RT\ln\gamma^{rec}_{B_\alpha Y_\beta Z_\gamma}, \tag{7.109}$$

$$\mu_{A_\alpha Y_\beta Z_\gamma} = \boldsymbol{\mu}^{o}_{A_\alpha Y_\beta Z_\gamma} + RT\ln(x^{\alpha}_{A} x^{\beta}_{Y}) + RT\ln\gamma^{rec}_{A_\alpha Y_\beta Z_\gamma}, \tag{7.110}$$

and

$$\mu_{B_\alpha X_\beta Z_\gamma} = \boldsymbol{\mu}^{o}_{B_\alpha X_\beta Z_\gamma} + RT\ln(x^{\alpha}_{B} x^{\beta}_{X}) + RT\ln\gamma^{rec}_{B_\alpha X_\beta Z_\gamma}. \tag{7.111}$$

The reciprocal solution is only ideal if $\gamma_i^{rec} = 1$, which means that there is no interaction between the atoms on the different sublattices and, therefore, the term $\Delta_r \boldsymbol{G}^{rec}$ equals zero. The activity coefficient, γ_i^{rec}, thus accounts for the *intersite* interaction.

The last two terms in Eqs. (7.108) through (7.111) can be replaced by the activity of the corresponding component in the reciprocal solution i.e.

$$a_{A_\alpha X_\beta Z_\gamma} = x^{\alpha}_{A} x^{\beta}_{X} \gamma^{rec}_{A_\alpha X_\beta Z_\gamma}, \tag{7.112}$$

$$a_{B_\alpha Y_\beta Z_\gamma} = x^{\alpha}_{B} x^{\beta}_{Y} \gamma^{rec}_{B_\alpha Y_\beta Z_\gamma}, \tag{7.113}$$

$$a_{A_\alpha Y_\beta Z_\gamma} = x^{\alpha}_{A} x^{\beta}_{Y} \gamma^{rec}_{A_\alpha Y_\beta Z_\gamma} \tag{7.114}$$

and

$$a_{B_\alpha X_\beta Z_\gamma} = x^{\alpha}_{B} x^{\beta}_{X} \gamma^{rec}_{B_\alpha X_\beta Z_\gamma} \tag{7.115}$$

or

$$a_{A_\alpha X_\beta Z_\gamma} = x_A^\alpha x_X^\beta \exp\left\{-x_B x_Y(\Delta_r \boldsymbol{G}^{rec})/RT\right\}, \tag{7.116}$$

$$a_{B_\alpha Y_\beta Z_\gamma} = x_B^\alpha x_Y^\beta \exp\left\{-x_A x_X(\Delta_r \boldsymbol{G}^{rec})/RT\right\}, \tag{7.117}$$

$$a_{A_\alpha Y_\beta Z_\gamma} = x_A^\alpha x_Y^\beta \exp\left\{x_B x_X(\Delta_r \boldsymbol{G}^{rec})/RT\right\} \tag{7.118}$$

and

$$a_{B_\alpha X_\beta Z_\gamma} = x_B^\alpha x_X^\beta \exp\left\{x_A x_Y(\Delta_r \boldsymbol{G}^{rec})/RT\right\}. \tag{7.119}$$

Example: In garnet, $(Mg,Ca)_3(Al,Cr)_2Si_3O_{12}$, mixing takes place over two different sites. Magnesium and calcium occupy an 8-fold coordinated dodecahedral site, whereas aluminium and chromium enter the 6-fold coordinated octahedral site. This composition, thus, represents a reciprocal solution with the following end-member components: pyrope, $Mg_3Al_2Si_3O_{12}$, uwarowite, $Ca_3Cr_2Si_3O_{12}$, grossular, $Ca_3Al_2Si_3O_{12}$ and korringite, $Mg_3Cr_2Si_3O_{12}$. The reciprocal reaction that relates the fourth dependent component to the three independent ones reads:

$$Mg_3Al_2Si_3O_{12} + Ca_3Cr_2Si_3O_{12} = Mg_3Cr_2Si_3O_{12} + Ca_3Al_2Si_3O_{12}.$$

If pyrope, uvarovite and grossular are chosen to describe the thermodynamic properties of the solid solution, the expression for the molar Gibbs free energy, $\overline{G}$, has the form:

$$\overline{G} = x_{Mg}^{[8]}\boldsymbol{\mu}^{\text{o}}_{Mg_3Al_2Si_3O_{12}} + x_{Cr}^{[6]}\boldsymbol{\mu}^{\text{o}}_{Ca_3Cr_2Si_3O_{12}}$$

$$+ (x_{Al}^{[6]} - x_{Mg}^{[8]})\boldsymbol{\mu}^{\text{o}}_{Ca_3Al_2Si_3O_{12}} + RT\Big\{3[x_{Mg}^{[8]}\ln x_{Mg}^{[8]}$$

$$+ x_{Ca}^{[8]}\ln x_{Ca}^{[8]}] + 2[x_{Al}^{[6]}\ln x_{Al}^{[6]} + x_{Cr}^{[6]}\ln x_{Cr}^{[6]}]\Big\} + x_{Mg}^{[8]}x_{Cr}^{[6]}\Delta_r\boldsymbol{G}^{rec},$$

where $\Delta_r \boldsymbol{G}^{rec}$ is the Gibbs free energy of the exchange reaction given above.

If, in addition to the interactions between atoms on different sites, the interactions between atoms on individual sites is to be considered, the term for the excess Gibbs free energy has to be extended as follows:

Consider the two-site reciprocal solid solution $(A,B)_\alpha(X,Y)_\beta Z_\gamma$ once again and assume that A-B and X-Y form separate solid solutions that can be modelled by the simple solution model (see Eq. (5.134)). The excess Gibbs free energy of mixing is then described by:

$$\Delta_m \overline{G}^{ex}_{(A_\alpha X_\beta Z_\gamma)} = \alpha(x_A x_B W^G_{A-B}) + \beta(x_X x_Y W^G_{X-Y}) + x_A x_X \Delta_r \boldsymbol{G}^{rec}, \tag{7.120}$$

where W^G_{A-B} and W^G_{X-Y} give the interaction energies between the atoms within the sites and α and β account for the number of the respective sites. There are four expressions of the form as given in Eq. (7.120) possible depending on the choice of the ternary end-members triple. In the above example, $A_\alpha X_\beta Z_\gamma$ is considered to be the dependent component. The other three possible equations read:

$$\Delta_m \overline{G}^{ex}_{(B_\alpha Y_\beta Z_\gamma)} = \alpha(x_A x_B W^G_{A-B}) + \beta(x_X x_Y W^G_{X-Y}) + x_B x_Y \Delta_r \boldsymbol{G}^{rec}, \tag{7.121}$$

$$\Delta_m \overline{G}^{ex}_{(A_\alpha Y_\beta Z_\gamma)} = \alpha(x_A x_B W^G_{A-B}) + \beta(x_X x_Y W^G_{X-Y}) - x_A x_Y \Delta_r \boldsymbol{G}^{rec} \tag{7.122}$$

and

$$\Delta_m \overline{G}^{ex}_{(B_\alpha X_\beta Z_\gamma)} = \alpha(x_A x_B W^G_{A-B}) + \beta(x_X x_Y W^G_{X-Y}) - x_B x_X \Delta_r \boldsymbol{G}^{rec}. \tag{7.123}$$

The activity coefficients for the four components are obtained by differentiating expressions given in Eqs. (7.120) through (7.123) with respect to composition.

$$RT\ln\gamma_{A_\alpha X_\beta Z_\gamma} = \alpha x_B^2 W^G_{A-B} + \beta x_Y^2 W^G_{X-Y} - x_B x_Y \Delta_r \boldsymbol{G}^{rec}, \tag{7.124}$$

$$RT\ln\gamma_{B_\alpha Y_\beta Z_\gamma} = \alpha x_A^2 W^G_{\text{A-B}} + \beta x_X^2 W^G_{\text{X-Y}} - x_A x_X \Delta_r \boldsymbol{G}^{rec}, \tag{7.125}$$

$$RT\ln\gamma_{A_\alpha Y_\beta Z_\gamma} = \alpha x_B^2 W^G_{\text{A-B}} + \beta x_X^2 W^G_{\text{X-Y}} + x_B x_X \Delta_r \boldsymbol{G}^{rec} \tag{7.126}$$

and

$$RT\ln\gamma_{B_\alpha X_\beta Z_\gamma} = \alpha x_A^2 W_{\text{A-B}}^G + \beta x_Y^2 W_{\text{X-Y}}^G + x_A x_Y \Delta_r G^{rec}. \tag{7.127}$$

From Eqs. (7.124) to (7.127) it follows that in the case of non-ideal mixing on sites the activity coefficient of a component in the reciprocal solution has the form:

$$\gamma_{ij} = \gamma_i^{\alpha} \cdot \gamma_j^{\beta} \cdot \gamma^{rec}, \tag{7.128}$$

where i and j designate the different sublattices and α and β give the number of the respective sites per formula unit.

Example: Consider the distribution of iron and magnesium between olivine and spinel that can be represented by the following exchange reaction:

$$MgAl_2O_4 + 1/2Fe_2SiO_4 = FeAl_2O_4 + 1/2Mg_2SiO_4 \tag{7.129}$$

with the equilibrium constant:

$$K_{P,T} = K_D \cdot K_\gamma, \tag{7.130}$$

where

$$K_D = \frac{x_{Fe}^{sp} \cdot x_{Mg}^{ol}}{x_{Mg}^{sp} \cdot x_{Fe}^{ol}} \tag{7.131}$$

and

$$K_\gamma = \frac{\gamma_{FeAl_2O_4}^{sp} \cdot (\gamma_{Mg_2SiO_4}^{ol})^{1/2}}{\gamma_{MgAl_2O_4}^{sp} \cdot (\gamma_{Fe_2SiO_4}^{ol})^{1/2}}. \tag{7.132}$$

Hence, the equilibrium condition for the exchange reaction reads :

$$\Delta_r G^{exch} + RT\ln\frac{x_{Fe}^{sp} \cdot x_{Mg}^{ol}}{x_{Mg}^{sp} \cdot x_{Fe}^{ol}} + RT\ln\frac{\gamma_{FeAl_2O_4}^{sp} \cdot (\gamma_{Mg_2SiO_4}^{ol})^{1/2}}{\gamma_{MgAl_2O_4}^{sp} \cdot (\gamma_{Fe_2SiO_4}^{ol})^{1/2}} = 0. \tag{7.133}$$

In this example, it is assumed that atomic mixing occurs only on tetrahedral sites and the chemical formula of spinel solid solution reads: $(Mg,Fe)Al_2O_4$. In order to use K_D as a geothermometer for natural spinel peridotites, however, the effect of chromium in spinel on the partitioning behavior of iron and magnesium between olivine and spinel must be considered. In other words, spinel must be treated as a reciprocal solution, $(Mg,Fe)(Al,Cr)_2O_4$, containing iron and magnesium on tetrahedral sites and aluminium and chromium on octahedral sites. The corresponding reciprocal reaction is given by:

$$MgAl_2O_4 + FeCr_2O_4 = FeAl_2O_4 + MgCr_2O_4. \tag{7.134}$$

If the mixing on the individual sites is non-ideal, the activity coefficients for spinel and hercynite consist of three different terms accounting for three different energetic contributions. Two of them are necessary to account for the interactions between the atoms within the sublattice sites (tetrahedral and octahedral), while the third one accounts for the interaction between the atoms on different sublattices. The activity coefficients can, thus, be represented by:

$$\gamma^{sp}_{MgAl_2O_4} = \gamma^{sp}_{Mg} \cdot (\gamma^{sp}_{Al})^2 \cdot \gamma^{rec}_{MgAl_2O_4} \tag{7.135}$$

and

$$\gamma^{sp}_{FeAl_2O_4} = \gamma^{sp}_{Fe} \cdot (\gamma^{sp}_{Al})^2 \cdot \gamma^{rec}_{FeAl_2O_4}. \tag{7.136}$$

In the case that mixing on both sublattices obeys the simple mixture model, the activity coefficients have the form:

$$RT\ln\gamma^{sp}_{MgAl_2O_4} = (x^{sp}_{Fe})^2 W^{sp}_{FeMg} + 2(x^{sp}_{Cr})^2 W^{sp}_{CrAl} + x^{sp}_{Fe} \cdot x^{sp}_{Cr} \cdot \Delta_r \boldsymbol{G}^{rec} \tag{7.137}$$

and

$$RT\ln\gamma^{sp}_{FeAl_2O_4} = (x^{sp}_{Mg})^2 W^{sp}_{FeMg} + 2(x^{sp}_{Cr})^2 W^{sp}_{CrAl} - x^{sp}_{Mg} \cdot x^{sp}_{Cr} \cdot \Delta_r \boldsymbol{G}^{rec}, \tag{7.138}$$

where $\Delta_r \boldsymbol{G}^{rec}$ is the Gibbs free energy of the reciprocal reaction given in Eq. (7.134). W^{sp}_{FeMg} and W^{sp}_{CrAl} are the interaction parameters for the tetrahedral and octahedral sites, respectively.

Substituting the expressions in Eqs. (7.137) and (7.138) for the last term in Eq. (7.133) yields:

$$\Delta_r G^{exch} + RT\ln\frac{x_{Fe}^{sp} \cdot x_{Mg}^{ol}}{x_{Mg}^{sp} \cdot x_{Fe}^{ol}} + RT\ln\frac{(\gamma_{Mg_2SiO_4}^{ol})^{1/2}}{(\gamma_{Fe_2SiO_4}^{ol})^{1/2}} \tag{7.139}$$

$$+ [(x_{Mg}^{sp})^2 - (x_{Fe}^{sp})^2] W_{FeMg}^{sp} - (x_{Mg}^{sp} \cdot x_{Cr}^{sp} + x_{Fe}^{sp} \cdot x_{Cr}^{sp}) \Delta_r G^{rec} = 0$$

or

$$\ln K_D = -\frac{\Delta_r G^{exch}}{RT} - \ln\frac{(\gamma_{Mg_2SiO_4}^{ol})^{1/2}}{(\gamma_{Fe_2SiO_4}^{ol})^{1/2}} - [(x_{Mg}^{sp})^2 \tag{7.140}$$

$$- (x_{Fe}^{sp})^2] W_{FeMg}^{sp}/RT + (x_{Mg}^{sp} \cdot x_{Cr}^{sp} + x_{Fe}^{sp} \cdot x_{Cr}^{sp}) \Delta_r G^{rec}/RT .$$

Eq. (7.140) can be used to determine the equilibration temperature of natural olivine- and spinel-bearing rocks.

Energetics of order-disorder

In the foregoing examples, one type of atom always occupied only one specific crystallorgraphic site. In garnet-solid solution, $(Mg,Ca)_3(Al,Cr)_2Si_3O_{12}$, for example, magnesium occupies only the dodecahedral and aluminium only the octahedral site. The same holds for calcium and chromium. In some crystal structures, however, an atom can occupy more than one crystallographic site and the question regarding its distribution over the different sites becomes relevant. If the size and the bonding character of the sites are similar, the distribution of the atom over the various sites can be completely random, and the atomic fractions of the respective atom will then be the same for each site. A crystal with this type of atomic distribution are referred to as *completely disordered.*

If there is a significant difference between the different crystallographic sites, an *ordering* of atoms can occur. In such a case, a given atom will preferentially occupy just one crystallographic site and the atomic fractions of various atoms over different sites will have different values. The degree of order is generally temperature dependent and it decreases with increasing temperatures and vice versa. On the other hand, the pressure dependence of ordering is small, because of small differences in volume between an ordered and disordered state. Perfect order is theoretically achieved only at absolute zero (i.e.0 K) and thus, a real crystal can never be perfectly ordered.

There are two types of long-range ordering described in the literature: *convergent* and *non-convergent*. In the case of convergent ordering, the difference between the sites disappears above some critical temperature and the partitioning of an atom between the site is the same. In the case of non-convergent ordering, the differences between the different sites remain at all temperatures and the partitioning of an atom between the sites is never equal.

An example for non-convergent ordering is Mg-Fe orthopyroxene, $(Mg,Fe)_2Si_2O_6$. Here, iron and magnesium are distributed over two octahedrally coordinated sites termed *M*1 and *M*2 and the chemical formula can be written as follows:

$$(Mg, Fe)_{M2}(Mg, Fe)_{M1}Si_2O_6.$$

Iron occupies preferentially the *M*2 sites, which are larger and more distorted than the *M*1 sites.

In order to discuss the energetic properties of the orthopyroxene solid solutions the reciprocal solution model can be used.

The four end-member components are:

$Mg_{M2}Mg_{M1}Si_2O_6$ enstatite,

$Fe_{M2}Fe_{M1}Si_2O_6$ ferrosilite,

$Fe_{M2}Mg_{M1}Si_2O_6$ ordered phase and

$Mg_{M2}Fe_{M1}Si_2O_6$ 'anti-ordered' phase.

The ordered and 'anti-ordered' phases are hypothetical end-member pyroxenes where iron occupies only the *M*2 and *M*1 site, respectively. Three components are independent while the fourth one is related to the other three by the reciprocal reaction, i.e.

$$Mg_{M2}Mg_{M1}Si_2O_6 + Fe_{M2}Fe_{M1}Si_2O_6 = Fe_{M2}Mg_{M1}Si_2O_6 + Mg_{M2}Fe_{M1}Si_2O_6. \quad (7.141)$$

The corresponding Gibbs free energy is given by:

$$\Delta_r G^{rec} = \overset{\circ}{\mu}_{Fe_{M2}Mg_{M1}Si_2O_6} + \overset{\circ}{\mu}_{Mg_{M2}Fe_{M1}Si_2O_6} - \overset{\circ}{\mu}_{Mg_{M2}Mg_{M1}Si_2O_6} - \overset{\circ}{\mu}_{Fe_{M2}Fe_{M1}Si_2O_6}. \quad (7.142)$$

According to Eq. (7.92) through Eq. (7.95), the excess Gibbs free energy associated with the reciprocal reaction in Eq. (7.141) can be expressed in four different ways depending on the choice of the components, namely:

$$\Delta_m \overline{G}^{ex}_{(Mg_{M2}Mg_{M1}Si_2O_6)} = -x_{Mg,\,M2}x_{Mg,\,M1}\Delta_r \boldsymbol{G}^{rec}, \tag{7.143}$$

$$\Delta_m \overline{G}^{ex}_{(Fe_{M2}Fe_{M1}Si_2O_6)} = -x_{Fe,\,M2}x_{Fe,\,M1}\Delta_r \boldsymbol{G}^{rec}, \tag{7.144}$$

$$\Delta_m \overline{G}^{ex}_{(Mg_{M2}Fe_{M1}Si_2O_6)} = x_{Mg,\,M2}x_{Fe,\,M1}\Delta_r \boldsymbol{G}^{rec} \tag{7.145}$$

and

$$\Delta_m \overline{G}^{ex}_{(Fe_{M2}Mg_{M1}Si_2O_6)} = x_{Fe,\,M2}x_{Mg,\,M1}\Delta_r \boldsymbol{G}^{rec}. \tag{7.146}$$

The chemical formula given in parenthesis designates the fourth dependent component.

The calculation procedure given in (7.96) through (7.99) leads to the following activity coefficients for the four components, i.e.:

$$RT\ln\gamma^{rec}_{Mg_{M2}Mg_{M1}Si_2O_6} = x_{Fe,\,M2}x_{Fe,\,M1}\Delta_r \boldsymbol{G}^{rec}, \tag{7.147}$$

$$RT\ln\gamma^{rec}_{Fe_{M2}Fe_{M1}Si_2O_6} = x_{Mg,\,M2}x_{Mg,\,M1}\Delta_r \boldsymbol{G}^{rec}, \tag{7.148}$$

$$RT\ln\gamma^{rec}_{Mg_{M2}Fe_{M1}Si_2O_6} = -x_{Fe,\,M2}x_{Mg,\,M1}\Delta_r \boldsymbol{G}^{rec} \tag{7.149}$$

and

$$RT\ln\gamma^{rec}_{Fe_{M2}Mg_{M1}Si_2O_6} = -x_{Mg,\,M2}x_{Fe,\,M1}\Delta_r \boldsymbol{G}^{rec}. \tag{7.150}$$

Assuming an ideal mixing behavior of atoms on individual sites, the chemical potentials of the four components read:

$$\mu_{Mg_{M2}Mg_{M1}Si_2O_6} = \boldsymbol{\mu}^{o}_{Mg_{M2}Mg_{M1}Si_2O_6} + RT\ln(x_{Mg,\,M2}\cdot x_{Mg,\,M1}) + x_{Fe,\,M2}x_{Fe,\,M1}\Delta_r \boldsymbol{G}^{rec}, \tag{7.151}$$

$$\mu_{Fe_{M2}Fe_{M1}Si_2O_6} = \mu^{o}_{Fe_{M2}Fe_{M1}Si_2O_6} + RT\ln(x_{Fe,\,M2} \cdot x_{Fe,\,M1}) + x_{Mg,\,M2}x_{Mg,\,M1}\Delta_r\boldsymbol{G}^{rec}, \tag{7.152}$$

$$\mu_{Mg_{M2}Fe_{M1}Si_2O_6} = \mu^{o}_{Mg_{M2}Fe_{M1}Si_2O_6} + RT\ln(x_{Mg,\,M2} \cdot x_{Fe,\,M1}) - x_{Fe,\,M2}x_{Mg,\,M1}\Delta_r\boldsymbol{G}^{rec} \tag{7.153}$$

and

$$\mu_{Fe_{M2}Mg_{M1}Si_2O_6} = \mu^{o}_{Fe_{M2}Mg_{M1}Si_2O_6} + RT\ln(x_{Fe,\,M2} \cdot x_{Mg,\,M1}) - x_{Mg,\,M2}x_{Fe,\,M1}\Delta_r\boldsymbol{G}^{rec}. \tag{7.154}$$

If the mixing behavior of atoms on individual sites is non-ideal, further terms must be added to the expressions given in Eqs. (7.151) through (7.154). Many workers (e.g. Thompson 1969, 1970; Sack 1980; Sack and Ghiorso 1989; Ganguly 1982; Yang and Ghose 1994; Kroll et al. 1994, 1997) who have addressed the mixing properties of orthopyroxenes, used a regular solution model. Following them, the chemical potentials obtain the form:

$$\mu_{Mg_{M2}Mg_{M1}Si_2O_6} = \mu^{o}_{Mg_{M2}Mg_{M1}Si_2O_6} + RT\ln(x_{Mg,\,M2} \cdot x_{Mg,\,M1}) + x^2_{Fe,\,M2}W^G_{M2} + x^2_{Fe,\,M1}W^G_{M1} + x_{Fe,\,M2}x_{Fe,\,M1}\Delta_r\boldsymbol{G}^{rec}, \tag{7.155}$$

$$\mu_{Fe_{M2}Fe_{M1}Si_2O_6} = \mu^{o}_{Fe_{M2}Fe_{M1}Si_2O_6} + RT\ln(x_{Fe,\,M2} \cdot x_{Fe,\,M1}) + x^2_{Mg,\,M2}W^G_{M2} + x^2_{Mg,\,M1}W^G_{M1} + x_{Mg,\,M2}x_{Mg,\,M1}\Delta_r\boldsymbol{G}^{rec}, \tag{7.156}$$

$$\mu_{Mg_{M2}Fe_{M1}Si_2O_6} = \mu^{o}_{Mg_{M2}Fe_{M1}Si_2O_6} + RT\ln(x_{Mg,\,M2} \cdot x_{Fe,\,M1}) + x^2_{Fe,\,M2}W^G_{M2} + x^2_{Mg,\,M1}W^G_{M1} - x_{Fe,\,M2}x_{Mg,\,M1}\Delta_r\boldsymbol{G}^{rec} \tag{7.157}$$

and

$$\mu_{Fe_{M2}Mg_{M1}Si_2O_6} = \mu^{o}_{Fe_{M2}Mg_{M1}Si_2O_6} + RT\ln(x_{Fe,\,M2} \cdot x_{Mg,\,M1}) + x^2_{Mg,\,M2}W^G_{M2} + x^2_{Fe,\,M1}W^G_{M1} - x_{Mg,\,M2}x_{Fe,\,M1}\Delta_r\boldsymbol{G}^{rec}, \tag{7.158}$$

where W^G_{M1} and W^G_{M2} are the interaction parameter for the $M1$ and $M2$ sites, respectively.

In order to calculate the Gibbs free energy of an orthopyroxene solid solution, the degree of order, that is the partitioning of iron (or magnesium) between $M1$ and $M2$, must be known.

The process of ordering is governed by the exchange reaction:

$$Mg_{M2}Fe_{M1}Si_2O_6 = Fe_{M2}Mg_{M1}Si_2O_6\,, \tag{7.159}$$

where iron and magnesium exchange their crystallographic positions. As for all other reactions, it holds that thermodynamic equilibrium is attained when the chemical potentials of the anti-ordered and the ordered component are equal, that is

$$\mu_{Mg_{M2}Fe_{M1}Si_2O_6} = \mu_{Fe_{M2}Mg_{M1}Si_2O_6} \tag{7.160}$$

or

$$\mu_{Fe_{M2}Mg_{M1}Si_2O_6} - \mu_{Mg_{M2}Fe_{M1}Si_2O_6} = 0. \tag{7.161}$$

Substituting the expressions given in Eqs. (7.157) and (7.158) for the chemical potentials $\mu_{Mg_{M2}Fe_{M1}Si_2O_6}$ and $\mu_{Fe_{M2}Mg_{M1}Si_2O_6}$, respectively yields:

$$\begin{aligned}
&\mu^{o}_{Fe_{M2}Mg_{M1}Si_2O_6} + RT\ln(x_{Fe,\,M2}\cdot x_{Mg,\,M1})\\
&+x^2_{Mg,\,M2}W^G_{M2} + x^2_{Fe,\,M1}W^G_{M1} - x_{Mg,\,M2}x_{Fe,\,M1}\Delta_r\boldsymbol{G}^{rec}\\
&-(\mu^{o}_{Mg_{M2}Fe_{M1}Si_2O_6} + RT\ln(x_{Mg,\,M2}\cdot x_{Fe,\,M1})\\
&+x^2_{Fe,\,M2}W^G_{M2} + x^2_{Mg,\,M1}W^G_{M1} - x_{Fe,\,M2}x_{Mg,\,M1}\Delta_r\boldsymbol{G}^{rec}) = 0
\end{aligned} \tag{7.162}$$

or

$$\begin{aligned}
RT\ln K_D &= -(\mu^{o}_{Fe_{M2}Mg_{M1}Si_2O_6} - \mu^{o}_{Mg_{M2}Fe_{M1}Si_2O_6})\\
&+(1-2x_{Fe,\,M2})W^G_{M2} - (1-2x_{Fe,\,M1})W^G_{M1}\\
&+(x_{Fe,\,M2} - x_{Fe,\,M1})\Delta_r\boldsymbol{G}^{rec}.
\end{aligned} \tag{7.163}$$

The term on the left side of Eq. (7.163) is the so-called intracrystalline distribution coefficient, K_D, that reads:

$$K_D = \frac{x_{Fe,\,M2} \cdot x_{Mg,\,M1}}{x_{Fe,\,M1} \cdot x_{Mg,\,M2}} \tag{7.164}$$

or, because

$$x_{Fe,\,M2} + x_{Fe,\,M1} = 1 \tag{7.165}$$

and

$$x_{Fe,\,M1} + x_{Mg,\,M2} = 1, \tag{7.166}$$

$$K_D = \frac{x_{Fe,\,M2} \cdot (1 - x_{Fe,\,M1})}{x_{Fe,\,M1} \cdot (1 - x_{Fe,\,M2})}. \tag{7.167}$$

In his discussion on the thermodynamics of order-disorder in pyroxenes, Thompson (1969) introduced an order parameter, Q, and a compositional parameter, r. The two variable are defined as follows:

$$Q = x_{Fe,\,M2} - x_{Fe,\,M1} \tag{7.168}$$

and

$$r = x_{Fe,\,M2} + x_{Fe,\,M1} - 1 = x_{Fs} - 1, \tag{7.169}$$

where x_{Fs} corresponds to the mole fraction of ferrosilite, $Fe_2Si_2O_6$, in orthopyroxene.

The relationships given in Eqs. (7.168) and (7.169) can be used to express the atomic fractions of iron and magnesium on $M1$ and $M2$, respectively, i.e.:

$$x_{Fe,\,M1} = \frac{r + 1 - Q}{2}, \tag{7.170}$$

$$x_{Fe,\,M2} = \frac{Q + 1 + r}{2}, \tag{7.171}$$

$$x_{Mg,\,M1} = \frac{1 + Q - r}{2} \tag{7.172}$$

and

$$x_{Mg,\,M2} = \frac{1-Q-r}{2}. \tag{7.173}$$

Replacing the atomic fractions in Eq. (7.163) by the expressions given in Eqs. (7.170) and (7.171) and considering that the difference $\overset{\mathrm{o}}{\mu}_{Fe_{M2}Mg_{M1}Si_2O_6}$ - $\overset{\mathrm{o}}{\mu}_{Mg_{M2}Fe_{M1}Si_2O_6}$ corresponds to the standard Gibbs free energy of the exchange reaction given in Eq. (7.159), the intracrystalline distribution coefficient, K_D, can be written in terms of composition, *r*, and the degree the of order, *Q*:

$$RT\ln K_D = -\Delta_r \boldsymbol{G}^{exch} + (W^G_{M2} - W^G_{M1})r + [\Delta_r \boldsymbol{G}^{rec} + (W^G_{M2} + W^G_{M1})]Q. \tag{7.174}$$

The molar Gibbs free energy of an orthopyroxene solid solution can, according to Shi et al. (1992), be written in terms of site occupancies, namely:

$$\begin{aligned}\overline{G} = {} & x_{Mg,\,M2}x_{Mg,\,M1}\overset{\mathrm{o}}{\mu}_{Mg_{M2}Mg_{M1}Si_2O_6} + x_{Fe,\,M2}x_{Fe,\,M1}\overset{\mathrm{o}}{\mu}_{Fe_{M2}Fe_{M1}Si_2O_6} \\ & + x_{Fe,\,M2}x_{Mg,\,M1}\overset{\mathrm{o}}{\mu}_{Fe_{M2}Mg_{M1}Si_2O_6} \\ & + x_{Mg,\,M2}x_{Fe,\,M1}\overset{\mathrm{o}}{\mu}_{Mg_{M2}Fe_{M1}Si_2O_6} + x_{Mg,\,M2}x_{Fe,\,M2}W^G_{M2} \\ & + x_{Mg,\,M1}x_{Fe,\,M1}W^G_{M1} + RT[x_{Mg,\,M2}\ln x_{Mg,\,M2} \\ & + x_{Fe,\,M2}\ln x_{Fe,\,M2} + x_{Mg,\,M1}\ln x_{Mg,\,M1} + x_{Fe,\,M1}\ln x_{Fe,\,M1}].\end{aligned} \tag{7.175}$$

Substituting Eqs. (7.170) through (7.173) into Eq. (7.175) and rearranging gives:

$$\begin{aligned}\overline{G} = {} & \frac{1}{2}(1+r)\overset{\mathrm{o}}{\mu}_{Fe_{M2}Fe_{M1}Si_2O_6} + \frac{1}{2}(1-r)\overset{\mathrm{o}}{\mu}_{Mg_{M2}Mg_{M1}Si_2O_6} \\ & + \frac{1}{2}Q(\Delta_r \boldsymbol{G}^{exch}) + \frac{1}{4}(1-r^2)(\Delta_r \boldsymbol{G}^{rec} + W^G_{M2} + W^G_{M1}) \\ & + \frac{1}{2}rQ(W^G_{M1} - W^G_{M2}) + \frac{1}{4}Q^2(\Delta_r \boldsymbol{G}^{rec} - W^G_{M1} - W^G_{M2}) \\ & + RT[x_{\mathrm{Mg,M2}}\ln x_{\mathrm{Mg,M2}} + x_{Fe,\,M2}\ln x_{Fe,\,M1} \\ & + x_{Mg,\,M1}\ln x_{Mg,\,M1} + x_{Fe,\,M1}\ln x_{Fe,\,M1}],\end{aligned} \tag{7.176}$$

where $\Delta_r \boldsymbol{G}^{exch}$ and $\Delta_r \boldsymbol{G}^{rec}$ represent the Gibbs free energy of exchange and reciprocal reaction, respectively, as defined in Eqs. (7.141) and (7.159).

Because

$$\frac{1+r}{2} = x_{Fs} \tag{7.177}$$

and

$$\frac{1-r}{2} = x_{En}, \tag{7.178}$$

Eq. (7.176) can be rewritten as:

$$\begin{aligned}\overline{G} = {} & x_{Fs}\overset{\text{o}}{\boldsymbol{\mu}}_{Fe_{M2}Fe_{M1}Si_2O_6} + x_{En}\overset{\text{o}}{\boldsymbol{\mu}}_{Mg_{M2}Mg_{M1}Si_2O_6} \\ & + x_{Fs}x_{En}(\Delta_r\boldsymbol{G}^{rec} + W^G_{M2} + W^G_{M1}) + \frac{1}{2}Q[(W^G_{M1} - W^G_{M2})r + \Delta_r\boldsymbol{G}^{exch}] \\ & + \frac{1}{4}Q^2(\Delta_r\boldsymbol{G}^{rec} - W^G_{M1} - W^G_{M2}) + RT[x_{Mg,\,M2}\ln x_{Mg,\,M2} \\ & + x_{Fe,\,M2}\ln x_{Fe,\,M2} + x_{Mg,\,M1}\ln x_{Mg,\,M1} + x_{Fe,\,M1}\ln x_{Fe,\,M1}],\end{aligned} \tag{7.179}$$

where x_{En} and x_{Fs} are the mole fractions of enstatite and ferrosilite in orthopyroxene, respectively.

The configurational entropy associated with the ordering corresponds to the difference between the entropies of partially ordered and completely disordered states, i.e.:

$$\begin{aligned}\Delta_{ord}\overline{S}^{conf} = {} & -R[x_{Mg,\,M2}\ln x_{Mg,\,M2} + x_{Fe,\,M2}\ln x_{Fe,\,M2} \\ & + x_{Mg,\,M1}\ln x_{Mg,\,M1} + x_{Fe,\,M1}\ln x_{Fe,\,M1}] \\ & + 2R[x_{Fs}\ln x_{Fs} + x_{En}\ln x_{En}].\end{aligned} \tag{7.180}$$

The 'two' in front of the second bracket in Eq. (7.180) accounts for the two crystallographically independent sites in orthopyroxene, which are thermodynamically equivalent in the case of complete disorder.

Introducing the relationship given in Eq. (7.180) into Eq. (7.179) yields:

$$\overline{G} = x_{Fs}\overset{\text{o}}{\mu}_{Fe_{M2}Fe_{M1}Si_2O_6} + x_{En}\overset{\text{o}}{\mu}_{Mg_{M2}Mg_{M1}Si_2O_6} \tag{7.181}$$
$$+ 2RT[x_{Fs}\ln x_{Fs} + x_{En}\ln x_{En}] + x_{Fs}x_{En}(\Delta_r\boldsymbol{G}^{\text{rec}} + W^G_{M2} + W^G_{M1})$$
$$+ \frac{1}{2}s[(W^G_{M1} - W^G_{M2})r + \Delta_r\boldsymbol{G}^{\text{exch}}] + \frac{1}{4}s^2(\Delta_r\boldsymbol{G}^{\text{rec}} - W^G_{M1} - W^G_{M2})$$
$$+ RT[x_{Mg,\,M2}\ln x_{Mg,\,M2} + x_{Fe,\,M2}\ln x_{Fe,\,M2} + x_{Mg,\,M1}\ln x_{Mg,\,M1}$$
$$+ x_{Fe,\,M1}\ln x_{Fe,\,M1}] - 2RT[x_{Fs}\ln x_{Fs} + x_{En}\ln x_{En}].$$

The first four terms in Eq. (7.181) give the Gibbs free energy of a completely disordered magnesium-iron pyroxene. The Gibbs free energy due to the ordering is, thus:

$$\overline{G}^{ord} = \frac{1}{2}s[(W^G_{M1} - W^G_{M2})r + \Delta_r\boldsymbol{G}^{exch}] \tag{7.182}$$
$$+ \frac{1}{4}s^2(\Delta_r\boldsymbol{G}^{rec} - W^G_{M1} - W^G_{M2}) + RT[x_{Mg,\,M2}\ln x_{Mg,\,M2}$$
$$+ x_{Fe,\,M2}\ln x_{Fe,\,M2} + x_{Mg,\,M1}\ln x_{Mg,\,M1} + x_{Fe,\,M1}\ln x_{Fe,\,M1}]$$
$$- 2RT[x_{Fs}\ln x_{Fs} + x_{En}\ln x_{En}].$$

The entropy due to the ordering, $\overline{S}^{ord}$, can be derived from Eq. (7.182) by differentiating with respect to temperature at constant composition and constant degree of order, i.e.:

$$\left(\frac{\partial \overline{G}^{ord}}{\partial T}\right)_{r,\,s} = -\overline{S}^{ord}. \tag{7.183}$$

The differentiation leads to:

$$\overline{S}^{ord} = \frac{1}{2}Q[(W^S_{M1} - W^S_{M2})r + \Delta_r\boldsymbol{S}^{exch}] \tag{7.184}$$
$$+ \frac{1}{4}Q^2(\Delta_r\boldsymbol{S}^{rec} - W^S_{M1} - W^S_{M2}) - R[x_{Mg,\,M2}\ln x_{Mg,\,M2}$$
$$+ x_{Fe,\,M2}\ln x_{Fe,\,M2} + x_{Mg,\,M1}\ln x_{Mg,\,M1} + x_{Fe,\,M1}\ln x_{Fe,\,M1}]$$
$$+ 2R[x_{Fs}\ln x_{Fs} + x_{En}\ln x_{En}],$$

where W^S_{M1}, W^S_{M2}, $\Delta_r\boldsymbol{S}^{exch}$ and $\Delta_r\boldsymbol{S}^{rec}$ are the entropic interaction parameters for $M1$ and $M2$ sites, the change in entropy associated with the exchange reaction and the change in entropy associated with the reciprocal reaction, respectively.

The enthalpy associated with the ordering is obtained by extracting the entropy contributions from Eq. (7.182). The result is:

$$\begin{aligned}\overline{H}^{ord} &= \frac{1}{2}Q[(W_{M1}^{H} - W_{M2}^{H})r + \Delta_r \boldsymbol{H}^{exch}] \\ &+ \frac{1}{4}Q^2(\Delta_r \boldsymbol{H}^{rec} - W_{M1}^{H} - W_{M2}^{H}).\end{aligned} \tag{7.185}$$

In Eq. (7.185), W_{M1}^{H} and W_{M2}^{H} designate the enthalpic interaction parameters for the $M1$ and $M2$ sites, respectively. $\Delta_r \boldsymbol{H}^{exch}$ gives the enthalpy of the exchange reaction and $\Delta_r \boldsymbol{H}^{rec}$ that of the reciprocal reaction.

The derivative of Eq. (7.182) with respect to pressure yields the volume of ordering and is given by:

$$\overline{V}^{ord} = \frac{1}{2}s[(W_{M1}^{V} - W_{M2}^{V})r + \Delta_r \boldsymbol{V}^{exch}] + \frac{1}{4}s^2(\Delta_r \boldsymbol{V}^{rec} - W_{M1}^{V} - W_{M2}^{V}). \tag{7.186}$$

In Eq. (7.186), W_{M1}^{V} and W_{M2}^{V} are the volumetric interaction parameters for $M1$ and $M2$ site, respectively. $\Delta_r \boldsymbol{V}^{exch}$ gives the change in volume associated with the exchange reaction and $\Delta_r \boldsymbol{V}^{rec}$ that associated with the reciprocal reaction.

If the entropy contribution is extracted from Eq. (7.181), the molar enthalpy, $\overline{H}$, of orthopyroxene is obtained, namely:

$$\begin{aligned}\overline{H} &= x_{Fs}\boldsymbol{H}_{Fe_{M2}Fe_{M1}Si_2O_6} + x_{En}\boldsymbol{H}_{Mg_{M2}Mg_{M1}Si_2O_6} \\ &+ x_{Fs}x_{En}(\Delta_r \boldsymbol{H}^{rec} + W_{M2}^{H} + W_{M1}^{H}) + \frac{1}{2}Q[(W_{M1}^{H} - W_{M2}^{H})r + \Delta_r \boldsymbol{H}^{exch}] \\ &+ \frac{1}{4}Q^2(\Delta_r \boldsymbol{H}^{rec} - W_{M1}^{S} - W_{M2}^{S}),\end{aligned} \tag{7.187}$$

where $\boldsymbol{H}_{Fe_{M2}Fe_{M1}Si_2O_6}$ and $\boldsymbol{H}_{Mg_{M2}Mg_{M1}Si_2O_6}$ designate the molar enthalpies of pure ferrosilite and pure enstatite, respectively.

Subtracting the enthalpy of a mechanical mixture consisting of the end-members ferrosilite and enstatite,

$$\overline{H}^{mm} = x_{Fs}\boldsymbol{H}_{Fe_{M2}Fe_{M1}Si_2O_6} + x_{En}\boldsymbol{H}_{Mg_{M2}Mg_{M1}Si_2O_6}, \tag{7.188}$$

yields the enthalpy of mixing or excess enthalpy, $\Delta_m\overline{H}^{ex}$, that is:

$$\Delta_m\overline{H}^{ex} = x_{Fs}x_{En}(\Delta_r\boldsymbol{H}^{rec} + W^H_{M2} + W^H_{M1}) + \frac{1}{2}Q[(W^H_{M1} - W^H_{M2})r + \Delta_r\boldsymbol{H}^{exch}] + \frac{1}{4}Q^2(\Delta_r\boldsymbol{H}^{rec} - W^H_{M1} - W^H_{M2}). \tag{7.189}$$

The molar volume of orthopyroxene is obtained by differentiating Eq. (7.181) with respect to pressure i.e.:

$$\overline{V} = x_{Fs}\boldsymbol{V}_{Fe_{M2}Fe_{M1}Si_2O_6} + x_{En}\boldsymbol{V}_{Mg_{M2}Mg_{M1}Si_2O_6} + x_{Fs}x_{En}(\Delta_r\boldsymbol{V}^{rec} + W^V_{M2} + W^V_{M1}) + \frac{1}{2}Q[(W^V_{M1} - W^V_{M2})r + \Delta_r\boldsymbol{V}^{exch}] + \frac{1}{4}Q^2(\Delta_r\boldsymbol{V}^{rec} - W^V_{M1} - W^V_{M2}), \tag{7.190}$$

where $\boldsymbol{V}_{Fe_{M2}Fe_{M1}Si_2O_6}$ and $\boldsymbol{V}_{Mg_{M2}Mg_{M1}Si_2O_6}$ are the molar volumes of the pure end-member components.

The excess volume of mixing is obtained by subtracting the ideal volume of mixing

$$\overline{V}^{id} = x_{Fs}\boldsymbol{V}_{Fe_{M2}Fe_{M1}Si_2O_6} + x_{En}\boldsymbol{V}_{Mg_{M2}Mg_{M1}Si_2O_6} \tag{7.191}$$

from Eq. (7.190). In this manner one arrives at

$$\Delta_m\overline{V}^{ex} = x_{Fs}x_{En}(\Delta_r\boldsymbol{V}^{rec} + W^V_{M2} + W^V_{M1}) + \frac{1}{2}Q[(W^V_{M1} - W^V_{M2})r + \Delta_r\boldsymbol{V}^{exch}] + \frac{1}{4}Q^2(\Delta_r\boldsymbol{V}^{rec} - W^V_{M1} - W^V_{M2}). \tag{7.192}$$

Example 1: A synthetic hypersthene containing 50 mol% ferrosilite shows the following iron site occupation: $x_{Fe,M1} = 0.222$ and $x_{Fe,M2} = 0.778$. After heating the crystal at 1173 K, the distribution changes to $x_{Fe,M1} = 0.318$ and $x_{Fe,M2} = 0.682$. Calorimetric measurements of the enthalpy, $\Delta_{dis}\overline{H}$, associated with the disordering process yielded 1.73 kJmol^{-1} (Cemič and Kähler 2000). What is the enthalpy of the exchange reaction, $\Delta_r\boldsymbol{H}^{exch}$?

The enthalpy of disordering equals the difference between the excess enthalpies of mixing before and after heating, i.e.

$$\Delta_{dis}\overline{H} = \Delta_m\overline{H}^{ex}(Q_a) - \Delta_m\overline{H}^{ex}(Q_b) = \overline{H}^{ord}(Q_a) - \overline{H}^{ord}(Q_b), \tag{7.193}$$

where s_b and s_a are the order parameters before and after heat treating, respectively.

$$\Delta_{dis}\overline{H} = \frac{1}{2}(Q_a - Q_b)[(W^H_{M1} - W^H_{M2})r + \Delta_r\boldsymbol{H}^{exch}] \\ + \frac{1}{4}(Q_a^2 - Q_b^2)(\Delta_r\boldsymbol{H}^{rec} - W^H_{M1} - W^H_{M2}). \tag{7.194}$$

For a pyroxene of composition $x_{Fs} = 0.5$ one has

$$r = 2 \times 0.5 - 1 = 0, \tag{7.195}$$

and the first term in the bracket of Eq. (7.194) vanishes. The enthalpy of disordering, $\Delta_{dis}\overline{H}$, is then given by:

$$\Delta_{dis}\overline{H} = \frac{1}{2}(Q_a - Q_b)\Delta_r\boldsymbol{H}^{exch} + \frac{1}{4}(Q_a^2 - Q_b^2)(\Delta_r\boldsymbol{H}^{rec} - W^H_{M1} - W^H_{M2}). \tag{7.196}$$

Solving Eq. (7.196) for $\Delta_r\boldsymbol{H}^{exch}$ yields:

$$\Delta_r\boldsymbol{H}^{exch} = \frac{1}{2}(Q_a + Q_b)(\Delta_r\boldsymbol{H}^{rec} - W^H_{M1} - W^H_{M2}) - \frac{2(\Delta_{dis}\overline{H})}{(Q_a - Q_b)}. \tag{7.197}$$

Kroll et al. (1997) give for the term $(\Delta_r\boldsymbol{H}^{rec} - W^H_{M1} - W^H_{M2})$ a value of - 4162 $Jmol^{-1}$. The order parameters before and after heat treating are:

$$Q_b = 0.778 - 0.222 = 0.556$$

and

$$Q_a = 0.682 - 0.318 = 0.364,$$

respectively.

Inserting these values together with the experimentally determined enthalpy of disordering, $\Delta_{dis}\boldsymbol{H}$, into Eq. (7.197), the enthalpy of the exchange reaction is calculated as follows:

$$\Delta_r \boldsymbol{H}^{exch} = \frac{1}{2}(0.364 + 0.556)(-4163\ \mathrm{Jmol}^{-1}) - \frac{2 \times 1730\ \mathrm{Jmol}^{-1}}{0.364 - 0.556} \quad (7.198)$$

$$= \underline{16106\ \mathrm{Jmol}^{-1}}.$$

Example 2: Chatillon-Colinet et al. (1983) measured calorimetrically the heat of mixing, $\Delta_m \overline{H}$, for a disordered orthopyroxene. The authors fitted a regular solution model to their experimental data and obtained

$$\Delta_m \overline{H} = x_{Fs} x_{En} W^H, \quad (7.199)$$

where $W^H = 7950\ \mathrm{Jmol}^{-1}$. This datum can be combined with the result in Example 1 to evaluate the enthalpy of the reciprocal reaction, $\Delta_r \boldsymbol{H}^{rec}$, and to determine the intra site interaction parameters W^H_{M1} and W^H_{M2}.

Because the enthalpy of mixing determined by Chatillon-Colinet et al. (1983) refers to a completely disordered pyroxene, the value of the order parameter, Q, is zero. In this case, Eq. (7.189) has the form:

$$\Delta_m \overline{H} = x_{Fs} x_{En} (\Delta_r \boldsymbol{H}^{rec} + W^H_{M2} + W^H_{M1}). \quad (7.200)$$

A comparison of Eqs. (7.199) and (7.200) shows that

$$\Delta_r \boldsymbol{H}^{rec} + W^H_{M2} + W^H_{M1} = W^H = \underline{7950\ \mathrm{Jmol}^{-1}}. \quad (7.201)$$

Combining Eq. (7.201) with $\Delta_r \boldsymbol{H}^{rec} - W^H_{M1} - W^H_{M2} = -4162\ \mathrm{Jmol}^{-1}$ (Kroll et al. (1997) gives:

$$\Delta_r \boldsymbol{H}^{rec} = \frac{1}{2}(7950 - 4162) = \underline{1894\ \mathrm{Jmol}^{-1}} \quad (7.202)$$

and

$$W^H_{M1} + W^H_{M2} = \underline{6056\ \mathrm{Jmol}^{-1}}. \quad (7.203)$$

Using Eq. (7.203) and the relationship:

$$W^H_{M1} - W^H_{M2} = 1796\ \text{Jmol}^{-1} \tag{7.204}$$

(Kroll et al., 1997) one obtains:

$$W^H_{M1} = \frac{6056\ \text{Jmol}^{-1} + 1796\ \text{Jmol}^{-1}}{2} = \underline{3926\ \text{Jmol}^{-1}} \tag{7.205}$$

and

$$W^H_{M2} = 6056\ \text{Jmol}^{-1} - 3926\ \text{Jmol}^{-1} = \underline{2130\ \text{Jmol}^{-1}}. \tag{7.206}$$

These results enables one to calculate the enthalpy of mixing for any orthopyroxene with any arbitrary distribution of iron over the $M1$ and $M2$ sites.

Example 3: In this example we want to calculate the entropy of disordering due to heating of an ordered orthopyroxene at 1173 K as described in Example 1. The calculation is carried out using Eq. (7.184) where:

$$\begin{aligned} \Delta_{ord}\bar{S} = {} & \tfrac{1}{2}(Q_a - Q_b)[(W^S_{M1} - W^S_{M2})r + \Delta_r S^{\text{exch}}] + \tfrac{1}{4}(Q_a^2 - Q_b^2) \\ & \times (\Delta_r S^{rec} - W^S_{M1} - W^S_{M2}) - R\Big\{ [x^{Q_a}_{Mg,\,M2} \ln x^{Q_a}_{Mg,\,M2} + x^{Q_a}_{Fe,\,M2} \ln x^{Q_a}_{Fe,\,M2} \\ & + x^{Q_a}_{Mg,\,M1} \ln x^{Q_a}_{Mg,\,M1} + x^{Q_a}_{Fe,\,M1} \ln x^{Q_a}_{Fe,\,M1}] - [x^{Q_b}_{Mg,\,M2} \ln x^{Q_b}_{Mg,\,M2} \\ & + x^{Q_b}_{Fe,\,M2} \ln x^{Q_b}_{Fe,\,M2} + x^{Q_b}_{Mg,\,M1} \ln x^{Q_b}_{Mg,\,M1} + x^{Q_b}_{Fe,\,M1} \ln x^{Q_b}_{Fe,\,M1}] \Big\}. \end{aligned} \tag{7.207}$$

In Eq. (7.207), the difference Q_a - Q_b designates the change in the order parameter associated with the process of Fe-Mg disordering.

Following Kroll et al. (1997), $\Delta_r S^{exch}$ has a value of - 2.719 $\text{Jmol}^{-1}\text{K}^{-1}$. The entropy of the reciprocal reaction, $\Delta_r S^{rec}$, and the entropic intra site interaction parameters W^S_{M1} and W^S_{M2} are approximately zero. Moreover, the compositional parameter, r, is zero, because the mole fraction of ferrosilite in the pyroxene is 0.5. Considering these facts the Eq. (7.207) simplifies to:

$$\Delta_{dis}\bar{S} = \frac{1}{2}(Q_a - Q_b)\Delta_r S^{exch} - R\Big\{ [x_{Mg,\,M2}^{Q_a} \ln x_{Mg,\,M2}^{Q_a} \quad (7.208)$$

$$+ x_{Fe,\,M2}^{Q_a} \ln x_{Fe,\,M2}^{Q_a} + x_{Mg,\,M1}^{Q_a} \ln x_{Mg,\,M1}^{Q_a} + x_{Fe,\,M1}^{Q_a} \ln x_{Fe,\,M1}^{Q_a}]$$

$$- [x_{Mg,\,M2}^{Q_b} \ln x_{Mg,\,M2}^{Q_b} + x_{Fe,\,M2}^{Q_b} \ln x_{Fe,\,M2}^{Q_b} + x_{Mg,\,M1}^{Q_b} \ln x_{Mg,\,M1}^{Q_b}$$

$$+ x_{Fe,\,M1}^{Q_b} \ln x_{Fe,\,M1}^{Q_b}]\Big\}.$$

Inserting the numerical data into Eq. (7.208) yields:

$$\Delta_{dis}\bar{S} = \frac{1}{2}(0.364 - 0.556)(-2.719\,\mathrm{Jmol^{-1}K^{-1}}) - 8.3144\,\mathrm{Jmol^{-1}K^{-1}} \quad (7.209)$$

$$\times \{2[0.318 \ln 0.318 + 0.682 \ln 0.682] - 2[0.222 \ln 0.222$$

$$+ 0.778 \ln 0.778]\} = \underline{1.86\ \mathrm{Jmol^{-1}K^{-1}}}.$$

The Gibbs free energy of disordering can be calculated using the Gibbs-Helmholtz equation, according to which, the following relationship holds:

$$\Delta_{dis}\bar{G} = \Delta_{dis}\bar{H} - T\Delta_{dis}\bar{S}. \quad (7.210)$$

Inserting the values for the enthalpy and entropy of disordering into Eq. (7.210) and solving for $\Delta_{dis}\bar{G}$ at $T = 1173$ K yields:

$$\Delta_{dis}\bar{G} = -1730\,\mathrm{Jmol^{-1}} - 1.86\,\mathrm{Jmol^{-1}K^{-1}} \times 1173\,\mathrm{K} = \underline{-3912\ \mathrm{Jmol^{-1}}}. \quad (7.211)$$

7.2 Problems

1. Consider the reaction:

$$Mg_3Si_4O_{10}(OH)_2 + 4MgSiO_3 \rightarrow Mg_7Si_8O_{22}(OH)_2.$$

- Calculate the equilibrium pressure of the reaction at 700°C assuming that the enthalpy, entropy and the volume of reaction are all pressure and temperature independent.

- Calculate the equilibrium pressure of the reaction at 700°C considering the temperature dependence of the enthalpy and entropy of reaction and the pressure and temperature dependence of the volume of reaction. Use the data given in Tab. 7.7.

Table 7.7 Thermodynamic data of the phases involved in the reaction: talc + enstatite → anthophyllite (Holland and Powell 1998)

Phase	$\Delta_f H_{298}$ [kJmol^{-1}]	S_{298} [Jmol^{-1}K^{-1}]	V_{298} [cm^{3}mol^{-1}]	α[K^{-1}] x10^5	β[Pa^{-1}] x10^{12}
Talc	-5896.92	260.00	136.25	2.04	21.94
Enstatite	-1545.13	66.25	31.31	2.32	9.84
Anthophyllite	-12068.59	536.00	265.40	2.75	15.04

$$C_{p,\,Mg_3Si_4O_{10}(OH)_2} = 622.2 - 63.855\times10^5 T^{-2} - 3916.3 T^{-1/2}$$

$$C_{p,\,MgSiO_3} = 178.1 - 1.495\times10^{-3} T - 2.9845\times10^5 T^{-2} - 1592.65 T^{-1/2}$$

$$C_{p,\,Mg_7Si_8O_{22}(OH)_2} = 1277.3 + 2.5825\times10^{-2} T - 97.046\times10^5 T^{-2} - 9074.7 T^{-1/2}$$

C_p[Jmol^{-1}K^{-1}]; (Holland and Powell 1998).

2. Consider the degassing reaction:

$$Mg_3Si_4O_{10}(OH)_2 + 5MgCO_3 \rightarrow 4Mg_2SiO_4 + 5CO_2 + H_2O.$$

- Calculate the various degassing temperatures at pressures 0.1 MPa. 10 MPa, 50 MPa, 100 MPa, 500 MPa and 1 GPa under the following assumptions: the enthalpy, entropy and volume of reaction do not depend on pressure and temperature, the gases behave ideally and their pressure equals the total pressure in the system.
- Draw the *P-T* diagram.
- Calculate the degassing temperature at 0.5 GPa for the case that the mole fraction of CO_2 equals 0.1. Use the data given in Tab. (7.7) and (7.8).

Table 7.8 Thermodynamic data of the phases involved in the reaction: talc + magnesite → forsterite + CO_2 + H_2O (Holland and Powell 1998)

Phase	$\Delta_f H_{298}$ [$kJmol^{-1}$]	S_{298} [$Jmol^{-1}K^{-1}$]	V_{298} [cm^3mol^{-1}]
Magnesite	- 1111.59	65.10	28.03
Forsterite	- 2171.85	95.10	43.06
CO_2	- 393.51	213.70	-
H_2O	- 241.81	188.80	-

3. A mechanical mixture of Ni and NiO is sealed into an evacuated quartz ampoule and annealed at 800°C.

- Write the chemical reaction that defines the oxygen fugacity in the quartz ampoule.
- Assume that the enthalpy and the entropy of reaction are pressure and temperature independent and that oxygen is an ideal gas and calculate the equilibrium oxygen pressure at 800°C.

$$\Delta_f H_{298,\,NiO} = -239.3\ \text{kJmol}^{-1},$$

$$S_{298,\,Ni} = 29.87\ \text{Jmol}^{-1}\text{K}^{-1},$$

$$S_{298,\,O_2} = 205.15\ \text{Jmol}^{-1}\text{K}^{-1}\ \text{and}$$

$$S_{298,\,NiO} = 37.99\ \text{Jmol}^{-1}\text{K}^{-1}$$

Robie and Hemingway (1995).

4. The distribution of iron and magnesium between olivine and garnet can be used to calculate the equilibrium temperature of olivine and garnet bearing rocks.

- Formulate the exchange reaction.
- Give the mathematical expression for the distribution coefficient, K_D, as a function of temperature and pressure assuming ideal mixing of cations in both phases.
- Give the mathematical expression for the distribution coefficient in the case that olivine solid solution is ideal and that of garnet non-ideal.
- Give the mathematical expression for the activity of $Mg_3Al_2Si_3O_{12}$ in the gar-

net solid solution assuming that the dodecahedral sites are occupied by Mg and Fe and the octahedral sites by Al and Cr. Further assume that the mixing on sites obeys the model of a simple solution.

Chapter 8 Geothermometry and geobarometry

The goal of geothermometry and geobarometry is to determine the P-T-conditions at which a rock equilibrated by using a so-called *geothermometer* and *geobarometer*.

Geothermometers are phase or reaction equilibria that depend strongly on temperature and not or only little on pressure, while geobarometers have significant pressure and negligible temperature dependencies. The fundamental thermodynamic relationship that constitutes the basis for geothermometry and geobarometry is given in Eq. (7.8), namely

$$\Delta_r \boldsymbol{G}^{\mathrm{o}} = -RT\ln K_{P,\,T}\,.$$

The temperature dependence of the thermodynamic equilibrium constant is given by:

$$\left(\frac{\partial \ln K_{P,\,T}}{\partial T}\right)_P = -\frac{\partial}{\partial T}\left(\frac{\Delta_r \boldsymbol{G}^{\mathrm{o}}}{RT}\right)_P = \frac{\Delta_r \boldsymbol{H}}{RT^2}, \tag{8.1}$$

and its pressure dependence by:

$$\left(\frac{\partial \ln K_{P,\,T}}{\partial P}\right)_T = -\frac{\partial}{\partial P}\left(\frac{\Delta_r \boldsymbol{G}^{\mathrm{o}}}{RT}\right)_T = -\frac{\Delta_r \boldsymbol{V}}{RT}. \tag{8.2}$$

The total differential of the logarithm of the thermodynamic constant, $\ln K_{P,T}$, is, therefore:

$$d\ln K_{P,\,T} = \left(\frac{\Delta_r \boldsymbol{H}}{RT^2}\right)dT - \left(\frac{\Delta_r \boldsymbol{V}}{RT}\right)dP. \tag{8.3}$$

At constant $K_{P,T}$ it holds that:

$$\left(\frac{\Delta_r H}{RT^2}\right) dT - \left(\frac{\Delta_r V}{RT}\right) dP = 0 \tag{8.4}$$

or

$$\left(\frac{dT}{dP}\right)_{K_{p,T}} = \frac{T\Delta_r V}{\Delta_r H}. \tag{8.5}$$

It is obvious from Eqs. (8.1), (8.2) and (8.5) that a reaction with a large $\Delta_r \boldsymbol{H}$ and a small $\Delta_r \boldsymbol{V}$ is suitable as a geothermometer and that a reaction with a large $\Delta_r \boldsymbol{V}$ and a small $\Delta_r \boldsymbol{H}$ as a geobarometer. (Note that $\Delta_r \boldsymbol{H}$ and $\Delta_r \boldsymbol{V}$ in Eqs. (8.1) through (8.6) are functions of pressure and temperature).

The principle behind geothermometry and geobarometry calculations is simple. Thermodynamic functions of state such as $\Delta_r \boldsymbol{H}$, $\Delta_r \boldsymbol{S}$, $\Delta_r \boldsymbol{V}$, $\Delta_r \boldsymbol{C}_p$, activities of the components, etc., which are measured using different calorimetric and x-ray methods or derived from experimental phase equilibrium studies, are combined to give the thermodynamic equilibrium constant, $K_{P,T}$, i.e.:

$$\ln K_{P,T} = -\frac{1}{RT}\left(\Delta_r \boldsymbol{H}_{298} + \int_{298}^{T} \Delta_r \boldsymbol{C}_p dT + \int_{10^5}^{P} \Delta_r \boldsymbol{V}(T) dP\right) + \frac{1}{R}\left(\Delta_r \boldsymbol{S}_{298} + \int_{298}^{T} \frac{\Delta_r \boldsymbol{C}_p}{T} dT\right). \tag{8.6}$$

Eq. (8.6) defines a surface in a three dimensional system having the coordinates P-$1/T$-$\ln K_{P,T}$. The intersection of this surface with a P-$1/T$ plane at any constant $\ln K_{P,T}$ gives a curve in a P-$1/T$ diagram (see Fig. 8.1).

If the activity coefficients of the components involved in the reaction are known, the value of the thermodynamic equilibrium constant, $K_{P,T}$, can be determined by measuring the compositions of the coexisting phases. The equilibrium pressure, as a function of temperature, for the determined equilibrium constant is then calculated and drawn in a P-T diagram. The coordinates of the curve represent the P-T conditions at which the phase assemblage may have equilibrated. In order to evaluate the relevant equilibrium pressure and temperature, an analogous P-T-curve of another phase assemblage from the same rock is required. The intersection of the two curves then defines uniquely the temperature and pressure of equilibration.

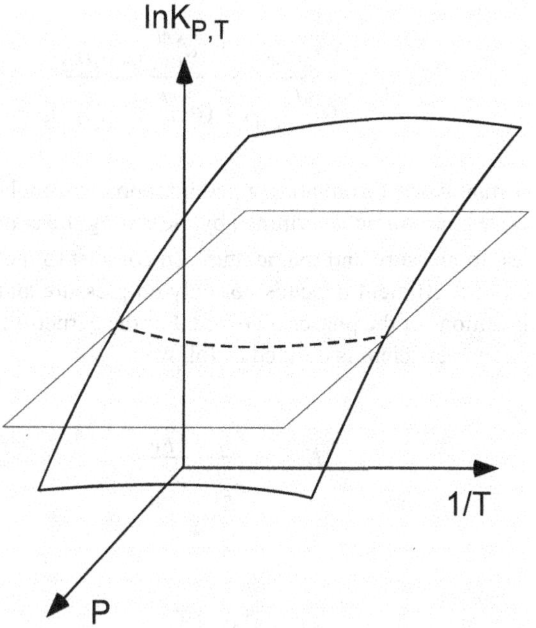

Fig. 8.1 Graphic representation of Eq. (8.6). The dashed curve gives the *P-T* coordinates for a constant value of $\ln K_{P,T}$.

8.1 Exchange geothermometers

Geothermometers and geobarometers are based on different types of reactions. One type is a so-called *exchange reaction*. This is a reaction that involves the exchange of two atoms between two different minerals. The volume changes associated with exchange reactions are generally small and the enthalpy changes relatively large and, therefore, show a strong temperature dependence and are largely pressure independent and thus make ideal geothermometers.

An example of an exchange *thermometer* that is based on the partitioning of iron and magnesium between garnet and biotite as presented in the foregoing chapter. Another example is the so-called *garnet-olivine geothermometer*. It was calibrated by O'Neill and Wood (1979) and is based on the equilibrium distribution of iron and magnesium between garnet and olivine. The corresponding exchange reaction reads:

$$2Mg_3Al_2Si_3O_{12} + 3Fe_2SiO_4 = 2Fe_3Al_2Si_3O_{12} + 3Mg_2SiO_4 \qquad (8.7)$$

and thermodynamic equilibrium constant given by:

$$K_{P,T} = \frac{\left(a^{ol}_{Mg_2SiO_4}\right)^3 \left(a^{grt}_{Fe_3Al_2Si_3O_{12}}\right)^2}{\left(a^{ol}_{Fe_2SiO_4}\right)^3 \left(a^{grt}_{Mg_3Al_2Si_3O_{12}}\right)^2}. \tag{8.8}$$

It is important when formulating a geothermometer to relate the distribution coefficient, K_D, which can be determined by measuring the compositions of the coexisting phases, to pressure and temperature. In contrast to the equilibrium constant, the distribution coefficient depends not only on pressure and temperature but also on the compositions of the phases involved. For the garnet-olivine geothermometer, the distribution coefficient is defined as follows:

$$K_D = \frac{x^{ol}_{Mg} \cdot x^{grt}_{Fe}}{x^{ol}_{Fe} \cdot x^{grt}_{Mg}}, \tag{8.9}$$

where

$$x^{ol}_{Mg} = \left(\frac{Mg}{Mg + Fe}\right)_{olivine} \tag{8.10}$$

and

$$x^{ol}_{Fe} = \left(\frac{Fe}{Mg + Fe}\right)_{olivine}. \tag{8.11}$$

In order to account for the effect of the calcium content on the distribution coefficient, O'Neill and Wood (1979) used a ternary Mg - Fe - Ca garnet in their experiments. The atomic fractions of magnesium and iron are, therefore, given by:

$$x^{grt}_{Mg} = \left(\frac{Mg}{Mg + Fe + Ca}\right)_{garnet} \tag{8.12}$$

and

$$x^{grt}_{\mathrm{Fe}} = \left(\frac{Fe}{Mg + Fe + Ca}\right)_{garnet}, \tag{8.13}$$

respectively.

Considering the fact that there are two thermodynamically equivalent sites in olivine and three in garnet, the relationship between the atomic fraction of a component and its activity reads:

$$a_{Mg_2SiO_4}^{ol} = (x_{Mg}^{ol} \cdot \gamma_{Mg}^{ol})^2, \tag{8.14}$$

$$a_{Fe_2SiO_4}^{ol} = (x_{Fe}^{ol} \cdot \gamma_{Fe}^{ol})^2, \tag{8.15}$$

$$a_{Mg_3Al_2Si_3O_{12}}^{grt} = (x_{Mg}^{grt} \cdot \gamma_{Mg}^{grt})^3 \tag{8.16}$$

and

$$a_{Fe_3Al_2Si_3O_{12}}^{grt} = (x_{Fe}^{grt} \cdot \gamma_{Fe}^{grt})^3. \tag{8.17}$$

Substituting expressions (8.14) through (8.17) for the activities in Eq. (8.8) yields:

$$K_{P,T} = \frac{(x_{Mg}^{ol} \cdot \gamma_{Mg}^{ol})^6 (x_{Fe}^{grt} \cdot \gamma_{Fe}^{grt})^6}{(x_{Fe}^{ol} \cdot \gamma_{Fe}^{ol})^6 (x_{Mg}^{grt} \cdot \gamma_{Mg}^{grt})^6} \tag{8.18}$$

or

$$K_D = K_{P,T}^{1/6} \cdot \left(\frac{\gamma_{Fe}^{ol} \cdot \gamma_{Mg}^{grt}}{\gamma_{Mg}^{ol} \cdot \gamma_{Fe}^{grt}}\right). \tag{8.19}$$

Taking logarithms and multiplying by the gas constant, *R*, and temperature, *T*, leads to:

$$RT\ln K_D = \frac{1}{6} RT\ln K_{P,T} + RT\ln\left(\frac{\gamma_{Mg}^{grt}}{\gamma_{Fe}^{grt}}\right) - RT\ln\left(\frac{\gamma_{Mg}^{ol}}{\gamma_{Fe}^{ol}}\right). \tag{8.20}$$

O'Neill and Wood (1979) assumed that olivine, as well as garnet, could be represented by a regular solution. Thus, the activity coefficients are given by:

$$RT\ln\gamma_{Mg}^{grt} = (x_{Fe}^{grt})^2 W_{FeMg}^{grt} + (x_{Ca}^{grt})^2 W_{CaMg}^{grt} + x_{Fe}^{grt}x_{Ca}^{grt}(W_{FeMg}^{grt} + W_{CaMg}^{grt} - W_{FeCa}^{grt}), \tag{8.21}$$

$$RT\ln\gamma_{Fe}^{grt} = (x_{Mg}^{grt})^2 W_{FeMg}^{grt} + (x_{Ca}^{grt})^2 W_{FeCa}^{grt} + x_{Ca}^{grt}x_{Mg}^{grt}(W_{FeMg}^{grt} + W_{FeCa}^{grt} - W_{CaMg}^{grt}), \tag{8.22}$$

$$RT\ln\gamma_{Mg}^{ol} = (x_{Fe}^{ol})^2 W_{FeMg}^{ol} \tag{8.23}$$

and

$$RT\ln\gamma_{Fe}^{ol} = (x_{Mg}^{ol})^2 W_{FeMg}^{ol}, \tag{8.24}$$

where W_{ij} are the interaction parameters.

Considering that the sum of $x_{Mg} + x_{Fe}$ is 1, from Eqs. (8.21) and (8.22) one obtains:

$$RT\ln\left(\frac{\gamma_{Mg}^{grt}}{\gamma_{Fe}^{grt}}\right) = (x_{Fe}^{grt} - x_{Mg}^{grt})W_{FeMg}^{grt} + x_{Ca}^{grt}(W_{CaMg}^{grt} - W_{FeCa}^{grt}) \tag{8.25}$$

and from Eqs. (8.23) and (8.24):

$$RT\ln\left(\frac{\gamma_{Mg}^{ol}}{\gamma_{Fe}^{ol}}\right) = (x_{Fe}^{ol} - x_{Mg}^{ol})W_{FeMg}^{ol}. \tag{8.26}$$

Disregarding the temperature and pressure dependence on the enthalpy, entropy and volume of reaction, the equilibrium constant is, according to Eqs. (7.8) and (7.16), given by:

$$\frac{1}{6}\ln K_{P,T} = -\frac{\Delta_r \boldsymbol{H}_{298}}{6RT} + \frac{\Delta_r \boldsymbol{S}_{298}}{6R} - \frac{\Delta_r \boldsymbol{V}_{298}}{6RT}(P - P_o). \tag{8.27}$$

Eq. (8.15) can, thus, be written as:

$$\ln K_D = -\frac{\Delta_r H_{298}}{6RT} + \frac{\Delta_r S_{298}}{6R} - \frac{\Delta_r V_{298}}{6RT}(P - P_o) + (x_{Fe}^{grt} - x_{Mg}^{grt})W_{FeMg}^{grt} \qquad (8.28)$$

$$+ x_{Ca}^{grt}(W_{CaMg}^{grt} - W_{FeCa}^{grt}) - (x_{Fe}^{ol} - x_{Mg}^{ol})W_{FeMg}^{ol}.$$

If the pressure and temperature dependence on the changes in enthalpy, entropy and volume of reaction are to be considered, the molar heat capacities, thermal expansion and compressibility coefficients of the end-member phases must be introduced into the calculation. In addition, the interaction parameters must be given as a function of temperature and pressure. This leads to complex expressions, but does not introduce any substantially new aspects and is, therefore, not presented here.

8.2 Solvus thermometry

Solvus thermometry is based on the distribution of a component between two coexisting phases occurring on the limbs of a miscibility gap. The boundary of the gap, which is referred to as the solvus, indicates the temperature dependent degree of miscibility between two structurally related phases. The composition of coexisting mineral pairs can, therefore, be used to estimate their temperature of equilibration. The location of a solvus in a *T*-*x*-diagram is generally determined experimentally. Based on the experimental results, the equilibrium conditions as a function of temperature, pressure and composition are then expressed by the usual thermodynamic formulas.

One important mineralogical system with a miscibility gap that has been well studied experimentally is that consisting of enstatite-diopside. The subsolidus equilibria in this mineral pair are governed by the partial immiscibility between Ca-rich diopside and Ca-poor enstatite. Fig. 8.2 shows the mutual solubility of the components as a function of temperature at 1.5 GPa (Lindsley and Dixon 1976).

Several thermodynamic models for the system enstatite-diopside have been presented in the literature. Here, we want to describe the model developed by Lindsley et al. (1981).

In a binary system containing two phases there are two equations that relate the chemical potentials of the components to one another. For coexisting enstatite and diopside, these relations are:

$$\mu_{Mg_2Si_2O_6}^{opx} = \mu_{Mg_2Si_2O_6}^{cpx} \qquad (8.29)$$

and

$$\mu_{CaMgSi_2O_6}^{opx} = \mu_{CaMgSi_2O_6}^{cpx}, \tag{8.30}$$

where $\mu_{Mg_2Si_2O_6}^{opx}$, $\mu_{Mg_2Si_2O_6}^{cpx}$, $\mu_{CaMgSi_2O_6}^{opx}$ and $\mu_{CaMgSi_2O_6}^{cpx}$ are the chemical potentials of the components $Mg_2Si_2O_6$ and $CaMgSi_2O_6$ in the phases orthopyroxene and clinopyroxene, respectively. Eqs. (8.29) and (8.30) correspond to the reactions:

$$(Mg_2Si_2O_6)^{opx} \rightarrow (Mg_2Si_2O_6)^{cpx} \tag{8.31}$$

and

$$(CaMgSi_2O_6)^{opx} \rightarrow (CaMgSi_2O_6)^{cpx}. \tag{8.32}$$

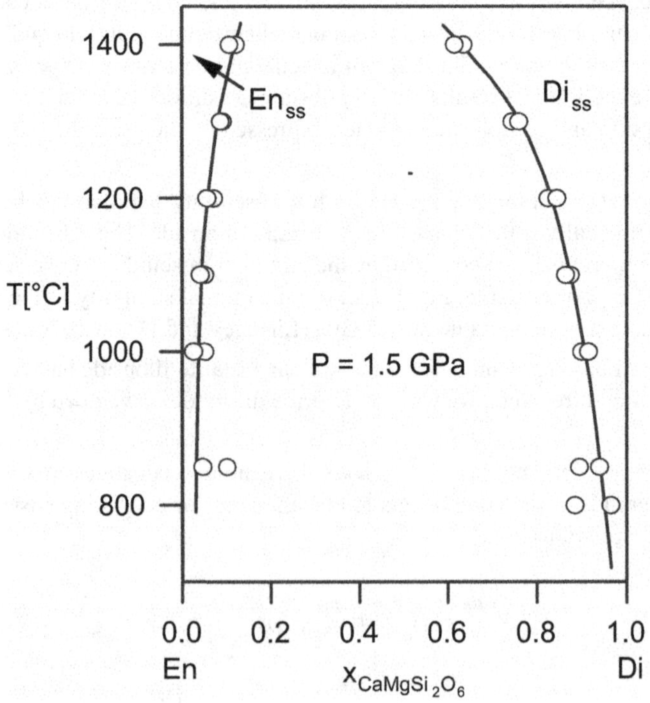

Fig. 8.2 *T-x* diagram of the system $Mg_2Si_2O_6$ - $CaMgSi_2O_6$ at P = 1.5 GPa (after Lindsley and Dixon 1976). En_{ss} = enstatite solid solution, Di_{ss} = diopside solid solution

If the chemical potentials are written in their extended form, Eqs. (8.29) and (8.30) read:

$$\mu^{0,opx}_{Mg_2Si_2O_6} + RT\ln x^{opx}_{Mg_2Si_2O_6} + RT\ln\gamma^{opx}_{Mg_2Si_2O_6} = \mu^{0,cpx}_{Mg_2Si_2O_6} + RT\ln x^{cpx}_{Mg_2Si_2O_6} + RT\ln\gamma^{cpx}_{Mg_2Si_2O_6} \quad (8.33)$$

and

$$\mu^{0,opx}_{CaMgSi_2O_6} + RT\ln x^{opx}_{CaMgSi_2O_6} + RT\ln\gamma^{opx}_{CaMgSi_2O_6} = \mu^{0,cpx}_{CaMgSi_2O_6} + RT\ln x^{cpx}_{CaMgSi_2O_6} + RT\ln\gamma^{cpx}_{CaMgSi_2O_6} \quad (8.34)$$

or

$$\mu^{0,cpx}_{Mg_2Si_2O_6} - \mu^{0,opx}_{Mg_2Si_2O_6} = RT\ln\frac{x^{opx}_{Mg_2Si_2O_6}}{x^{cpx}_{Mg_2Si_2O_6}} + RT\ln\frac{\gamma^{opx}_{Mg_2Si_2O_6}}{\gamma^{cpx}_{Mg_2Si_2O_6}} \quad (8.35)$$

and

$$\mu^{0,cpx}_{CaMgSi_2O_6} - \mu^{0,opx}_{CaMgSi_2O_6} = RT\ln\frac{x^{opx}_{CaMgSi_2O_6}}{x^{cpx}_{CaMgSi_2O_6}} + RT\ln\frac{\gamma^{opx}_{CaMgSi_2O_6}}{\gamma^{cpx}_{CaMgSi_2O_6}}. \quad (8.36)$$

The differences between the standard potentials in Eqs. (8.35) and (8.36) correspond to the Gibbs free energy of phase transformation for pure $Mg_2Si_2O_6$ and pure $CaMgSi_2O_6$, respectively. Hence, it holds that

$$\Delta_{tr}\boldsymbol{G}_{Mg_2Si_2O_6} = \Delta_{tr}\boldsymbol{H}_{Mg_2Si_2O_6} - T\Delta_{tr}\boldsymbol{S}_{Mg_2Si_2O_6} + \Delta_{tr}\boldsymbol{V}_{Mg_2Si_2O_6}(P - P_o) \quad (8.37)$$

$$= RT\ln\frac{x^{opx}_{Mg_2Si_2O_6}}{x^{cpx}_{Mg_2Si_2O_6}} + RT\ln\frac{\gamma^{opx}_{Mg_2Si_2O_6}}{\gamma^{cpx}_{Mg_2Si_2O_6}}$$

and

$$\Delta_{tr}\boldsymbol{G}_{CaMgSi_2O_6} = \Delta_{tr}\boldsymbol{H}_{CaMgSi_2O_6} - T\Delta_{tr}\boldsymbol{S}_{CaMgSi_2O_6} \tag{8.38}$$

$$+ \Delta_{tr}\boldsymbol{V}_{CaMgSi_2O_6}(P - P_o) = RT\ln\frac{x^{opx}_{CaMgSi_2O_6}}{x^{cpx}_{CaMgSi_2O_6}} + RT\ln\frac{\gamma^{opx}_{CaMgSi_2O_6}}{\gamma^{cpx}_{CaMgSi_2O_6}}.$$

Orthopyroxene is considered to be a regular and clinopyroxene a subregular solid solution. The activity coefficients of the components have, therefore, the form:

$$RT\ln\gamma^{opx}_{Mg_2Si_2O_6} = (x^{opx}_{CaMgSi_2O_6})^2 W^{G,\,opx}, \tag{8.39}$$

$$RT\ln\gamma^{opx}_{CaMgSi_2O_6} = (x^{opx}_{Mg_2Si_2O_6})^2 W^{G,\,opx}, \tag{8.40}$$

$$RT\ln\gamma^{cpx}_{Mg_2Si_2O_6} = (x^{cpx}_{CaMgSi_2O_6})^2 [W^{G,\,cpx}_{EnDi} + 2(W^{G,\,cpx}_{DiEn} - W^{G,\,cpx}_{EnDi})x^{cpx}_{Mg_2Si_2O_6}] \tag{8.41}$$

and

$$RT\ln\gamma^{cpx}_{CaMgSi_2O_6} = (x^{cpx}_{Mg_2Si_2O_6})^2 [W^{G,\,cpx}_{DiEn} + 2(W^{G,\,cpx}_{EnDi} - W^{G,\,cpx}_{DiEn})x^{cpx}_{CaMgSi_2O_6}]. \tag{8.42}$$

Inserting expressions (8.39) through (8.42) into Eq. (8.37) and (8.38), respectively, yields:

$$\Delta_{tr}\boldsymbol{H}_{Mg_2Si_2O_6} - T\Delta_{tr}\boldsymbol{S}_{Mg_2Si_2O_6} + \Delta_{tr}\boldsymbol{V}_{Mg_2Si_2O_6}(P - P_o) \tag{8.43}$$

$$= RT\ln\frac{x^{opx}_{Mg_2Si_2O_6}}{x^{cpx}_{Mg_2Si_2O_6}} + (x^{opx}_{CaMgSi_2O_6})^2 W^{G,\,opx}$$

$$-(x^{cpx}_{CaMgSi_2O_6})^2 [W^{G,\,cpx}_{EnDi} + 2(W^{G,\,cpx}_{DiEn} - W^{G,\,cpx}_{EnDi})x^{cpx}_{Mg_2Si_2O_6}]$$

for reaction (8.31) and

$$\Delta_{tr}\boldsymbol{H}_{CaMgSi_2O_6} - T\Delta_{tr}\boldsymbol{S}_{CaMgSi_2O_6} + \Delta_{tr}\boldsymbol{V}_{CaMgSi_2O_6}(P - P_o) \tag{8.44}$$

$$= RT\ln\frac{x^{opx}_{CaMgSi_2O_6}}{x^{cpx}_{CaMgSi_2O_6}} + (x^{opx}_{Mg_2Si_2O_6})^2 W^{G,\,cpx}$$

$$-(x^{cpx}_{Mg_2Si_2O_6})^2[W^{G,\,cpx}_{DiEn} + 2(W^{G,\,cpx}_{EnDi} - W^{G,\,cpx}_{DiEn})x^{cpx}_{CaMgSi_2O_6}].$$

for reaction (8.32).

Using the compositions of 23 orthopyroxene-clinopyroxene pairs taken from their experiments, Lindsley et al. (1981) determined the solution parameters and obtained the following values:

$$\Delta_{tr}\boldsymbol{G}_{Mg_2Si_2O_6}[\mathrm{Jmol}^{-1}] = 3561 - 1.91T[\mathrm{K}] + 0.355\times10^{-6}P[\mathrm{Pa}],$$

$$\Delta_{tr}\boldsymbol{G}_{CaMgSi_2O_6}[\mathrm{Jmol}^{-1}] = -21178 + 8.16T[\mathrm{K}] - 0.908\times10^{-6}P[\mathrm{Pa}],$$

$$W^{G,\,opx} = 25000\ \mathrm{Jmol}^{-1},$$

$$W^{G,\,cpx}_{EnDi}[\mathrm{Jmol}^{-1}] = 25484 + 0.812\times10^{-6}P[\mathrm{Pa}]\ \text{and}$$

$$W^{G,\,cpx}_{DiEn}[\mathrm{Jmol}^{-1}] = 31216 - 0.061\times10^{-6}P[\mathrm{Pa}].$$

Inserting these data into Eqs. (8.43) and (8.44) and solving for T, yields:

$$T = \Bigl\{3561 + 0.355\times10^{-6}P + 25000(x^{opx}_{Mg_2Si_2O_6}) \tag{8.45}$$

$$-(x^{cpx}_{CaMgSi_2O_6})^2\Bigl\{(25484 + 0.812\times10^{-6}P)$$

$$+2[(31216 - 0.061\times10^{-6}P) - (25484 + 0.812\times10^{-6}P)]x^{cpx}_{Mg_2Si_2O_6}\Bigr\}\Bigr\}/$$

$$[1.91 - R\ln(x^{opx}_{Mg_2Si_2O_6}/x^{cpx}_{Mg_2Si_2O_6})]$$

and

$$T = \Bigg\{-21178 - 0.908\times10^{-6}P + 25000x^{opx}_{CaMgSi_2O_6} \tag{8.46}$$

$$-(x^{cpx}_{Mg_2Si_2O_6})^2\Big\{(31216 - 0.061\times10^{-6}P)$$

$$+2[(25484 + 0.812\times10^{-6}P) - (31216 - 0.061\times10^{-6}P)]x^{cpx}_{Mg_2Si_2O_6}\Big\}\Bigg\}/$$

$$[-8.16 - R\ln(x^{opx}_{CaMgSi_2O_6}/x^{cpx}_{CaMgSi_2O_6})],$$

respectively.

Eqs. (8.45) and (8.46) can be used to estimate the equilibrium temperature of enstatite-diopside pairs at any pressure.

8.3 Solid-solid reactions

Some geobarometers and geothermometers are based on reactions that produce or consume of phases. Although these reactions generally are both temperature and pressure dependent, they are preferentially used as geobarometers. This is because solid-solid reactions often result in large volume changes. The application of these geobarometers usually requires a chemical analysis of the coexisting phases in order to account for the effect of solid solutions. This can be, however, also an advantage because an increased number of components increases the variance of the system.

8.3.1 Reactions in one-component system

The most widely used system to estimate the equilibrium conditions of metamorphic rocks is Al_2SiO_5. The system comprises the polymorphs kyanite, andalusite and sillimanite (see Fig. 7.2) and the thermodynamic equilibria are described by the following three reactions:

$$(Al_2SiO_5)^{ky} = (Al_2SiO_5)^{and},$$

$$(Al_2SiO_5)^{and} = (Al_2SiO_5)^{sill} \quad \text{and}$$

$$(Al_2SiO_5)^{ky} = (Al_2SiO_5)^{sill}.$$

The significance of these reactions for the estimation of P-T conditions in metamorphic rocks was recognized very early. A generation of geoscientists has tried to determine experimentally the location of the univariant reaction (transformation) curves in *P-T*-space. The results of their investigations are, however, inconsistent. These discrepancies or problems are related to the small Gibbs free energies of transitions and to the variation in the composition and structural state of the synthetic versus natural phases. Thus, several phase diagrams have been published. Field petrologists prefer the triple point at 0.376 GPa and 501°C as determined by Holdaway (1971), which is consistent with the calorimetrically determined thermodynamic properties of natural andalusite (Anderson et al. 1977). Another phase diagram that is cited often, gives the triple point at 0.55 GPa and 622°C (Richardson et al. 1969). The usefulness of this system as a geobarothermometer is related to the frequent occurrence of the three polymorphs in metapelitic rocks and the fact that some isograd reactions correspond to polymorphic transformations. Al_2SiO_5 phases are used sometimes to calibrate other geothermometers and geobarometers.

8.3.2 Reactions in multicomponent systems

One key geobarometer for medium-grade pelitic rocks is based on the assemblage plagioclase, garnet, Al-silicate and quartz. The reaction describing thermodynamic equilibrium between the phases reads:

$$3CaAl_2Si_2O_8 = Ca_3Al_2Si_3O_{12} + 2Al_2SiO_5 + SiO_2. \tag{8.47}$$

where anorthite and grossular are components in plagioclase and garnet solid solutions, respectively. Aluminium silicate can occur as kyanite or sillimanite. The reaction was first proposed as a geobarometer by Gent (1976). Several experimental studies have been made on the end-member reaction (e.g., Hays 1976;Harriya and Kennedy 1968;Schmid et al. 1978;Goldsmith 1980;Koziol and Newton 1988). Fig. 8.3 shows the results of Koziol and Newton (1988).

The pressure dependence of the reaction is given by:

$$P[Pa] = \frac{-\Delta_r \boldsymbol{H} + T\Delta_r \boldsymbol{S} - RT\ln K_{P,T}}{\Delta_r \boldsymbol{V}} + 1\times10^5, \tag{8.48}$$

where $\Delta_r \boldsymbol{H}$, $\Delta_r \boldsymbol{S}$ and $\Delta_r \boldsymbol{V}$ are the enthalpy, entropy and volume of reaction, respectively. $K_{P,T}$ is the thermodynamic equilibrium constant that reads:

$$K_{P,T} = \frac{a^{grt}_{Ca_3Al_2Si_3O_{12}} \cdot (a^{ky/sill}_{Al_2SiO_5})^2 a^{qtz}_{SiO_2}}{(a^{fsp}_{CaAl_2Si_2O_8})^3}, \tag{8.49}$$

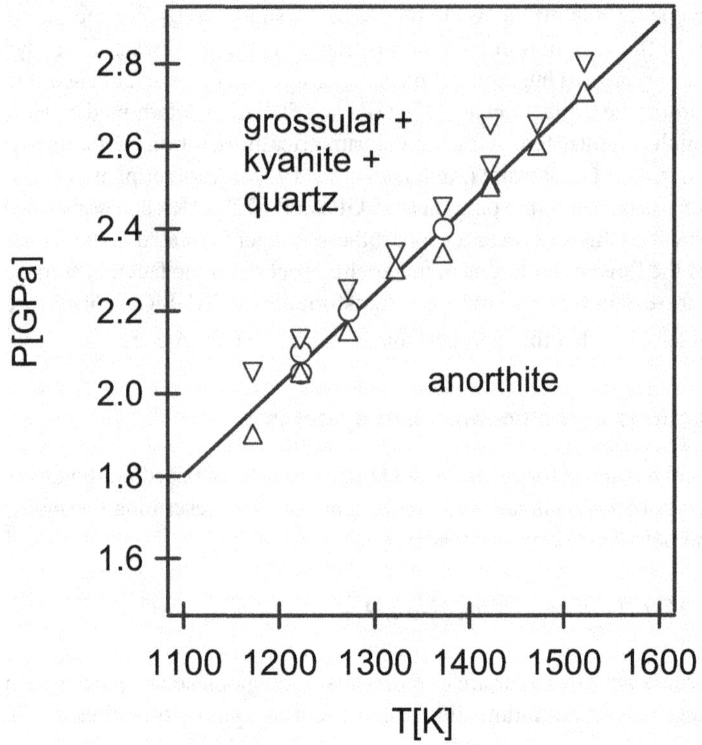

Fig. 8.3 Experimental results for the reaction $3an = grt + 2ky + qtz$ (Koziol and Newton 1988). The reaction curve is calculated using $P[\text{GPa}] = 0.0022T[\text{K}] - 0.62$ from McKenna and Hodges (1988).

where the superscript *ky/sill* indicates the two possible modifications of aluminium silicate.

All phases except garnet and plagioclase are pure. Therefore, their activities equal 1 and the expression for the equilibrium constant reduces to:

$$K_{P,T} = \frac{a^{grt}_{Ca_3Al_2Si_3O_{12}}}{(a^{fsp}_{CaAl_2Si_2O_8})^3} = \frac{(x^{grt}_{Ca})^3 \cdot (\gamma^{grt}_{Ca})^3}{[x^{fsp}_{Ca}(1/4(1+x^{fsp}_{Ca})^2/4)\gamma^{fsp}_{Ca}]^3} . \tag{8.50}$$

The term $[1/4(1+x^{fsp}_{Ca})^2/4]$ is the entropy correction factor that takes into account the aluminum avoidance principle, which says that adjacent aluminium containing tetrahedra are energetically unfavorable and are, therefore, avoided.

Berman (1990) developed a model for the activity coefficients of quaternary Ca-Mg-Fe-Mn garnets, whereby the activity coefficient for grossular reads:

$$\begin{aligned}3RT\ln\gamma^{grt}_{Ca_3Al_2Si_3O_{12}} &= W_{112}(2x_1x_2 - 2x_1^2x_2) + W_{122}(x_2^2 - 2x_1x_2^2) \\ &+ W_{113}(2x_1x_3 - 2x_1^2x_3) + W_{133}(x_3^2 - 2x_1x_3^2) \\ &+ W_{114}(2x_1x_4 - 2x_1^2x_4) + W_{144}(x_4^2 - 2x_1x_4^2) \\ &+ W_{223}(-2x_2^2x_3) + W_{233}(-2x_2x_3^2) \\ &+ W_{224}(-2x_2^2x_4) + W_{244}(-2x_2x_4^2) + W_{334}(-2x_3^2x_4) \\ &+ W_{344}(-2x_3x_4^2) + W_{123}(x_2x_3 - 2x_1x_2x_3) \\ &+ W_{124}(x_2x_4 - 2x_1x_2x_4) + W_{124}(x_3x_4 - 2x_1x_2x_4) \\ &+ W_{134}(x_3x_4 - 2x_1x_3x_4) + W_{234}(-2x_2x_3x_4).\end{aligned} \tag{8.51}$$

In Eq. (8.51), W_{ijk} are the ternary interaction parameters and the numbers in the subscripts designate the components: 1 = $Ca_3Al_2Si_3O_{12}$, 2 = $Mg_3Al_2Si_3O_{12}$, 3 = $Fe_3Al_2Si_3O_{12}$ and 4 = $Mn_3Al_2Si_3O_{12}$.

Tab. 8.1 gives the values for the interaction parameters.

Table 8.1 Interaction parameters for garnet solid solutions (Berman 1990)

Parameter	W^H[$Jmol^{-1}$]	W^S[$Jmol^{-1}K^{-1}$]	W^V[cm^3mol^{-1}]
112	21560	18.79	1.0
122	69200	18.79	1.0
113	20320	5.08	1.7
133	2620	5.08	0.9
223	230	-	0.1
233	3720	-	0.6
123	58825	23.87	2.65
124	45424	18.79	1.0
134	11470	5.08	1.3
234	1975	-	0.35

The activity coefficient for $CaAl_2Si_2O_8$ in a ternary feldspar solid solution is given by Fuhrman and Lindsley (1988) and has the following form:

$$
\begin{aligned}
RT\ln\gamma^{fsp}_{CaAl_2Si_2O_8} &= W_{OrAb}[x_{Ab}x_{Or}(1/2 - x_{An} - 2x_{Ab})] \\
&+ W_{AbOr}[x_{Ab}x_{Or}(1/2 - x_{An} - 2x_{Or})] \\
&+ W_{OrAn}[2x_{Or}x_{An}(1 - x_{An})] + x_{Ab}x_{Or}(1/2 - x_{An}) \\
&+ W_{AnOr}[x^2_{Or}(1 - 2x_{An}) + x_{Ab}x_{Or}(1/2 - x_{An})] \\
&+ W_{AbAn}[2x_{Ab}x_{An}(1 - x_{An}) + x_{Ab}x_{Or}(1/2 - x_{An})] \\
&+ W_{AnAb}[x^2_{Ab}(1 - 2x_{An}) + x_{Ab}x_{Or}(1/2 - x_{An})] \\
&+ W_{OrAbAn}[x_{Or}x_{Ab}(1 - 2x_{An})],
\end{aligned}
\tag{8.52}
$$

where An = $CaAl_2Si_2O_8$, Ab = $NaAlSi_3O_8$ and Or = $KAlSi_3O_8$. Note that Eq. (8.52) represents just another form of Eq. (5.172) that is given on page 210.

The interaction parameters required in Eq. (8.52) are given in Tab. 8.2. The standard thermodynamic functions including heat capacities, thermal expansion and compressibility coefficients that are required to calculate the pressure according to Eq. (8.48) can be taken from Berman (1988) or Holland and Powell (1990).

Table 8.2 Interaction parameters for ternary feldspars (Fuhrman and Lindsley 1988)

Parameter	W^H[$Jmol^{-1}$]	W^S[$Jmol^{-1}K^{-1}$]	W^V[cm^3mol^{-1}]
Ab-Or	18810	10.3	3.94
Or-Ab	27320	10.3	3.94
Ab-An	28226	-	-
An-Ab	8471	-	-
Or-An	52468	-	-
An-Or	47396	-	-1.20
OrAbAn	8700	-	-10.94

8.4 Reactions involving gaseous phases

Systems involving gaseous phases are generally less useful in geothermometry or geobarometry studies because a knowledge of fugacity of the gas involved in the metamorphic reaction is required. This information is, however, rarely available. In order to circumvent the problem, it is often assumed that the pressure of the participating gas equals the total pressure. This simplification can, however, introduce significant error in the determination of equilibrium conditions, especially if the gaseous phase consists of several different species. In such cases, the fugacities of each separate species should be known. Occasionally, other independent reactions in the same sample can be used to determine the composition of the fluid phase and the fugacities can be calculated.

One of the widely used geothermometers involving oxygen as the gaseous phase was proposed by Buddington and Lindsley (1964). It is based on coexisting magnetite-ilmenite pairs and it allows a determination of equilibrium temperature and oxygen fugacity. Both minerals, magnetite and ilmenite are binary solid solutions occurring in the systems Fe_3O_4-Fe_2TiO_4 and Fe_2O_3-$FeTiO_3$, respectively. Thermodynamic equilibrium is defined by a temperature-dependent exchange reaction and an oxidation reaction (Spencer and Lindsley 1981). The exchange reactions reads:

$$Fe_3O_4^{mt} + FeTiO_3^{ilm} = Fe_2TiO_4^{mt} + Fe_2O_3^{ilm} \tag{8.53}$$

and the oxidation reaction reads:

$$4Fe_3O_4^{mt} + O_2 = 6Fe_2O_3^{ilm}. \tag{8.54}$$

The superscripts '*mt*' and '*ilm*' designate the cubic magnetite-ulvöspinel and the rhombohedral hematite-ilmenite solid-solution phase, respectively.

For the system at equilibrium it holds:

$$\Delta_r \boldsymbol{G}^{exch} = -RT\ln\frac{a_{Fe_2TiO4}^{mt} \cdot a_{Fe_2O_3}^{ilm}}{a_{Fe_3O_4}^{mt} \cdot a_{FeTiO_3}^{ilm}} \tag{8.55}$$

and

$$\Delta_r \boldsymbol{G}^{ox} = -RT\ln\frac{(a_{Fe_2O_3}^{ilm})^6}{(a_{Fe_3O_4}^{mt})^4} + RT\ln f_{O_2}. \tag{8.56}$$

In the ilmenite structure, Fe and Ti are ordered in alternate layers perpendicular to the *c*-axis. The layers are termed *A* and *B*, respectively. Spencer and Lindsley (1981) assume that Fe^{2+} occupies exclusively the *A* layers and Ti^{4+} exclusively *B* layers and that Fe^{3+} mixes randomly in both layers showing no preference neither for *A* nor for *B*. The activity of ilmenite is, therefore, given by:

$$a_i^{ilm} = (x_i^{ilm})^2 (\gamma_i^{ilm})^2 . \tag{8.57}$$

Magnetite, Fe_3O_4, and ulvöspinel, Fe_2TiO_4, have an inverse spinel structure. For the magnetite-ulvöspinel solid solution, Spencer and Lindsley (1981) adopt a so-called ħmolecular mixing'model. Octahedrally coordinated Ti^{4+} is linked to octahedrally coordinated Fe^{2+} and each octahedrally coordinated Fe^{2+} is linked to octahedrally coordinated Fe^{3+}. The activity of magnetite can, therefore, be written as:

$$a_i^{mt} = x_i^{mt} \cdot \gamma_i^{mt} . \tag{8.58}$$

Substituting the activities in Eqs. (8.55) and (8.56) by expressions (8.57) and (8.58), respectively, lead to:

$$\Delta_r \boldsymbol{G}^{exch} = -RT\ln\frac{x_{Fe_2TiO4}^{mt} \cdot (x_{Fe_2O_3}^{ilm})^2}{x_{Fe_3O_4}^{mt} \cdot (x_{FeTiO_3}^{ilm})^2} - RT\ln\frac{\gamma_{Fe_2TiO4}^{mt} \cdot (\gamma_{Fe_2O_3}^{ilm})^2}{\gamma_{Fe_3O_4}^{mt} \cdot (\gamma_{FeTiO_3}^{ilm})^2} \tag{8.59}$$

and

$$\Delta_r \boldsymbol{G}^{ox} = -RT\ln\frac{(x_{Fe_2O_3}^{ilm})^{12}}{(x_{Fe_3O_4}^{mt})^4} - RT\ln\frac{(\gamma_{Fe_2O_3}^{ilm})^{12}}{(\gamma_{Fe_3O_4}^{mt})^4} + RT\ln f_{O_2}. \tag{8.60}$$

Spencer and Lindsley (1981) express the activity coefficients of the components in terms of a regular solution model. They have the form:

$$RT\ln\gamma_i = (1-x_i)^2[W_{ij}^G + 2(W_{ji}^G - W_{ij}^G)x_i], \tag{8.61}$$

where the subscripts *i* and *j* designate the components of the binary system *i-j*. Thus, the activity coefficient of the end-member component Fe_2TiO_4 reads:

$$RT\ln\gamma_{Fe_2TiO4}^{mt} = (1-x_{Fe_2TiO4}^{mt})^2 [W_{usp-mt}^G + 2(W_{mt-usp}^G - W_{usp-mt}^G)x_{Fe_2TiO4}^{mt}], \tag{8.62}$$

and that of Fe_2O_3 in the rhombohedral ilmenite phase:

$$RT\ln\gamma^{ilm}_{Fe_2O_3} = (1 - x^{ilm}_{Fe_2O_3})^2 [W^G_{hem-ilm} + 2(W^G_{ilm-hem} - W^G_{hem-ilm})x^{ilm}_{Fe_2O_3}]. \tag{8.63}$$

The activity coefficients of magnetite and ilmenite have similar forms.

Inserting the activity coefficients into Eqs. (8.59) and (8.60) and replacing the Gibbs free energy of reaction by the enthalpy and entropy of reaction yields:

$$\Delta_r \boldsymbol{H}^{exch} - T\Delta_r \boldsymbol{S}^{exch} = -RT\ln\frac{x^{mt}_{Fe_2TiO4} \cdot (x^{ilm}_{Fe_2O_3})^2}{x^{mt}_{Fe_3O_4} \cdot (x^{ilm}_{FeTiO_3})^2} \tag{8.64}$$

$$- (1 - x^{mt}_{Fe_2TiO4})^2 [W^G_{usp-mt} + 2(W^G_{mt-usp} - W^G_{usp-mt})x^{mt}_{Fe_2TiO4}]$$

$$- 2(1 - x^{ilm}_{Fe_2O_3})^2 [W^G_{hem-ilm} + 2(W^G_{ilm-hem} - W^G_{hem-ilm})x^{ilm}_{Fe_2O_3}]$$

$$+ (1 - x^{mt}_{Fe_3O_4})^2 [W^G_{mt-usp} + 2(W^G_{usp-mt} - W^G_{mt-usp})x^{mt}_{Fe_3O_4}]$$

$$+ 2(1 - x^{ilm}_{FeTiO_3})^2 [W^G_{ilm-hem} + 2(W^G_{hem-ilm} - W^G_{ilm-hem})x^{ilm}_{FeTiO_3}]$$

and

$$\Delta_r \boldsymbol{H}^{ox} - T\Delta_r \boldsymbol{S}^{ox} = -RT\ln\frac{(x^{ilm}_{Fe_2O_3})^{12}}{(x^{mt}_{Fe_3O_4})^{4}} + RT\ln f_{O_2} \tag{8.65}$$

$$-12(1-x^{ilm}_{Fe_2O_3})^2[W^G_{hem-ilm} + 2(W^G_{ilm-hem} - W^G_{hem-ilm})x^{ilm}_{Fe_2O_3}]$$

$$+4(1-x^{mt}_{Fe_3O_4})^2[W^G_{mt-usp} + 2(W^G_{usp-mt} - W^G_{mt-usp})x^{mt}_{Fe_3O_4}].$$

The temperature dependence of the interaction parameters is given by:

$$W^G_{ij} = W^H_{ij} - TW^S_{ij}. \tag{8.66}$$

Using the relationship in (8.66), the equilibrium temperature can be calculated as follows:

$$T = (-\Delta_r \boldsymbol{H}^{exch} + A_1 W^H_{usp-mt} + A_2 W^H_{mt-usp} \tag{8.67}$$
$$+ A_3 W^H_{hem-ilm} + A_4 W^H_{ilm-hem})/(R\ln K^{exch}_D - \Delta_r \boldsymbol{S}^{exch}$$
$$+ A_1 W^H_{usp-mt} + A_2 W^H_{mt-usp} + A_3 W^H_{hem-ilm} + A_4 W^H_{ilm-hem}),$$

where

$$A_1 = -3(x^{mt}_{Fe_3O_4})^2 + 2x^{mt}_{Fe_3O_4}, \tag{8.68}$$

$$A_2 = 3(x^{mt}_{Fe_3O_4})^2 - 4x^{mt}_{Fe_3O_4} + 1, \tag{8.69}$$

$$A_3 = -6(x^{ilm}_{Fe_2O_3})^2 + 8x^{ilm}_{Fe_2O_3} - 2, \tag{8.70}$$

$$A_4 = 6(x^{ilm}_{Fe_2O_3})^2 - 4x^{ilm}_{Fe_2O_3} \tag{8.71}$$

and

$$K_D^{exch} = \frac{x_{Fe_2TiO4}^{mt} \cdot (x_{Fe_2O_3}^{ilm})^2}{x_{Fe_3O_4}^{mt} \cdot (x_{FeTiO_3}^{ilm})^2}. \tag{8.72}$$

The oxygen fugacity is then given by:

$$\begin{aligned} \ln f_{O_2} &= \ln K_D^{ox} + \frac{\Delta_r \boldsymbol{H}^{ox}}{RT} - \frac{\Delta_r \boldsymbol{S}^{ox}}{R} \\ &+ \Big\{ 12(1 - x_{Fe_2O_3}^{ilm})^2 [W_{hem-ilm}^G + 2(W_{ilm-hem}^G \\ &- W_{hem-ilm}^G) x_{Fe_2O_3}^{ilm}] - 4(1 - x_{Fe_3O_4}^{mt})^2 [W_{mt-usp}^G \\ &+ 2(W_{usp-mt}^G - W_{mt-usp}^G) x_{Fe_3O_4}^{mt}] \Big\} / RT, \end{aligned} \tag{8.73}$$

where

$$\ln K_D^{ox} = \frac{(x_{Fe_2O_3}^{ilm})^{12}}{(x_{Fe_3O_4}^{mt})^4}. \tag{8.74}$$

Spencer and Lindsley (1981) assume ideal mixing behavior for the magnetite-ulvöspinel solid solution above 800°C. Thus , the values for W_{mt-usp}^G and W_{usp-mt}^G are zero for temperatures higher than 800°C. Other solution parameters were determined as follows:

$$\begin{aligned} W_{usp-mt}^H &= 64835 \text{ Jmol}^{-1} & W_{usp-mt}^S &= 60.296 \text{ Jmol}^{-1}\text{K}^{-1} \\ W_{mt-usp}^H &= 20798 \text{ Jmol}^{-1} & W_{mt-usp}^S &= 19.652 \text{ Jmol}^{-1}\text{K}^{-1} \\ W_{ilm-hem}^H &= 102374 \text{ Jmol}^{-1} & W_{ilm-hem}^S &= 71.095 \text{ Jmol}^{-1}\text{K}^{-1} \\ W_{hem-ilm}^H &= 36818 \text{ Jmol}^{-1} & W_{hem-ilm}^S &= 7.7714 \text{ Jmol}^{-1}\text{K}^{-1} \\ \Delta_r H^{exch} &= 27799 \text{ J} & \Delta_r S^{exch} &= 4.192 \text{ JK}^{-1}. \end{aligned}$$

The enthalpy and the entropy of the oxidation reaction can be calculated using the data given in the usual thermodynamic tables (e.g. Robie and Hemingway 1995).

Solutions to problems

Chapter 01

1. $\dfrac{\partial^2 z}{\partial x \partial y} = \dfrac{\partial^2 z}{\partial y \partial x} = -\dfrac{a}{y^2}$

2. $x_{MgO}^{fo} = 0.667, \quad x_{SiO_2}^{fo} = 0.333$

   $x_{MgO}^{en} = 0.500, \quad x_{SiO_2}^{en} = 0.500$

4. $x_{MgO}^{crd} = 0.222, \quad x_{Al_2O_3}^{crd} = 0.222, \quad x_{SiO_2}^{crd} = 0.556$

   $wt\%(\mathrm{MgO}) = 13.78, \quad wt\%(\mathrm{Al_2O_3}) = 34.86, \quad wt\%(\mathrm{SiO_2}) = 51.36$

7. $x_{Al_2SiO_5}^{ky} = 0.7, \quad x_{SiO_2}^{qtz} = 0.3$

Chapter 02

1. $V_{1MPa} = 8.913\ \mathrm{dm^3 mol^{-1}}$

2. $a = \frac{9}{8} R T_c V_c = 0.339\ \mathrm{Jm^3 mol^{-2}}, \quad a = 3 P_c V_c = 0.208\ \mathrm{Jm^3 mol^{-2}}$

   - $b = \dfrac{V_c}{3} = 18.67\ \mathrm{cm^3}, \quad b = \dfrac{RT_c}{8P_c} = 30.49\ \mathrm{cm^3}$

   - The formulas are based on simplified assumptions.

3. $V = 1.31 \times 10^{-4}\ \mathrm{m^3 mol^{-1}}$

4. $V_E = 579.84\ \mathrm{Å^3}$,

- $V = 34.918 \times 10^{-6}\ \mathrm{m^3 mol^{-1}} = 34.918\ \mathrm{cm^3 mol^{-1}}$

5. $V = 1.139\ \mathrm{cm^3}$

6. $\Delta V^{and/sill} = -\ 2.047\ \mathrm{cm^3 mol^{-1}}$

7. $V^{di}_{CaMgSi_2O_6} = 66.043\ \mathrm{cm^3 mol^{-1}}$, $V^{CaTs}_{CaAl_2SiO_6} = 63.603\ \mathrm{cm^3 mol^{-1}}$

- $V^{di}_{CaMgSi_2O_6} = 65.960\ \mathrm{cm^3 mol^{-1}}$, $V^{di\infty}_{CaMgSi_2O_6} = 65.523\ \mathrm{cm^3 mol^{-1}}$
- $V^{CaTs}_{CaAl_2SiO_6} = 63.416\ \mathrm{cm^3 mol^{-1}}$, $V^{CaTs\infty}_{CaAl_2SiO_6} = 63.083\ \mathrm{cm^3 mol^{-1}}$
- $V^{di,\ ex}_{CaMgSi_2O_6} = -\ 0.083\ \mathrm{cm^3 mol^{-1}}$, $V^{CaTs,\ ex}_{CaAl_2SiO_6} = -\ 0.187\ \mathrm{cm^3 mol^{-1}}$
- $V^{di,\ ex,\ \infty}_{CaMgSi_2O_6} = -\ 0.520\ \mathrm{cm^3 mol^{-1}}$, $V^{CaTs,\ ex,\ \infty}_{CaAl_2SiO_6} = -\ 0.520\ \mathrm{cm^3 mol^{-1}}$
- $\Delta_m \overline{V}^{ex} = (1 - x^{cpx}_{CaTs}) \cdot x^{cpx}_{CaTs} \times (-\ 0.52)$

8. $\Delta_r V = 67.9\ \mathrm{cm^3}$

Chapter 03

1. $W_{0.1MPa} = 5.45 \times 10^{-3}\ \mathrm{J}$

- $W_{2.5\ GPa} = 63.5\ \mathrm{J}$ or

- $W_{2.5\ GPa} = 64.0$ J

- The formulas are based on simplified assumptions.

2. $\nu_E = 1.589 \times 10^{13}$ sec^{-1}

- $U_{E,300} = 456.476$ kJmol^{-1}

- $C_{p,300} = 612.52$ kJmol^{-1}
- $C_{p,500} = 845.36$ kJmol^{-1}

3. $\Delta H = -6.815 \times 10^6$ kJ

4. subregular

- $H^{ex}_{geh} = -0.811$ kJmol^{-1}, $H^{ex}_{ak} = 7.239$ kJmol^{-1}

- $H^{ex,\infty}_{geh} = -24.288$ kJmol^{-1}, $H^{ex,\infty}_{ak} = 0.502$ kJmol^{-1}

- $\Delta_m \overline{H}^{ex} = 1.6$ kJmol^{-1}

5. $\Delta_r H_{298} = 12.27$ kJ, $\Delta_r H_{1073} = 9.26$ kJ

6. $\Delta_r H_{298} = -21.74$ kJ

- $\Delta_r H_{1073} = -36.664$ kJ

7. $\Delta_{tr} \boldsymbol{H}^{\alpha/\beta}_{975} = 29.970$ kJmol^{-1}, $\Delta_{tr} \boldsymbol{H}^{\beta/\gamma}_{975} = 6.82$ kJmol^{-1},
   $\Delta_{tr} \boldsymbol{H}^{\alpha/\beta}_{298} = 32.769$ kJmol^{-1}, $\Delta_{tr} \boldsymbol{H}^{\beta/\gamma}_{298} = 9.40$ kJmol^{-1}

- $\Delta_f \boldsymbol{H}^{\beta}_{298} = -2140.2 \text{ kJmol}^{-1}$, $\Delta_f \boldsymbol{H}^{\gamma}_{298} = -2130.8 \text{ kJmol}^{-1}$

Chapter 04

1. $\boldsymbol{S}^{herc}_{1273} = 338.537 \text{ Jmol}^{-1}\text{K}^{-1}$

2. $S_{tot} > 0 = 10 \text{ JK}^{-1}$

3. $\boldsymbol{S}^{conf}_{Mg_2Al_3AlSi_5O_{18}} = 47.61 \text{ Jmol}^{-1}\text{K}^{-1}$

- $\boldsymbol{S}^{conf}_{rand} = 60.68 \text{ Jmol}^{-1}\text{K}^{-1}$

4. $\bar{S}^{conf}_{(Ca,\,Mg)(Mg,\,Al)(Al,\,Si)O_6} = 4.22 \text{ Jmol}^{-1}\text{K}^{-1}$

5. $\Delta_r \boldsymbol{S}_{800} = 701.2 \text{ JK}^{-1}$

Chapter 05

1. $\Delta_m \bar{G} = -15.237 \text{ kJmol}^{-1}$

2. $\Delta\mu = 56.652 \text{ kJmol}^{-1}$

3. $\Delta_m \bar{G}^{ex} = (1 - x_{mu}) \cdot x^2_{mu} W^G_{pg-mu} + (1 - x_{mu})^2 \cdot x_{mu} W^G_{mu-pg}$

- $\Delta_m \bar{H}^{ex} = 1.86 \text{ kJmol}^{-1}$

- $\Delta_m \bar{V}^{ex} = 0.429 \text{ cm}^3\text{mol}^{-1}$, $\Delta_m \bar{S}^{ex} = 0.14 \text{ JK}^{-1}$

- $\gamma_{pg} = 5.152, \quad a_{pg} = 0.773$
- $\gamma_{mu} = 1.028, \quad a_{mu} = 0.874$

- $\Delta_m \overline{G} = -1.114 \text{ kJmol}^{-1}$

4. $\mu^{gr}_{Ca_3Al_2Si_3O_{12}} = -6976.149 \text{ kJmol}^{-1}$

5. $\Delta_r G = 55.46 \text{ kJmol}^{-1}$

- The educts are more stable than the products at 0.3 GPa and 600 K. Therefore, the reaction can not proceed in the given direction.

Chapter 06

1. $P_{tr} = 6.61$ GPa

2. $T_c = 1203 \text{ K} = 930°\text{C}$

- $x_1 = 0.2814, \quad x_2 = 0.7186$

3. $T_{fus} = 1808$ K

- $\gamma^{an}_{CaAl_2Si_2O_8} = 1.049$

4. $x^{\alpha}_{Fe_2SiO_4} = 0.6150, \quad x^{\gamma}_{Fe_2SiO_4} = 0.8644$

Chapter 07

1. $P_{eq} = 0.47$ GPa

- $P_{eq} = 0.68$ GPa

2.

P x 10^{-6} [Pa]	T[K]
0.1	519
10.0	659
50.0	723
100.0	750
500.0	778
1000	737

- $T_{x_{CO_2} = 0.1} = 699$ K

3. $P_{O_2} = 3.73 \times 10^{-9}$ Pa

4. $3Mg_2SiO_4 + 2Fe_3Al_2Si_3O_{12} \rightarrow 3Fe_2SiO_4 + 3Mg_3Al_2Si_3O_{12}$

- $$\ln K_D = -\frac{\Delta_r \boldsymbol{H}}{6RT} + \frac{\Delta_r \boldsymbol{S}}{6R} - \frac{\Delta_r \boldsymbol{V}}{6RT}(P - 10^5)$$

- $$\ln K_D = -\frac{\Delta_r \boldsymbol{H}}{6RT} + \frac{\Delta_r \boldsymbol{S}}{6R} - \frac{\Delta_r \boldsymbol{V}}{6RT}(P - 10^5) + RT\ln\gamma_{Mg}^{grt} - RT\ln\gamma_{Fe}^{grt}$$

- $$a_{Mg_3Al_2Si_3O_{12}}^{grt} = x_{Mg}^3 \cdot x_{Al}^2 \left(\exp\left\{ x_{Fe}^2 \frac{W^G}{RT} \right\} \right)^3 \times \left(\exp\left\{ x_{Cr}^2 \frac{W^G}{RT} \right\} \right)^2 \times \exp\left\{ -x_{Fe} x_{Cr} \frac{\Delta_r \boldsymbol{G}^{rec}}{RT} \right\}$$

References

Ambrose D (1994) Critical constants, boiling points, and melting points of selected compounds. In Handbook of chemistry and physics. Lide RD ed. CRC-Press, Boca Raton Ann Arbor London, Tokyo, p 6-54 - 6-64

Anderson PAM, Newton RC, Kleppa OJ (1971) The enthalpy change of the andalusite-sillimanite reaction and the Al_2SiO_5 diagram Am J Sci 277: 585-593

Aranovich LY, Newton RC (1999): Experimental determination of CO_2-H_2O activity-composition relations at 600-1000°C and 6-14 kbar by reversed decarbonation and dehydration. Am Mineral 84: 1332-1332

Berman RG (1990) Mixing properties of Ca-Mg-Fe-Mn garnets. Am Mineral 77: 328-344

Berman RG, Brown TH (1985) Heat capacity of minerals in the system Na_2O-K_2O-CaO-MgO-FeO-Fe_2O_3-Al_2O_3-SiO_2-TiO_2-CO_2: representation, estimation, and high temperature extrapolation. Contrib Miner Petrol 89: 168-183

Boettcher AL (1970) The system CaO-Al_2O_3-SiO_2-H_2O at high pressures and temperatures J Petrol 11: 337-379

Bosenick A, Geiger CA (1997) Powder x-ray diffraction study of synthetic pyrope-grossular garnets between 20 and 295 K J Geophys Res 102: 22,649-22,657

Bosenick A, Geiger CA, Cemič (1996) Heat capacity measurements of synthetic pyrope-grossular garnets between 320 and 1000 K by differential scanning calorimetry. Geochim Cosmochim Acta 60: 3215-3227

Buddington AF, Lindsley DH (1964) Iron-titanium oxide minerals and synthetic equivalents J Petrol 5: 310-357

Burnham CW, Holloway JR, Davis NF (1969a) The specific volume of water in the range 1000 to 8900 bars, 20° to 900°C. Am J Sci 267A: 277-280

Burnham CW, Holloway JR, Davis NF (1969b) Thermodynamic properties of water to 1,000°C and 10,000 bars.
Geol Soc Am Spec Pap 132: p 1-96

Cahn JW (1961) On spinodal decomposition
Acta Metall 9: 795-801

Cahn JW (1962) Coherent fluctuations and nucleation in isotropic solids.
Acta Metall 10: 907-913

Cahn JW (1968) Spinodal decomposition.
Trans Metall Soc AIME 242: 166-180

Carlson HC, Colburn AP (1942) Vapor-liquid equilibria of non-ideal solutions. Utilization of theoretical method to extended data.
Ind Eng Chem 34: 581-589

Carnahan NF, Starling KE (1969) Equation of state for non attracting rigid spheres.
J Chem Phys 51: 635-636

Cemič L (1983) Chemische Aktivitäten in mineralogischen Systemen: Theorie und ihre Anwendung auf das System ZnS-FeS.
Fortschr Miner 61: 169-191

Cemič L, Kähler W (2000) Calorimetric determination of the enthalpy of Mg-Fe ordering in orthopyroxene.
Phys Chem Minerals 27: 220-224

Chatterjee ND (1991) Applied mineralogical thermodynamics.
Springer-Verlag, Berlin Heidelberg New York London Paris Tokyo Hong Kong Barcelona

Chatterjee ND, Froese E (1975) A thermodynamic study of pseudobinary join muscovite - paragonite in the system
$KAlSi_3O_8$-$NaAlSi_3O_8$-Al_2O_3-SiO_2-H_2O.
Am Mineral 60: 985-993

Chatillon-Colinet C, Newton RC, Perkins III D, Kleppa OJ (1983): Thermochemistry of $(Fe_{2+},Mg)SiO_3$ orthopyroxene.
Geochim Cosmochim Acta 47: 1597-1603

Cheng W, Ganguly J (1994) Some aspects of multicomponent excess free energy models with subregular binaries.
Geochim Cosmochim Acta 58: 3763-3767

de Santis R, Breedveld GJF, Prausnitz JM (1974) Thermodynamic properties of aqueous gas mixtures at advanced pressures.
Ind Eng Chem Proc Des Dev 13:374-377

Dove MT, Powell RM (1989) Neutron diffraction study of the tricritical orientational order/disorder phase transition in calcite at 1260 K.
Phys Chem Minerals 16: 503-507

Faßhauer DW, Cemič (2001) Heat of formation of petalite, $LiAlSi_4O_{10}$

Phys Chem Minerals 28: 531-533

Froese E, Gunter AE (1976) A note on the pyrrhotite sulfur vapor equilibrium. Econ Geol 71: 1589-1594

Ferry JM, Spear FS (1978) Experimental calibration of the partitioning of Fe and Mg between biotite and garnet. Contrib Mineral Petrol 66: 113-117

Fuhrman ML, Lindlsey DH (1988) Ternary feldspar modeling and thermometry. Am Mineral 73: 201-215

Ganguly J (1982) Mg-Fe order-disorder in ferromagnesian silicates. II Thermodynamics, kinetics, and geological implications. In: Saxena SK (ed) Advances in Physical Geochemistry, Vol 2. Springer-Verlag, New York, 58-99

Ganguly J, Saxena SK (1984) Mixing properties of alumosilicate garnets: Constraints from natural and experimental data, and applications to geothermo-barometry. Am Mineral 69: 88-97

Ganguly J, Saxena SK (1987) Mixtures and mineral reactions. Springer-Verlag, Berlin Heidelberg New York London Paris Tokyo

Geiger CA, Feenstra A (1997) Molar volumes of mixing of almandine-pyrope and almandine-spessartine garnets and the crystal chemistry and thermodynamic-mixing properties of the alumosilicate garnets. Am Mineral 82: 571-581

Geiger CA, Newton RC, Kleppa OJ (1987) Enthalpy of mixing of synthetic almandine-grossular and almandine-pyrope garnet from high temperature solution calorimetry. Geochim Cosmochim Acta 51: 1755-1763

Gent ED (1976) Plagioclase-garnet-Al_2SiO_5-quartz: a potential geothermometer-geobarometer. Am Mineral 61:710-714

Ghiorso MS (1984) Activity/composition relations in the ternary feldspars. Contrib Mineral Petrol 87: 282-296

Goldsmith JR (1980) The melting and breakdown reactions of anorthite at high pressures and temperatures. Am Mineral 65: 272-284

Grevel KD Chatterjee ND (1992) A modified Redlich-Kwong equation of state for H_2 - H_2O fluid mixtures at high pressures and at temperatures above 400°C Eur J Mineral 4: 1303-1310

Grevel KD (1993): Modified Redlich-Kwong equations of state for CH_4 and CH_4-H_2O.

N Jb Miner Mh 10: 462-480

Guggenheim EA (1937) A theoretical basis of Raoult's law. Trans Faraday Soc 33: 151-159

Guggenheim EA (1967): Thermodynamics, Elsevir/North Holland, Amsterdam New York

Haas JL, Fisher JR (1976) Simultaneous evaluation and correlation of thermodynamic data. Am J Sci 276: 525-545

Halbach H, Chatterjee ND (1982) An empirical Redlich-Kwong-type equation of state of water to 1000°C and 200 kbar. Contrib Miner Petrol 79: 337-345

Hardy HK (1953) A "sub-regular" solution model and its application to some binary alloy systems. Acta Metall 1: 202-209

Harriya Y, Kennedy GC (1968) Equilibrium study of anorthite under high pressure and high temperature. Amer J Sci 266: 193-203

Hays JF (1967) Lime-alumina-silica. Carnegie Institution of Washington Year Book 65: 234-239

Hazen RM, Finger LW (1978) Crystal structures and compressibilities of pyrope and grossular to 60 kbar. Amer Mineral 63: 297-303

Hildebrand JH (1929) Solubility, XII, Regular solutions. J Amer Chem Soc 51: 66-80

Holland TJB (1981) Thermodynamics of simple mineral systems. In: Newton R, Navrotsky A, Wood BJ (eds) Thermodynamics of minerals and melts. Springer-Verlag, Berlin Heidelberg New York p 19-34

Holland TJB, Powell R (1990) An enlarged and updated internally consistent thermodynamic data set for phases with uncertainties and correlations: the system K_2O-Na_2O-CaO-MgO-MnO-FeO-Fe_2O_3-Al_2O_3-TiO_2-SiO_2-C-H_2-O_2. J metamorphic Geol 16: 309-343

Holland T, Powell R (1991) A compensated-Redlich-Kwong (CORK) equation for volumes and fugacities of CO2 and H2O in the range 1 bar to 50 kbar and 100 -1600°C. Contrib Mineral Petrol 109: 265-273

Holdaway MJ (1971) Stability of andalusite and the aluminium silicate phase diagram. Amer J Sci 271: 97-131

Holdaway MJ, Mukhopadhyay B, Dyar MD, Guidotti CV, Dutrow BL (1997) Garnet-biotite geothermometry revised: New Margules parameters and a natural specimen data set from Maine.
Amer Mineral 82: 582-595

Holloway JR (1977) Fugacity and activity of molecular species in supercritical fluids. In: Fraser DG (ed) Thermodynamics in geology. Reidel, Dordrecht, p 161-181

Hovis GL (1986) Behavior of alkali feldspars: Crystallographic properties and characterization of composition and Al-Si distribution.
Amer Mineral 71: 869-890

Hovis GL (1988) Enthalpies and volumes related to K-Na mixing and Al-Si order/disorder in alkali feldspars.
J Petrol 29: 731-763

Huckenholz HG, Hölzel E, Lindbauer W (1975) Grossularite, its solidus and liquidus relations in the $CaO-Al_2O_3-SiO_2-H_2O$ system up to 10 kbar.
N Jb Mineral Abh 124: 1-46

Kerrick DM, Jacobs GK (1981) A modified Redlich-Kwong equation for H_2O, CO_2, and H_2O-CO_2 mixtures at elevated pressures and temperatures.
Am J Sci 281:735-767

Kieffer SW (1985): Heat capacity and entropy: systematic relations to lattice vibrations.
In: Kieffer SW, Navrotsky A (eds) Reviews in Mineralogy 14, Mineralogical Society of America, p 65-126

King MB (1969) Phase equilibrium in mixtures, Vol 9, International series of monographs in chemical engineering, London Pergamon press

Koziol AM, Newton RC (1988) Redetermination of the anorthite breakdown reaction and improvement of the plagioclase-garnet-Al_2SiO_5-quartz geobarometer.
Am Mineral 73: 216-223

Kroll H, Ribbe PH (1983) Lattice parameters, composition, and Al,Si order in alkali feldspars. In: Ribbe PH (ed) Feldspar mineralogy, Mineralogical Society of America Reviews in Mineralogy, 2 (2nd edition)

Kroll H, Schlenz H, Phillips MW (1994) Thermodynamic modelling of non-convergent ordering in orthopyroxenes: a comparison of classical and Landau approaches.
Phys Chem Minerals 21: 555-560

Kroll H, Lueder T, Schlenz H, Kirfel A, Vad T (1997) The Fe2+, Mg distribution in orthopyroxene: A critical assessment of its potential as a geospeedometer.
Eur J Mineral 9: 705-733

Lindsley DH, Dixon SA (1976) Diopside-enstatite equilibria at 850°C, 5 to 35 kb Am J Sci 276: 1285-1301

Lindsley DH, Grover JE, Davidson PM (1981) The thermodynamics of $Mg_2Si_2O_6$ - $CaMgSi_2O_6$ join: A review and an improved model
In Saxena SK (ed) Advances in physical geochemistry, Springer-Verlag, New York, 149-175

Mäder UK, Berman RG (1991) An equation of state for carbon dioxide to high pressure and temperature.
Am Mineral 76: 1547-1559

Maier CG, Kelly KK (1932) An equation for the representation high temperature heat content data.
J Am Chem Soc 54: 3243-3246

Margules M (1895) Sitzber Akad Wiss Wien, Math naturw Klasse II 104: 1243-1278

McKenna LW, Hodges KV (1988) Accuracy versus precision in locating reaction boundaries: Implications for the garnet-plagioclase-aluminium silicate-quartz geobarometer:
Am Mineral 73: 1205-1208

Mukhopadhyay B, Basu S, Holdaway (1993) A discussion of Margules type formulation for multicomponent solutions with a generalized approach.
Geochim. Cosmochim Acta 57: 277-283

Newton RC, Charlu TV, Kleppa OJ (1977) Thermochemistry of high pressure garnets and clinopyroxenes in the system $CaO-MgO-Al_2O_3-SiO_2$.
Geochimica et Cosmochimica Acta 41: 369-377

Newton RC, Geiger CA, Kleppa OJ, Brousse C (1986) Thermochemistry of binary and ternary garnet solid solutions.
IMA Abstr Prog p 186

O'Neill HStC, Wood BJ (1979) An experimental study of Fe-Mg partitioning between garnet and olivien and its calibration as a geothermometer.
Contrib Mineral Petrol 70: 59-70

Orville PM (1967) Unit-cell parameters of the microcline-low albite and sanidine-high albite solid solution series.
Am Mineral 52: 55-86

Pavese A, Diella V, Pischeda V, Merli M (2001) Pressure-volume-temperature equation of state of andradite and grossular, by high-pressure and temperature powder diffraction.
Phys Chem Minerals 28: 242-248

Ramberg H, DeVore DGW (1951) The distribution of Fe^{2+} and Mg in coexisting olivines and pyroxenes.
J Geol 59: 193-210

Redlich O, Kister AT (1948) Algebraic representation of thermodynamic properties and the classification of solutions.
Industr Engng Chem 40: 345-348

Redlich O, Kwong JNS (1949) On thermodynamics of solutions. V. An equation of state. Fugacities of gaseous solutions.
Chem Rev 44: 233-244

Redfern SAT, Salje E, Navrotsky A (1989) High-temperature enthalpy at the orientational order-disorder transition in calcite: implications for the calcite/aragonite phase equilibrium.
Contrib Minerl Petrol 101: 479-484

Richardson SW, Giblert MC, Bell PM (1969): Experimental determination of kyanite-andalusite and andalusite-sillimanite equilibria: the aluminium silicate triple point.
Amer J Sci 267: 259 -272

Richet P, Leclerc F, Benoist L (1993) Melting of forsterite and spinel, with implications for the glass transition of Mg_2SiO_4 liquid.
Geophys Res Letters 20: 1675-1678

Richet P, Mysen BO, Ingrin J (1998) High-temperature x-ray diffraction and Raman spectroscopy of diopside and pseudowollastonite.
Phys Chem Minerals 25: 401-41

Robie RA, Hemingway BS (1995) Thermodynamic properties of minerals and related substances at 298.15 K and 1 bar (10^5 Pascals) pressure and at high temperatures.

Robie RA, Hemingway BS, Takei H (1982) Heat capacities and entropies of Mg_2SiO_4, Mn_2SiO_4 and Co_2SiO_4 between 5 and 380 K.
Amer Mineral 67: 470-482

Robin P-YF (1974) Stress and strain in cryptoperthite lamellae and the coherent solvus in alkali feldspars.
Amer Mineral 59: 1299-1318

Robinson GP, Haas JL Jr (1983) Heat capacity, relative enthalpy, and calorimetric entropy of silicate minerals: an empirical method of prediction.
Amer Mineral 68:541-553

Sack RO (1980) Some constraints on the thermodynamic mixing properties of Fe-Mg-orthopyroxenes and olivines
Contrib Mineral Petrol 71:257-269

Sack RO, Ghiorso MS (1989) Importance of considerations of mixing properties in establishing an internally consistent thermodynamic database: thermochemistry of minerals in the system Mg_2SiO_4-Fe_2SiO_4-SiO_2.
Contrib Mineral Petrol 102: 42-68

Saxena SK (1973) Thermodynamics of rock-forming crystalline solutions.

Springer-Verlag, Berlin Heidelberg New York

Saxena SK, Ribbe PH (1972) Activity-composition reactions in feldspars. Contrib Mineral Petrol 37: 131 -138

Saxena SK, Tazzoli V, Domeneghetti MC (1987) Kinetics of Fe^{2+}-Mg distribution in aluminous orthopyroxenes. Phys Chem Mineral 15: 140-147

Seck HA (1971) Koexistierende Alkalifeldspäte und Plagioklase im System $NaAlSi_3O_8$-$KAlSi_3O_8$-$CaAl_2Si_2O_8$-H_2O bei Temperaturen von 650 bis 900°C. N Jb Miner Abh 115: 315-345

Schmid R, Ceressy G, Wood BJ (1978) Experimental determination of univariant equilibria using divariant solid solution assemblages. Am Mineral 63: 511-515

Shi P, Saxena SK, Sundman B (1992) Sublattice solid solution model and its application to orthopyroxene $(Mg,Fe)_2Si_2O_6$. Phys Chem Mineral 18: 393-405

Spear FS (1993) Metamorphic phase equilibria and pressure-temperature-time paths, MSA monograph. Washington

Spencer KJ, Lindsley DH (1981) A solution model for coexisting iron-titanium oxides. Amer Mineral 66. 1186-1201

Stebbins JF, Carmichael ISE (1984) The heat of fusion of fayalite. Amer Mineral 69: 292-297

Suzuki I, Anderson OL, Sumino Y (1983) Elastic properties of single-crystal forsterite Mg_2SiO_4, up to 1200 K. Phys Chem Minerals 10: 38-46

Thompson JB (1967) Thermodynamic properties in simple solutions. In: Abelson PH (ed.) Researches in geochemistry, vol 2, John Wiley and Sons, New York p 340-361

Thompson JB (1969) Chemical reactions in crystals. Am Mineral 54: 341-375

Thompson JB (1970) Chemical reactions in crystals: Corrections and clarification. Am Mineral 55: 528-532

Thompson JB (1982) Composition space: an algebraic and geometric approach. Rev Mineral 10: 1-31

Thompson JB, Waldbaum DR (1969): Mixing properties of sanidine crystalline solutions: III. Calculation based on two-phase data. Am Mineral 54: 811-838

Tribaudino M, Prencipe M, Bruno M, Levy D (2000) High-pressure behavior of

Ca-rich C2/c clinopyroxenes along the join diopside-enstatite ($CaMgSi_2O_6$-Mg_2Si_2O).
Phys Chem Minerals 27: 656-664

Toledano J-C, Toledano P (1987) The landau theory of phase transitions
World Scientic, Singapore, New Jersey, Hong Kong pp 451

Welch MD, Crichton WA (2002): Compressibility of chlinochlore to 8 GPa at 298 K and comparison with micas.
Eur J Mineral 14: 561-565

Wohl K (1953) Thermodynamic evaluation of binary and ternary liquid systems.
Chem Eng Progress 49: 218-219

Wood BJ, Nicholls J (1978) The thermodynamic properties of reciprocal solid solutions.
Contrib Mineral Petrol 66: 389-400

Yang H, Ghose S (1994) In-situ Fe-Mg order-disorder studies and thermodynamic properties of orthopyroxene $(Mg,Fe)_2Si_2O_6$
Am Mineral 79: 633-643

Yang H, Downs RT, Finger LW, Hazen RM, Prewitt CT (1997) Compressibility and crystal structure of kyanite, Al_2SiO_5, at high pressure.
Am Mineral 82: 467-474

Zhang J, Celestian A, Parise JB, Xu H, Heany PJ (2002) A new polymorph of eucryptite ($LiAlSiO_4$), ε-eucryptite, and thermal expansion of α- and ε-eucryptite at high pressure.
Am Mineral 87: 566-571

Subject index

E

F

G

H

I

K

L

M

N

O

P

R

S

T

U

V

W

Z